THE PREPPER'S WATER MANUAL:

An Illustrated Resource Guide For Smart Preppers And Owners Of Self-Sufficient And Off-The-Grid Homesteads

Harvest Rainwater - Drill For Water - Drive A Well Point - Dig A Well - Build A Distillation Unit - Design A Filtration System - Manage Your Well - Store Your Water - Avoid Waterborne Disease - Learn Why The Weather Matters

Daniel Schoeman

Art production and Book design by A.D. Schoeman.

Cover design and Illustration by A.D. Schoeman.

All illustrations done by A.D. Schoeman and the Inkscape Openclipart project.

Note To Readers.
The information provided in this book is designed to provide helpful information on the subjects of water management. This book is a reference guide presented solely for educational and entertainment purposes. The author and publisher are not offering it as professional services advice. While best efforts have been used in preparing this book, the author and publisher make no representations or warranties of any kind and assume no liabilities of any kind with respect to the accuracy or completeness of the contents and specifically disclaim any implied warranties of merchantability or fitness of use for a particular purpose. No warranties or guarantees are expressed or implied by the publisher's choice to include any of the content in this volume. Neither the publisher nor the individual author shall be liable for any physical, psychological, emotional, financial, or commercial damages, including, but not limited to, special, incidental, consequential or other damages. Our views and rights are the same: You are responsible for your own choices, actions, and results.
References are provided for informational purposes only and do not constitute endorsement of any websites or other sources. Readers should be aware that the websites listed in this book may change.

Copyright © 2024 Daniel Schoeman
All rights reserved.
ISBN-13: 978-0-7961-5250-3
Second Edition

DEDICATION

This book is dedicated to my father, a man honest and wise.

To my mother, whose light shines brightest.

CONTENTS

1. **Foreword** 11
2. **How To Use This Guide** 12
3. **New Projects And Common Sense** 15
4. **Introduction:** Water. Why It's An Issue 17
5. **Chapter 1:** Understanding The Origins Of Water 23
   - The Water Cycle 25
   - Conservation 27
6. **Chapter 2:** Finding A Water Source For Your Land 31
   - Public Water 32
   - Alternative Sources 34
7. **Chapter 3:** Storing Water For An Emergency . . 39
8. **Chapter 4:** Collection & Storage Of Rainwater . 51
   - Rainwater Collection 52
   - The Conveyance System 65
     - Rooftops 65
     - Gutters 69
     - Downspouts 84
     - Conduits 95
     - Storage Tanks 101
   - How To Install Fittings On A Water Tank 121
   - Installing A First Flush Diverter 129

- Vents 132
- Single Tank Installation 139
- Linking Multiple Tanks From The Top 149
- Linking Multiple Tanks From The Bottom 157
- Multiple IBC Totes Installation 169
- Rain Barrel Installation 176
- Chicken Waterer System Installation 188
- Rainwater Pumps 194
- How To Install A Pump 205
- Irrigation Installation 207
- Maintenance Of A Rainwater System 211

9. **Chapter 5:** Extracting Water From The Ground 223
- Well Basics 224
- The Soil 225
- The Water Table .226
- How To Drill & Construct A Well: PVC Pipe . . .230
- How To Dig & Construct A Well: Auger 246
- How To Dig & Construct A Well: Well Point . . .252
- How To Dig & Construct A Well 258
- Summary 269

10. **Chapter 6:** The Basics Of Wells & Water Pumps 271
- Pump Basics 272
- Pitless Adapters .274

- The Basics Of A Shallow Well Jet Pump 276
- The Basics Of A Deep Well Jet Pump 277
- The Basics Of A Submersible Well Pump 278
- Connecting A Solar Water Pump 280
- General Pump Maintenance 284
- Home Water Treatment Setup 286

11. **Chapter 7:** The Basics Of Managing A Well . . .289
    - Protect Your Well 290
    - Visual Inspections 291
    - Disinfecting A Well 293

12. **Chapter 8:** Cleaning Water For Survival 301
    - Filtration Vs. Purification 304
    - Comparing Disinfection Options 307
    - Boiling Water 307
    - Faucet/Sink/Desktop/Pitcher Filters 308
    - Reverse Osmosis 309
    - Reservoirs & Removable Gravity-Fed Filters . .310
    - Ceramic Multi Candle Filter 311
    - Carbon Filters 312
    - Distillation Systems 313
    - UV Treatment Systems 314
    - Ozone Purification 315
    - Water Softeners 317
    - Land-Based Marine Filters 317
    - Solar Disinfection 318

13. **Chapter 9:** Disinfecting Water With Chemicals 323
    - Water Disinfection 324
    - Chlorine Bleach 325
    - Granular Calcium Hypochlorite 327
    - Disinfection Tablets 327

14. **Chapter 10:** Designing A Filtration System . . .331
    - A Homemade Filtration System 332
    - Alternative: Bio-filter 335
    - Gravity-Fed Filtration System 337

15. **Chapter 11:** Designing A Distillation System . .341
    - Distillation .342
    - Solar Still Distillation 346

16. **Chapter 12:** Testing Water: Disease And . . . . . . Contamination 357
    - Testing Odors & Colors In Your Water 362
    - Water Related Diseases & Contaminants In Wells 365
    - How To Manage Waste 371
    - Discussing Waterborne Diseases 377
    - Treating Waterborne Diseases 379

17. **Chapter13:** Weather Patterns & Systems 389
    - El Nino 390
    - La Nina 392

18. **Chapter 14:** Plumbing Basics 395
    - How To Apply PVC Cement 396

- How To Thread Plumber's Tape. 396
- How To Install PVC Union Coupling For Maintenance 397
- Spigot Vs. Tap 398
- When To Use Reinforced Flexible Housing . . . 398
- Metric To Imperial Pipe Sizes 399
- Metric Conversions 399
- Types Of PVC 399
- Pipe Thread Acronyms 399
- Slip Socket 400
- Threaded Socket 400
- Sockets And Spigots 401
- Female And Male Threaded 401
- Add A PVC Ball Valve & Garden Hose To A Bulkhead 402
- How To Add A PVC Ball Valve To A PVC Pipe. . 402
- How To Add A Faucet To A PVC Pipe 403

18. **Resources** 404
19. **Index** 406
20. **About The Author** 413

Dear Reader

"Thanks for purchasing this book.

If you found the information in this book helpful, then please consider

leaving an honest review on your favorite store.

It would be greatly appreciated."

*For When The Well Is Dry, We
Will Know The Worth Of Water.*
   - Benjamin Franklin

Foreword

Welcome to a conversation about water.

Initially, I wrote this manual as a guide to assist me during a potential, water-related emergency of my own. Almost two decades of traveling through South-East Asia, Australasia and living in Africa for another 25 years, has definitely left me with an awareness of the value of water as a resource for sanitation, medicine, farming, and sustaining life. Although very informative and rewarding, the research also revealed that water distribution and weather cycles were starting to cause a scarcity of clean, drinkable water and this was becoming a major topic of discussion, even at government level.

It led me to an alarming conclusion:

At this moment in human history, the probability of us humans running out of potable water is actually very high.

Fear has the capacity to cripple and it is my hope that the information in this book will empower the reader; the aim is to cultivate a strong belief in the individual's own ability, and to motivate each of us to reach out to the community and to bring like-minded people together.

I trust that this knowledge will serve as a tool to give you the peace of mind to, like the camel, weather the storm no matter what nature throws at you.

Every chapter of this book discusses a certain aspect of water management. The reader will have to adapt the information to his or her real-life environment and individual water-related needs. When dealing with water, don't forget that water is a very simple medium and that it just wants to flow. Whether at the bottom of a well or on top of a roof, water needs to flow without being obstructed. Keep this philosophy in mind when drilling, connecting pump lines, dealing with contaminants and when installing filters in a water feed line.

Allow a steady flow of living water.

How To Use This Guide

This book is a guide about how to manage water on your property, and it is designed to help you, the DIY enthusiast. To make the most of this guide, it is recommended that you read it from cover to cover.

- You will get ideas for new setup designs or installations.
- You will find a step-by-step, illustrated guide for installations.
- You will find information needed for maintenance or repairs.
- If you are interested in a specific topic, then you can skip to the sections you are most interested in.
- If you are new to plumbing and DIY, read this book from cover to cover.

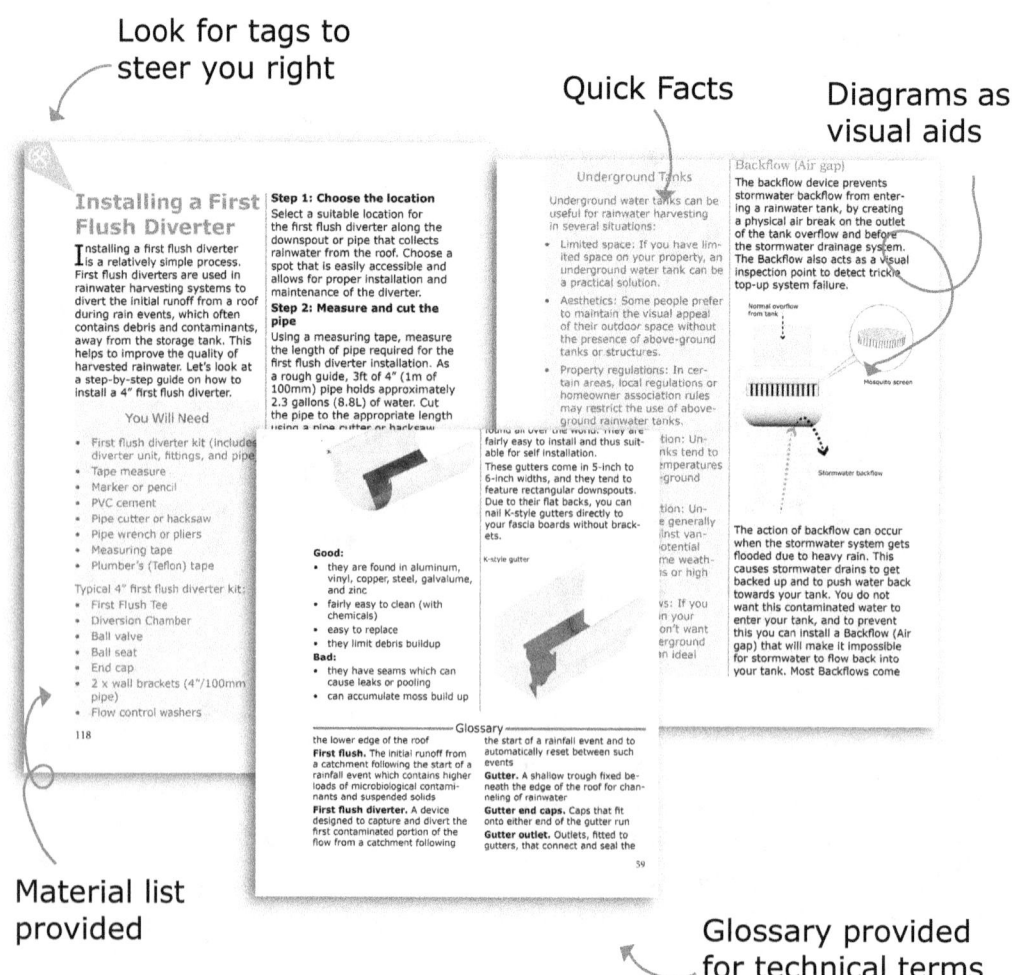

Look for tags to steer you right

Quick Facts

Diagrams as visual aids

Material list provided

Glossary provided for technical terms

12

Tags

Each unit will have a series of tags to help you find the right topic. They include:

These tags simply represent a specific topic that forms part of a unit. Looking up specific sections within a chapter is much easier this way.

Installation Sections

For completeness sake, we added the complete installation process to the "installation sections", which means that previously covered installations might be repeated when part of a larger project.

Example: *Gutter Installation* is an installation section on its own, but it might be repeated as part of the complete *Tank Installation*.

The Practical Side Of Things

This guide is designed to offer practical solutions for any rainwater-related challenges you may encounter. However, it's important to acknowledge that individual needs, environments, skill levels, and other factors vary greatly. While the information provided is comprehensive and will guide you through the entire process, it is impossible to address every unique situation that DIY enthusiasts may face. In the event of a specific problem not covered in the guide, feel free to improvise and apply your own wisdom and knowledge. The primary goal of this book is to empower readers to develop technical expertise, become adept problem-solvers, and acquire new skills that foster self-reliance and confidence.

Some tips to keep in mind:
- ◇ Safety awareness is of utmost importance.
- ◇ Do read the section on Plumbing Basics. Water and plumbing is synonymous.
- ◇ Know your tools. Selecting the right tool for the job is half the battle won.

Basic Tool List

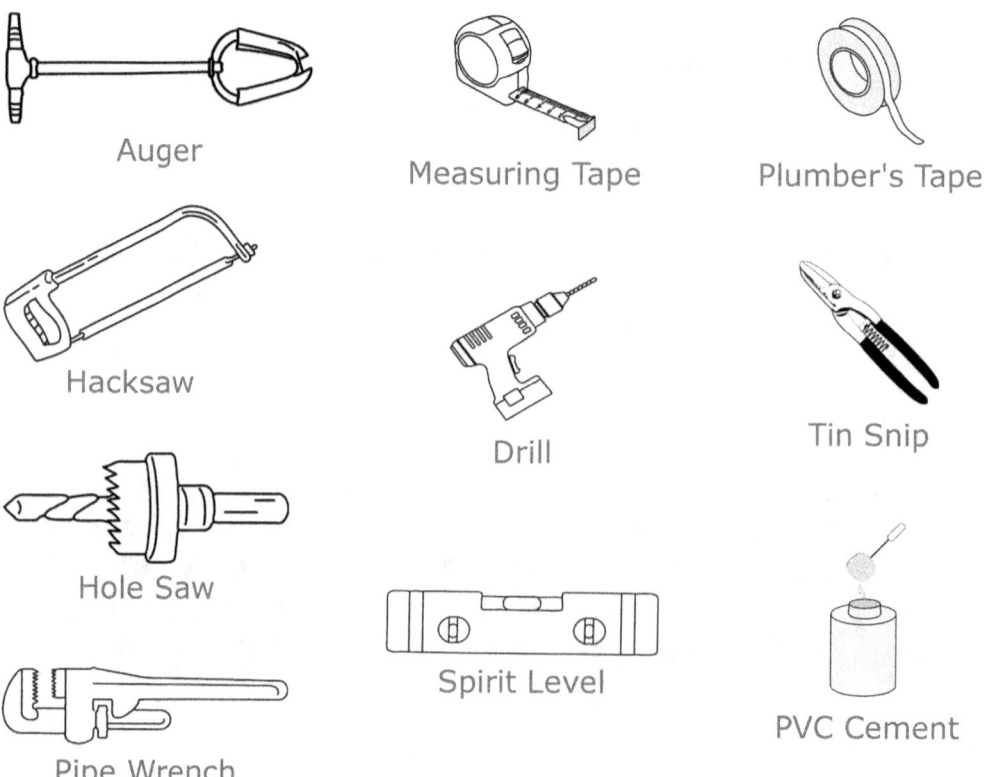

Auger

Measuring Tape

Plumber's Tape

Hacksaw

Drill

Tin Snip

Hole Saw

Spirit Level

PVC Cement

Pipe Wrench

New Projects And Common Sense

- **Safety** is everything! Whether you dig, install, buy, modify or remove, remember to use your common sense and to put safety first. Safety applies not just to you, but to children, visitors, workers, livestock, pets and wild animals.

- Follow local **laws** and regulations and avoid unwanted attention. Keep everything legal and obtain permits where needed.

- Buy **quality**. If it sounds too cheap, it probably is. Avoid tanks, filters and pumps imported from countries known for their cheap prices. Why take a chance on something that's function is to sustain you?

  Example: Use Schedule 40 PVC pipe and avoid the thinner, cheaper selections. Every storage container, no matter how big or small, must be food grade quality or have a food grade liner on the inside. Don't use any container that was previously used for chemical storage.

- Know how to **repair and maintain** all systems related to water on your property. Keep spare parts.

- Always have a backup **plan**. Pumps use electricity. What will you do when there's no power? Be prepared for any worst-case scenario.

- **Test** your water! Whether from a lake, river, well or rooftop, always have it tested. If unsure about the presence of pathogens, boil it. Never assume anything. Always have a water test kit at hand. They are reliable, inexpensive and simple to use. When in doubt, test first!

- Make nice with your neighbors. **Communities** create a network of trust and cooperation. Try to find like-minded people in the community and reach out to them. This is not just about getting, but also about giving.

- **Research**. Don't take resources for granted. Use the websites provided by government agencies that are responsible for environmental protection, disease control and water management.

Safety

Wear appropriate protective gear: Prioritize personal protective equipment like safety goggles, ear protection (if applicable), and sturdy footwear to ensure your safety during installation.

Control Electrical Hazards: If your project involves electrical components, take necessary precautions and ensure the system is properly grounded.

Proper Safety Care: Before installing the system, evaluate the installation site. Identify any potential hazards such as uneven terrain, overhead power lines, or obstructions. Clear the area of debris or objects that could pose a risk during the installation process.

Ladders: If your installation involves working at heights, such as installing gutters or downspouts, use proper ladder safety techniques. Ensure the ladder is stable, positioned on solid ground, and follow proper climbing procedures.

Preventative Maintenance: Utilize the correct tools and equipment for the installation process. Inspect your tools to ensure they are in good working condition and follow the manufacturer's instructions for their safe use. After installation, establish a routine maintenance schedule and conduct regular inspections of the system.

Introduction

Water. Why It's An Issue.

Growing up in the 70's and 80's, I still remember what a pleasure it was to run into a neighbor's yard and to grab a fistful of strawberries, a peach or even a fig. It wasn't about the thrill of taking something that didn't belong to me, but more about having access to, what I believed to be, the absolute best fruit in the neighborhood. This particular neighbor was well-known in the community and considered quite the innovator, a real "can do" man who had a solution for every problem. He allowed us kids to come and go as we please, pretending not to see, just as long as we didn't do any harm to the network of pipes, pumps and barrels that crisscrossed his yard. His backyard creations were surely works of art, way ahead of their time, but as kids we were ignorant and unaware of the ingenuity and craftiness that lay before our eyes. I often think back to those days. At the time, many families owned their own swimming pools where neighbors, friends and family congregated over weekends. If you didn't have a pool, there were water holes, rivers and lakes to explore. The greatest convenience was, of course, to grab a glass and fill it up with water straight from the faucet; it tasted good and during the hot summer months I drank at least five glasses per day. As a child, I did not realize that I was living in a world of abundance, a landscape rich in water. Safe to say that I took the whole experience for granted, not realizing that I was part of a generation who still had the peace of mind that there was enough to go around. This abundance was designed and created by a system of common sense and the unselfish aim of serving and providing for the community. Something from those days must have stuck with me, because many years later I became obsessed with self-sufficiency, organic farming and prepping for scenarios where we run out of resources.

The importance of water is realized universally. No matter where in the world, we all envision a life with an abundance of water. Water means growth and it is the foundation of life, found inside the cells of all living organisms and responsible for maintaining health at the cellular level. It is the one commodity we cannot do without. Found in a liquid, solid and gaseous states it plays a vital part in the survival of every species on planet earth. Not a day goes by that we humans don't use gallons of water, whether on ourselves or on our homestead. Water nourishes, replenishes, cleans, and treats; it's a base in our cooking, is used for cleaning and nourishes plants and livestock. Not only is water used for drinking but it is also responsible for sanitation and removal of waste material.

> Not a day goes by that we humans don't use gallons of water, whether on ourselves or on our homestead.

Water-born diseases are known to decimate populations and can quickly spread to adjacent areas. An absence of water causes droughts, crops to fail and livestock to perish.

Nowadays, more and more of us are starting to experience new stages of dissatisfaction when it comes to our water quality and source. The situation is dismal in certain parts of the world as weather systems, economic development and the exploitation of water as a resource are all creating conflict and disapproval. Corporations are continuously exploiting our water resources and sadly, sacrificing the well-being of the community has become the norm, all in the quest of making a buck at any cost. In some areas pesticides, chemicals and antibiotics contaminate runoffs and the result is tap water that's not fit for hu-

Start To Pay Attention

Getting serious about water means that you are aware of the issues that face you and your household. To become an in-house activist, start by calculating your family's monthly water consumption. Inspect faucets and pipes for leaks beforehand and start thinking how you can save water, especially in the bathroom.

Hint! Toilets use about 40% of your household water!

man consumption. Bottled water seemed like a good idea not so long ago, till we calculated the costs and realized that most companies do not deliver "mineral water" but mere tap water. Natural disasters and extreme weather events are on the rise. Earthquakes, hurricanes, tsunamis, flooding and droughts all pose a threat to disrupt water supply lines or to contaminate the water sources that we rely on. Even countries with an abundance of freshwater, take it for granted; they assume that this commodity will just exist indefinitely.

Although water covers about two thirds of the planet, finding it in a fresh and drinkable form is no simple task. Actually, nearly one billion people on this planet still do not have access to it and in some parts of the world people even spend their entire day searching for semi-drinkable water. Recently the UN Deputy Secretary General declared that "The human right to water is a historic, noble, even sacred mission and cause."

The problem with finding water is not a simple one and one driving force behind water scarcity is overpopulation. Simply put: More people means that more water is required for waste removal, manufacturing, sanitation, construction, food production, etc. Water treatment infrastructure has to adapt to the needs of the people and more

We can safely say that water plays a vital part in the survival of man and to live a self-sufficient lifestyle, we will need to know more than the average man about water.

money will have to be invested in water research and treatment facilities. The future definitely holds some problematic issues for us humans and the current prediction is, that in the next 20 years, the quantity of fresh water available will decrease by 30%. If this statistic becomes a reality, it has the makings of a catastrophe and it also has the potential to disrupt the daily routines of millions of people. It certainly holds true that we humans have neglected the safeguarding of our water supply, and we have created a mountain of problems out of a resource that should have been free and available to all in the first place.

This book is about water. Whether you live in the inner city, suburbs or out in the country, you can bet that water matters and that it always will. It now makes more sense than ever to secure a water source on your property and to store as much as possible for emergencies. Your water needs must be met, even before you think about owning a homestead, going off-grid or living a self-sufficient lifestyle. If you are an individual who has stopped taking resources for granted, and if you want to give yourself a decent chance at having access to clean water at all times, then this book is for you. To give ourselves a fair shot at surviving any water crisis, we need to know what we are dealing with. Knowledge can provide us with the know-how to weather this crisis and to survive in comfort.

Over the next few chapters we'll discuss the ins and outs of water. We will look at weather patterns and how to locate, collect, treat and store water. There is a "can do" attitude in all of us and this is exactly what's needed to get us back on course again. With some knowledge, a good dose of common sense and the right attitude, you can ride out the storm and be wiser for it. The aim is not to isolate, but to create a community of like-minded people who are passionate about survival and who want to educate themselves for the greater good of man.

Let's take a look at the various topics covered in this book.

- Understanding water. If we want to be prepared for whatever disaster situation, we will have to become highly knowledgeable about water. This means understanding how the water cycle works and how it influences our environment. Extreme weather events need to be investigated to avoid becoming a victim.
- Storage. Knowing how to store water is very important. The container you use must be suitable for long-term storage.
- Rainwater Harvesting. This is where you start saving on your water bill and where you can fill tanks for the long emergency.
- Cleaning water. Water filtration and purifying is essential. If you are able to clean water, no matter the source, you stand a much better chance at having potable water.
- Wells and boreholes. Having a well in your backyard just takes a big load off your shoulders. It is a precious resource that needs careful management.
- Testing your water. When, how and where to test is an issue, especially after floods or suspected contamination.

Start To Figure Things Out

There is no substitute for water. You cannot rely on soda, juice or sports drinks for nourishment. All liquids are not the same when it comes to water absorption. A liquid with sugars and additives have to be processed in the body before the actual water is extracted and absorbed. This means that your body has to work harder and use more energy for this task. Water is the real deal.

The Benefits

There are many reasons why we as land owners eventually decide to look to the skies or ground for help in the form of natural water. Most people are satisfied with their municipal water, but the concept of harvesting something that is "free" and provided by nature wakes up the survivalist in all of us, and inevitably we start to look for alternative ways to sustain our homesteads. By not collecting rainwater, you are allowing this precious resource to go to waste and to flow into the ground or down the road into the storm-water drains. These days more and more people are investing in rainwater harvesting systems and some of the main reasons are:

Economics

1. Investing in a self-sustained system reduces pressure on **future water infrastructure cost**.
2. An immediate reduction in your **monthly water-related expenses**.
3. Having an additional supply of stored water (calculated) can offset your **insurance premiums** in areas prone to wildfires.

Survival

1. **Quality.** Depending on where you live, rainwater or groundwater can be of a better quality that municipally treated water.
2. **Lack of main water line.** Living in rural areas can be very hard and lack of municipal water can make a property uninhabitable.
3. **Poor groundwater** resources in your area.
4. **Looking for a backup** source of water. Water restrictions due to droughts, overuse, or damage to the local water infrastructure can bring a homestead to its knees.

Environmental

1. Rainwater is an excellent option for **plants and wildlife**.
2. Rainwater can feed **dams and ponds** on your property.
3. A reduction in **storm-water runoff** which in turn limits the spread of chemicals, metals and pesticides in some areas.
4. Rainwater can be used to **fight wildfires**.
5. Harvesting rainwater **slows erosion** in dry environments.

Chapter 1

Understanding The Origins Of Water

The blue planet. As seen from space, our planet is an attractive blue ball, wrapped in water and covered in a web of clouds. It is easy to become mesmerized by the appeal that the oceans, rivers and lakes have, for those are the very places where humans chose to settle and to start civilizations. Innately, every human being is aware of the possibilities that water can create; it is the very source of life that we have come to depend and to rely on. For billions of years, tiny drops have been migrating in an endless cycle round and round the earth. Minute little molecules have traveled as droplets, from the oceans to the clouds, across the land, and down into rivers and lakes and into the ground. This constant movement of droplets, from solid to liquid to gaseous states, is what's known as the Water Cycle. Our fauna and flora have come to depend on these droplets that circulate the planet, to the extend that nothing on earth can survive without it. When we look at the water cycle, we quickly realize something very important. Water, or its absence, represents the absolute most important resource known to man, and it has been a part of our planet from the beginning of time. The water in my glass, has been traveling nonstop, since before the age of the dinosaurs. It's fallen as a drop of rain in the Sahara, as a torrential monsoon in India, a flash-flood in Arizona or become part of a glacier in Canada. It has never ceased to exist; it has merely taken on a different state of existence. Water cycles exist in a state of balance and provide us with the fundamental basics that we need to understand the dynamics of droughts, floods and extreme weather patterns. As landowners, we have to become experts at understanding the inner workings of the water cycle on our planet. If there is one thing that space exploration has taught us, it's that a planet without water is completely uninhabitable.

When we look at our planet we have to ask ourselves, why do only some planets have water and where does this water come from? Currently there are two theories related to the origins of water on planet Earth. The older and more established theory states that water was possibly carried all the way across the universe through comets and meteorites. A newer, and also very plausible theory states that water has been a part of this planet from the beginning. Whatever the case, it is correct to assume that the vast majority of the water on planet earth has been here for millions of years. Of concern for Preppers, and owners of self-sufficient homesteads, is making sure that we

Water Cycle

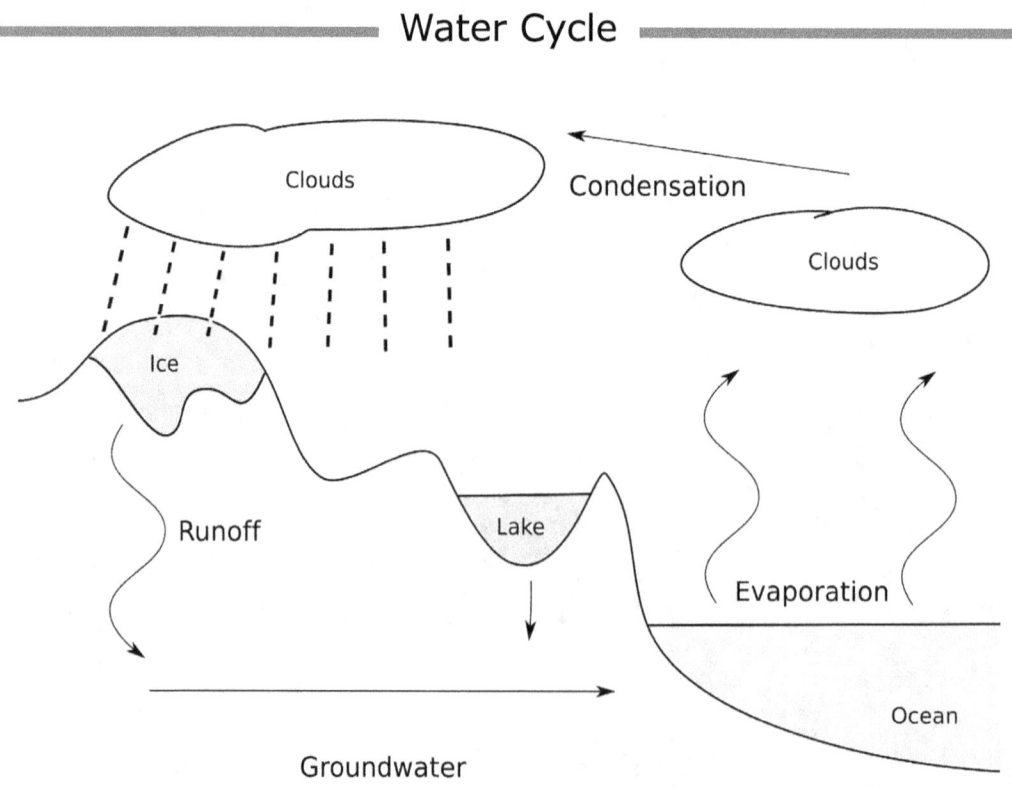

Liquid water > Evaporation > Condensation > Precipitation > Runoff/ flow back to the original state

understand the "water cycle" and how it applies to us and our immediate environment.

The Water Cycle

We are all familiar with the classic water cycle (a.k.a. the hydrological cycle) that we learned about as youngsters.

The cycle starts with the evaporation of water from the surface of the ocean.

The moist air rises up and it starts to cool down.

This water vapor condenses and it forms clouds.

Clouds filled with moisture are moved all around the globe and the moisture returns to the surface as precipitation (rain). Not only does the rain form lakes and rivers, but the rain also replenishes groundwater deep inside the earth.

(Take note that this is the very system that we mimic in water distillation and filtration systems.)

What makes the water cycle really interesting is the fact that we call this a "closed hydrological system". It is a closed system, because the amount of water on the planet does not fluctuate. Yes, it can be broken down and reformed through various chemical processes, but there is no significant input or output of water to or from the planet. Where does it go, you might ask, in times of drought? The simple answer is:

Somewhere else.

In theory, when there is a drought in one part of the world, there will be floods in other parts. Abundance over yonder, means scarcity over here. It is a balanced system that is in constant flux. Water does not disappear, it just moves with the weather systems and returns as precipitation when conditions permit.

Let's look at a breakdown of where, and in what forms, water is found:

96% of water accumulates on

Global Water Crisis

- Droughts caused by extreme weather events are destroying agriculture and displacing people. Waves of migrants are fleeing areas ravaged by droughts.
- Population growth has increased the need for water and also released more contaminants into existing groundwater sources.
- The demand for more food is putting stress on groundwater reserves.
- Weak infrastructure is not able to cope with the demand and most countries are in dire need of new technology to solve the water crisis.

the earth's crust. This is the water present in reservoirs like the oceans, lakes, seas and rivers.

2% is found in a solid state in the ice caps, glaciers and mountain caps.

1.7% is found as groundwater inside the crust.

0.001% is found in the atmosphere and air around us as clouds, falling rain and water vapor.

At first glance, the above-mentioned statistics might seem pretty decent. The percentage (96%) of water found on the surface crust seems to favor man's chances of survival. But how much of this water is actually potable freshwater?

The answer is quite disturbing. Only about 2.5% of this water is freshwater. Add to that the fact that most of this freshwater is frozen as ice, and suddenly we are looking at a figure of only 0.65 %, which represents the earth's total water supply that is neither salty nor frozen.

The moral of the story?

Right now, water is absolutely the most precious resource on earth!

The tiny percentage of 0.65, represents the possibility of life on our planet and we humans are scrambling to safeguard ourselves from running out. The National Geographic Society calls it " Water Wars" and states that underground water is pumped so aggressive-

> The aggressive pumping of our groundwater supply is putting stress on aquifers which can have catastrophic consequences.

ly around the globe that land is sinking, civil wars are being waged and agriculture is being transformed. Right now, the city of Beijing is sinking at a rate of 4 inches per year, due to the depletion of the groundwater underneath.

The water cycle teaches us that we cannot pump water from wells more rapidly than they can be replenished. We rely on precipitation runoff to replenish our groundwater. This is a very important aspect of the water cycle, since more than 30% of the planet's fresh water (in liquid form) is groundwater. We all want to share in this resource, but first we will have to familiarize ourselves with the location, depth and quality of groundwater in our area. Water is not an infinite resource and to avoid complications, we will have to learn how to manage and conserve this resource. You might need **X** amount of water for your land, but if every neighbor in your area is extracting water from the same aquifer, things can take a turn for the worse. Imagine a bottle with 15 straws in it. The bottle will be empty in seconds. If, during one summer, we have fewer rain events that usual, that groundwater aquifer will not be replenished and we could potentially sit with a dry well. This is the reality as more and more people are drilling wells and extracting groundwater.

Conservation

There is a good chance, that you have very good water just a few feet underneath your land and if so, you will be eager to drill a well. That's your right to do so, but be aware that besides just drilling a well, you also might have to change your lifestyle and your opinions about sustainability. Traditionally, homestead owners have always required large quantities of water for irrigation and livestock. If we look at the present evidence as seen through the water cycle, we quickly start to realize that the perceived abundance of water is an illusion.
In certain areas the situation is deteriorating rapidly and we will

Water and Droughts

Surface water is freshwater found above ground that accumulates in nature through precipitation. Once this water penetrates the soil, it gets filtered down into aquifers that store groundwater. A large portion of the freshwater on earth is lost through runoff to the oceans. When a period of abnormally dry weather endures in a region, it can create an imbalance that causes a drought.

We can identify four typical drought scenarios.

- **Rain events** - When there is a shortage in local precipitation.

- **Moisture content of soil** - If conditions arise that causes the soil to lack the moisture necessary to sustain a particular crop.

- **Water levels** - When the surface water and groundwater levels are
both below normal.

- **Supply** - When the demand for water cannot be met through the supply infrastructure.

Understanding The Origins Of Water

have to adapt and evolve with it. This could mean that some of us will have to start thinking about scaling down. If the supply cannot cover the demand, we will have to change on our end. This is already happening all over the world in many first world countries. Farmers are looking at alternative agriculture practices that can still secure a high yield, but that utilize less land and less water. This is definitely possible and actually recommended by the experts. Make the right decision and start thinking about water conservation on your property.

Consider some of the following options:

- Scaling down on irrigation. Drip irrigation reduces evaporation and saves huge amounts of water. Timers can schedule irrigation for the cooler parts of the day.
- Create a water-rich vegetation with low water needs.
- Recycle your house's gray water. This is something that every human on the planet should be doing. Why waste water that can feed your garden?
- Go organic. Organic methods help retain soil moisture and have a higher yield.
- Harvest, capture and store water. Rainwater is naturally captured and stored in ponds. Using rainwater for livestock and irrigation is a great way to lessen the impact on the local groundwater.
- Retain the moisture of the soil by using compost and mulch. Extreme weather events are causing a definite rise in temperatures in certain parts of the world. By using compost and mulch, you can retain some of the moisture and reduce water evaporation.
- Educate yourself about anything water conservation related. How to rotate crops, how to protect crops, increase yield, grow local indigenous crops, etc.
- The water-related problems of the day, affect not just the individual, but the whole community of land owners. We have to investigate and part of the process is to look at the problem from all angles.

Before considering a well or bore-hole, design a system for water conservation that will make you less reliant on the groundwater.

A major component of the water cycle is the weather. We'll discuss that in Chapter 13.

Saving Means More

Saving water is very rewarding in the long run. It means that you will put less stress on your water supply and stored water will last longer. Some simple measures to save water and minimize waste:

- When brushing your teeth, turn off the tap. You can save up to 1.5 gal (6L) of water a minute.
- Take shorter showers. You can easily save more than 1.5 gal (6L) a minute. Showers are more water-efficient than bathing.
- Check and double-check all pipes, shower heads and faucets for leaks.
- Replace older toilets, shower-heads and faucets with efficient "water-saving" products.
- Place a water displacement device in your toilet tank. This reduces water usage with every flush.
- Design and implement a gray water system. Shower and washing machine runoff can be reused.
- Create a water-smart garden or no-waste irrigation set-up for crops.
- For small gardens, use a watering can instead of a hose.
- Educate your family about awareness. Wash dishes by hand, don't flush every time, plug the tub when you shower, etc.
- Install a water meter. The evidence of water fees will encourage saving.

Try to calculate your household's water consumption to see where you can save.

A typical domestic water consumption graph:

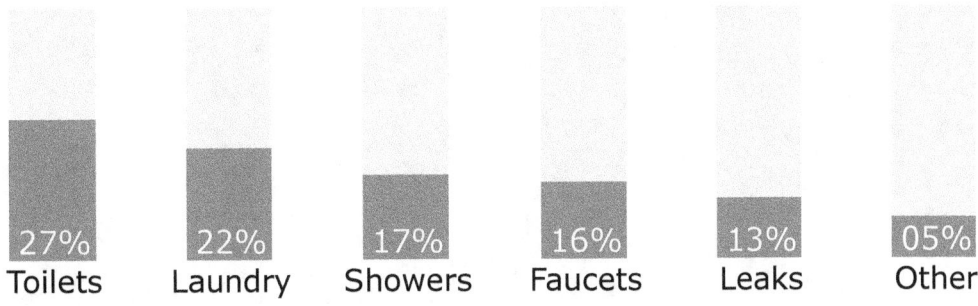

| 27% | 22% | 17% | 16% | 13% | 05% |
| Toilets | Laundry | Showers | Faucets | Leaks | Other |

Chapter 2

Finding A Water Source For Your Land

It's only when we try to find water ourselves, that we realize just how hard it can be. When looking at buying rural land, finding a dependable water source should be your first concern. In looking for water, you have three basic options:
Public or municipal water is water that's provided by the government in service to the people. Certain rural areas may not be serviced by a municipal water system. **Surface water** includes water from a river, stream, lake, or spring.
Groundwater is water pumped from a well.
Do not even think about buying a property if your water source is not guaranteed. Don't let emotion get the better of you. Make sure you do your research, for in certain areas potable water is scarce and to make it work, you'll have to become an expert at water management.

I'll go so far as to say that, when buying a property, finding water is not an afterthought, it is the decider. Factors like surface water, groundwater, depth of the water table, local rainfall and community preparedness, all have to be taken into account. You also have to consider the possibility of contamination through runoff from adjacent properties. Living downhill from a farm that uses harmful pesticides means that during a rain event your property will be affected adversely. Areas prone to flooding are more susceptible to pathogen contamination due to stagnant water formation that causes microbe proliferation. This can cause a myriad of problems down the line. If you want to avoid these issues, you will have to make a thorough assessment of the water situation not only on your land, but also in the surrounding communities

A Water Source Must Be Found

Having access to water, whether from a public municipal line or from a well or spring, will define your property. Everything that you learn about the process of attaining water rights will be helpful in the long run. It will provide you with the knowledge to secure your water rights, and with he peace of mind that comes with it.

This also applies to current landowners who might be dissatisfied with their water situation, for whatever reason. Unfortunately, in many areas you will need government approval, in the form of permits and licenses, before you can drill a well, harvest rain from the roof or develop a spring running through your property. No matter which direction you choose, the first and most obvious step will be to examine the information provided by your local government. You can go downtown to the department of water management or to a water utility office. Alternatively, you can also view the information provided online, by various government agencies. Let's take a look at the different options.

Public Water

Having access to the public water supply is considered a great benefit, since you are almost always guaranteed a steady flow from the faucet. When buying a new property, this kind of information will be provided by the seller, sales agent or on a government provided land listing. It should clearly state whether the land is served by a public water supply line. On paper this sounds wonderful, but the reality is that your water has a long way to travel between source and faucet, which means that many variables can influence the quality of the water before it flows out your kitchen faucet. It will be worth your while to investigate the potability of your water and if contaminants are found to be present, to be aware of the options at your disposal. You will have to do some investigation yourself to determine what forms of water sources are available. A property that's within the city limits will definitely have access to a public main water line. This is provided through a network of pipes that's managed by the government or a government appointed entity. Many rural properties are also connected to the public water line. In some areas the water is supplied by a local community well. This is also considered a form of "public water" and similarly you will have to apply for the right to join this well. There are of course properties without public water or community wells. This means that you will have to improvise to find an alternative source of water.

The Main Water Line

Find out where the main water line is located and where the original source is located. Inquire as to the quality of the source and also about the water pipes and if asbestos cement pipes are still in circulation. You want to avoid contaminants, or at the very least, be aware of them. Concerning connecting to the main water line, make sure that you get comprehensive answers concerning the following:

- Be legal

Make sure you know which permits and legal documentation are required to connect to this main water line.

- Fees

Check the fees to connect to the main water line. In certain countries the government will connect your water free of charge. Check with the local government office to inquire after fees and whether you need a water meter or not. Water meters are very expensive.

- Regulations

Make sure that you get answers regarding the following:

- What regulations will affect your usage of the water?
- Who should install the connection line?
- Does the installation require government inspection or regulation?

Local Community Water

In more remote areas it's not uncommon for the local communities to pool their resourc-

Before Buying

- Investigate extreme weather patterns in the area and the probability of droughts and floods.
- Research regulations concerning rainwater harvesting and having a well.
- Find the expected annual rainfall for the area.
- Talk to neighbors about their experiences.
- Check the depth of the local water table.
- Investigate all surface water souces.

Finding A Water Source For Your Land

es together and to create a local water well. These wells are operated and maintained by the community members themselves and are considered a "public" water source for the local area. When inquiring about a local community water source, make sure about the following:

- The water quality
Since no government agencies are involved you have to make sure that the water is 100% potable. Take a sample and have it tested at a government-certified laboratory. If your situation is more urgent, and you need an immediate answer, test the water yourself with a DIY water test kit. These kits are simple to use and very reliable.

- Life expectancy
How long is the water source (well) expected to produce water? This is not a simple question. Understand that the groundwater is dependent on precipitation. If the community extracts water more rapidly than it can be naturally replenished, then you will end up with a dry well. This is where extreme weather events can destroy communities with unexpected droughts.

- Hidden costs
What are the hidden costs, maintenance, infrastructure, etc. Can you access the water line and how much does it cost? Do you need a water meter (expensive)? What are the local rates.

Finding water relies on you becoming an expert at everything related to water. This means research and prepping.

Alternative Sources

Having a community or government provided water source is obviously a great option. You should still look into finding an alternative water source on your land. If you live in a real remote area, this might be your only option. Using surface water from a spring or lake, or groundwater from a well, are both great options to reduce your water bill and to provide you with water in a time of need. To prospective land owners, this means that you will have to make a thorough assessment of your land.

● **Step 1. Walk your land**
Get a general feel for the vegetation and wildlife present. Try to locate areas with lush vegetation compared to their surroundings. Shallow groundwater can be found by knowing the landscape. Vegetation with water-loving plants, like willows and aspen trees will be an indicator of groundwater at shallow depths. Conditions for large quantities of shallow groundwater are more favorable under valleys than under hills. Areas

where water is at the surface in a depression as springs, seeps, or swamps reflect the presence of groundwater, although not necessarily in large quantities or of usable quality. Make notes of all surface water present or evidence of past. If you find a tiny rivulet or stream flowing on your land, track it back to its source. A spring can be a great source of water. Start celebrating if the source provides a year-round, steady flow of clean water, and if it is located within your property boundaries. Call in an expert to help with capping the spring.

- **Step 2. Talk to the locals**

Finding water and knowing about water is something every landowner is willing to discuss and to offer advice about. Questions you should be asking:

a. How deep is the water table and are there springs in the area?
b. What's the water quality like?
c. Does the ground consist of clay, rock or sand?
d. What's the expected yearly local precipitation and which are the wettest months of the year?

- **Step 3. Test for presence of water**

If you are completely on your own, in the boonies, then you need to lower a pipe and look for groundwater yourself. This is why you have to do research

Spring Water

Springs are usually fed by shallow groundwater. Some springs seep out of banks, slopes or hills and typically groundwater emerges from one defined discharge in the earth's surface. This is called a **concentrated spring**. Others "seep" from the ground over a large area and has no defined discharge point. This is a **seepage spring**.

- Determine if it's year-round water. Look for signs like plant growth with roots growing towards the water source. Is there a worn-out channel for water to flow, indicating flow over a long period of time? Check the temperature of the water throughout the day. A constant temperature indicates consistent flow; the water should be slightly cooler than the air.
- Take a sample to test it.
- Determine the flow rate. Take a pipe and see how long it takes to fill a bucket (E.g. measure 5 gallons). Best to do it during the dry season, when the water table is at it lowest. Springs used for drinking water supplies should yield at least 2 gallons per minute throughout the entire year.
- Call an expert to cap your spring, or in an emergency, place a black flexible PVC pipe with a screen filter inside the source and secure and cover it without disrupting the original flow channel. Store the water in a holding tank.

first. If the locals provided you with hope that there is groundwater at drill-able depth, or if you saw unusual lush growth in one area of your land, then you will have to test. Consult an expert to assist you.

- **Step 4. Secure your water rights**

You have to secure the rights to your land. This means, depending on where you live, a visit to your local water office. Your water rights consist of the right to use, a reasonable quantity of public water, for a certain period of time, as it occurs on your property.

- **Step 5. Research**

Use the Internet, the local library, or visit a local government office. You must be aware of the water cycles and of the weather systems that affect precipitation in your area. This will give you a fairly clear indication of what to expect over the coming months. Once you are familiar with a certain area, you can use a system of deduction to figure out what is causing a water scarcity or drought.

Lakes are fed through rivers, which run down from the mountains.

The mountains depend on rain. It is crucial for any Prepper to know the topography and environment that he or she lives in.

- **Step 6. Talk to a licensed well driller or to a hydrologist**

Get a hydrologist to obtain information on the wells in the target area. The locations, depth to water, amount of water pumped, and types of rocks penetrated by wells also provide information on groundwater.

Licensed well drillers will not only inform you about drilling for surface water, but also about how to use submersible pumps to extract water from springs and lakes.

Questions To Ask A Licensed Well Driller

- The cost per depth drilled?
- What local regulations apply to drilled wells?
- What permits are required?
- The probability of finding water and the expected depth?
- The soil conditions in the area. Rock, clay, sand or gravel?
- Once you've struck water, what is the life expectancy of the well?
- How will weather cycles like El Nino and La Nina affect the water table and the yield of the well?
- For future scenarios, what could possible contamination causes be?

Your Water Rights

In general, your water right consists of the right to use:

1. a reasonable quantity of public water,
2. for a certain period of time,
3. occurring at a certain place.

Governments enforce legislation that's designed to protect the use and enjoyment of water that travels in streams, rivers, lakes, and ponds, gathers on the surface of the earth, or collects underground. If you have the right to divert water from a river, stream, lake, or spring, you have a surface water right. If you have the right to pump water from your well, then you have a groundwater right.

In areas with an abundance of surface water, these rights are mostly accepted, as insisted by the landowner.

In areas prone to droughts and water shortages, things can get more complicated because of competition, and you will have to apply for a permit before you can utilize the water resource.

When applying for water rights, the office with jurisdiction will consider the following:

- The protection of the environment.
- Preventing waste of precious water resources.
- The fair distribution of water resources.

The main focus is to ensure that the public interest is served.

Note that you do not own the water, you just own the rights to the use of it.

Finding A Water Source For Your Land

Chapter 3

Storing Water For An Emergency

Storing drinking water is essential for daily use and also for emergency situations. The supply and quality of water to your house can be disrupted by the following:

- A natural disaster like a hurricane, typhoon, earthquake, tsunami or even by a solar flare.

- These days extreme weather events are causing a multitude of water related problems like droughts, floods and unwanted algal blooms in lakes.

- Man-made disasters like chemical spills can bring communities to their knees.

- Wells that feed the community or your property can simply run dry.

Any of the above mentioned emergencies can cause serious problems for your homestead.

When this happens, you will possibly have to deal with the following:

- The disruption of water supply lines could mean that you and your family are without water, till maintenance crews come to the rescue. You still have to perform your daily duties. A grown man or woman working all day without enough water can lead to dehydration, heat stroke or heat cramps.

- The resulting water restrictions can be very hard on homesteads and can cause serious damage to crops and death to livestock.

- When panic sets in, the masses will besiege Asupermarkets and convenience stores. Water will be the first thing on everyone's minds. History has shown us that this is true; it's happened before and it will happen again.

Water In Containers

Water can be stored in various containers and select locations. Prepare storage in and around your home for sitting out an emergency. Also prepare containers for when you go mobile in a hurricane scenario. If you are prepping for future scenarios, plan carefully and make decisions concerning quantities, container material, types of containers and storage location.

- When electricity to your home is disrupted, the water pumps stop working and this means that water stops flowing out of the faucets. Do you have backup generators for your pumps? Are you prepared for a solar flare that can render all electrical appliances useless?

Nothing is guaranteed anymore and instead of waiting for things to change, why not take the initiative and start prepping. If you live on a homestead out in the country you must have water stored away for a time of need. The question that always comes up is, "Just how much water do you need?".

Take note of the following:

1. An average family of four needs around 400 gallons (380 liters) of water per day to perform their daily activities.
2. On a more serious note, just for drinking purposes in a dire emergency, a healthy person needs a gallon per day.
3. Children, sick people, pregnant women and people living in tropical zones will require more water.
4. Pets require around 1 gallon of water every three days.

To give yourself a decent chance of surviving a disaster, you should store at least 3 months' worth of water. To put it into perspective:

You will need to store large quantities of water!

Get your thinking cap on and consider every possible scenario that might affect your water supply. To get started, we will look at ways to store water in and around the house. Make sure that whatever product or container you buy, that it is a quality product that will not let you down.

Various Options

When using storage containers found in and around the house consider the following:

- **Glass** is a safe container that's easy to sterilize.

Glass must be kept away from

sunlight at all costs. If not, you will sit with algal growth in your water.

Glass breaks, so store containers carefully and do not stack them. Glass should not be used in areas prone to earthquakes.

- Food-grade **Stainless Steel** containers are safe to use. They are easy to clean, do not break and do not leach chemicals.

- Food-grade **Plastic** Containers are really not the best option, but are by far the most convenient. Remember that any translucent container will attract algal growth when exposed to sunlight.

Avoid any container with **Bisphenol A** and **Phthalates**. The recycling numbers found at the bottom of containers mean the following:

1. PET/ PETE. These are safe when new, but (if you have a choice) remember not to re-use water and soda bottles.
2. HDPE. Considered safe.
3. Vinyl (PVC). Unsafe. Absolutely unsafe! Avoid at all costs.
4. LDPE. Considered safe
5. Polypropylene (PP). Considered safe.
6. Polystyrene. Absolutely unsafe!
7. Various unmarked plastics. Absolutely unsafe!

Remember that all stored water, no matter the container, should

Store your water in a cool, dry and dark place. Don't expose plastic bottles to sunlight or to any heat source like a water heater or generator.

---

be rotated at least once every six months. This is not just for quality, but also for taste.

Depending on size, a container filled with water can be very heavy. If you are not young anymore and if you live on your own, then moving big containers will be problematic. Make sure you select containers that suit your needs and also your physical ability to move them around. **A small platform on wheels, like a dolly or pushcart, can be used for hauling heavy containers filled with water.**

Before using any container for water storage:

1. Wash the container inside out with dish washing soap and rinse throughly with water.
2. Mix one teaspoon of unscented household bleach with a ¼ gallon (0.95 Liter) of water.
3. Shake the container and thoroughly expose it to the bleach.
4. Wash around the opening where lids and screw on caps are used. Also wash the caps or lids.
5. Rinse thoroughly and spray

Storing Water For An Emergency

clean with a pressured hose. Once your containers are clean, you can fill them up with water. If your budget allows, get some water preserver concentrate to add to your water. This should especially be considered for untreated well-, spring- or rainwater. This will extend the life of your water to at least 5 years. Write the date on the container and replace the water every six

The CDC Recommends The Following After A Natural Disaster Has Occurred In Your Area

To clean clear tap water:

- Use unscented household liquid bleach.

- Add ⅛ teaspoon (0.75 ml) of bleach to 1 gallon (3.79L) of water.

- Mix well and wait 30 minutes or more before drinking.

To clean cloudy tap water:

- Use unscented household liquid bleach.

- Add ¼ teaspoon (1.5 ml) of bleach to 1 gallon (3.79L) of water.

- Mix well and wait 30 minutes or more before drinking.

(1 Gallon = 16 cups)

A family of four require around 100 gallons (380 liters) of water per day to perform their daily activities.

months from then on out.

Besides containers, you also have to think about location. Try to store your water in a cool, dark place. It's also not a good idea to store all of your water in one location. A natural disaster can cause a building to collapse, a fire to start or simply just render one part of your property inaccessible. Consider the following locations:

— Garage
— Barn or Shed
— Any secondary structure on your dwelling.
— Basement
— Attic or loft
— Under the stars, under protective covering.
— Storage rooms and storage containers.
— Closets and cupboards.
— Under beds.
— Cars, trucks and trailers. Being mobile with a supply of water in the back greatly increases your chances of survival.
— Secret location for dire emergencies. If, for whatever reason, you feel that there is a need to hide a cache of water underground, in a tree

house, a cave or on a different property, then make arrangements to do so. This will be a practical option in a flight scenario.

Depending on your level of alertness, you will do well to be prepared for both a "sit and wait" and a "get out of Dodge" scenario. You can never be too prepared!

Bottled Water

Keeping a supply of bottled water is a simple option and very practical. Ready-to-drink water will come in very handy in case of an emergency.

• Always keep bottled water at hand. Leave it in the back of your car, in the kitchen, or at the front door, and make sure that all family members are aware of the location. Bottled water is easy to transport, fairly cheap and readily available.

• Bottled water come in all sizes. Small bottles are light and fit into small spaces. They are great for storing all around the house and in vehicles. Larger containers like water dispensers can also come in handy. A five gallon water dispenser is great for everyday use, but the jugs can also be used in an emergency. Again, make sure that you are physically able to move these containers and if not, scale down and select smaller containers.

• It is best to check the brand before you buy. Government approved and commercially produced water is best. Avoid water with added "nutrients", ions, flavorings, fluoride or with the so-called "perfect Ph". Basically just stick to plain old water and a brand that you are familiar with. The aim is to store it indefinitely and chemical additives will just shorten the shelf-life of the water.

• Store your bottled water in a cool, dry and dark place. Don't expose plastic bottles to sunlight or to any heat source like a water heater or generator. Sunlight encourages bacterial growth and that is the last thing you want in your water.

• Don't stack your bottles too high. Water is heavy. Plastic is not indestructible and cracks and leaks can occur.

• Concrete floors are not suitable for storage of plastic water containers. Make shelves for your storage containers and keep them off the ground to avoid exposure to flooding, spilling and other domestic accidents.

• Rotate your bottled water often.

Boxed Bladders

You can also store your water in boxed bladders. This is a great option if you expect to be on the move during an emergency. They fit easily into trucks, cars and trailers.

• These bladders do come in different sizes, but the 5 gal (19L) ones are the most popular.

• They are food grade quality and BPA-free.

• The boxes are structurally strong and they keep the bladders in position which makes transportation easy.

• Look for ones with a fill hose and quality faucet. You do not want the container to leak after just a month in storage.

• They should come with a water preserver concentrate.

As an alternative, some people recommend using the bladders found in boxed wine containers. These are obviously not designed for water storage and also not structurally strong or durable enough to be transported. However, if in a true emergency, and you need to fill up a container with water, then consider this as an option.

Stackable Containers

Similar to the boxed bladders, having stackable containers is a great option if you expect to be mobile during an emergency. These are basically just quality containers that are convenient to move, refill and stack.

• Typically, these containers come in 5 gal (19L) sizes, but they are stacked as an "8 pack". That means 40 gal of stacked water.

• They are made of high density Polyethylene (HDPE), food grade quality and BPA-free.

• They are structurally strong and are very easy to stack. This makes moving your containers around, a simple task.

• They come with handles which

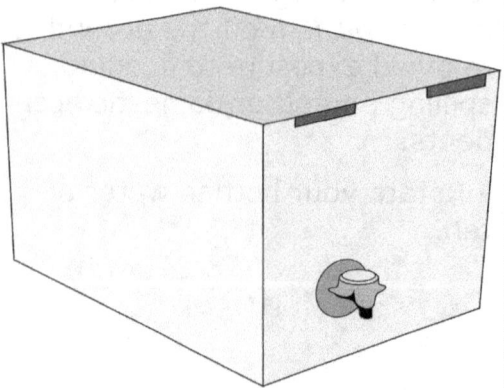

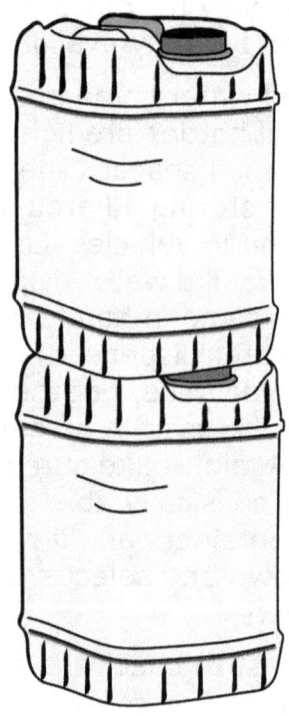

simplifies transportation.

• Included is a fill hose and faucet. Make sure that the quality of the faucet is up to standard.

These containers are very practical and they will provide a very realistic solution to your storage needs.

Water Bricks

Similar to stackable containers, water bricks are easy to move, stack and store.

• They typically come in 3.5 gal (13.2L) bricks and are also stacked as an "8 pack", which gives you 28 gal (106L) of stored water. Smaller 1.5 gal (5.7L) bricks are also available.

• What makes them unique is that they come with ridges that "lock" the bricks together and provide a very stable platform for water storage.

• They are made of high density Polyethylene (HDPE), food grade quality and BPA-free.

• Due to their design, they are great space-savers.

• They are considered very convenient due to the sturdy carry grips.

• Some brands have a wide opening which means that besides water, they can also be used for food storage or to keep items dry.

Water Barrels

These days you can buy 55 gallon (208L) barrels that are specifically designed for emergency water storage.

Before you consider using one of these, make sure that it is clean on the inside and that the "plastic smell" has been washed away. You can use a bit of bleach, or simply wash it out with vinegar and baking soda.

Typically, expect to find:

• The barrels are made of high density Polyethylene (HDPE), food grade quality and BPA-free.

• These containers come with treatment solutions, a hand pump and a siphon.

The barrels are heavy when full, so put a dolly underneath for transportation before filling it up. Avoid storing these barrels on concrete and prepare

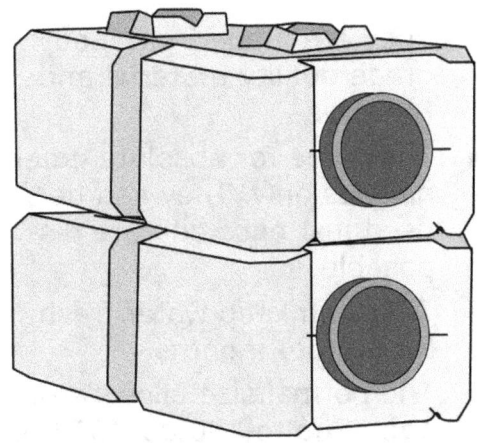
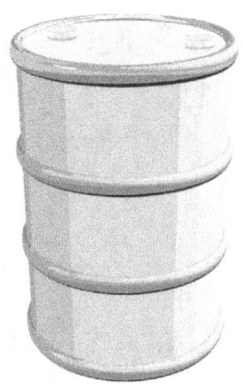

a little platform made out of wood or cardboard just to be on the safe side. You can run the water through a charcoal filter before drinking it.

Having six of these barrels will be enough water for a family of four, for a three month emergency.

They can be placed all around your property and are not as big and cumbersome as large, traditional rainwater tanks.

Storage Tanks

You can also look at larger options for in-house storage. High-capacity water storage tanks are becoming more and more popular. They are very similar to the standard water tanks that are used for rainwater harvesting.

• They are made of high density Polyethylene (HDPE), food grade quality and BPA-free.

• They come in a variety of sizes from 150-600 gallons (567-2270L).

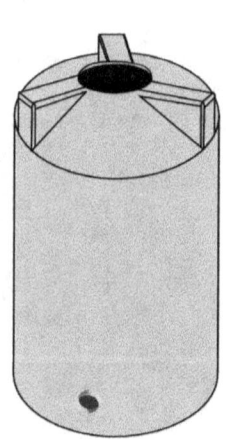

• Some brands offer small, 150gal (567L) tanks that are stackable and that fit through a standard doorway.

• They also come with spigots, inlets and outlets that make rotation of water a simple process. Try to get a brand with a spigot that connects to a standard garden hose. This makes draining and rotation a very simple process.

• They come with a water preserver concentrate.

Some of the larger models are very heavy once full, so plan carefully where you want to place them.

Water Liners For Bathtubs

If you are aware of an impending disaster in your area and you haven't had time to put emergency measures into place, then an "emergency water pod" might be the answer. Water pods are plastic liners that you put in the bathtub and fill with water from the faucet before the emergency strikes. You can buy them in different sizes and each bladder-like pod comes with a siphon. They are perfect for hurricane, earthquake or tsunami scares.

• They are made from food grade quality material and BPA-free.

• These are for absolute emergencies only. They can be used just once and are disposable.

• They can keep water fresh for up to 4 months.

• The normal size allows storage of up to 100 gal (378.5L)

Some Helpful Hints

- Most of the mentioned products come with water treatment tablets. These tablets purify the water and also improve the taste.

- Large, heavy containers become mobile with a dolly or pushcart. Water is heavy. Don't wait till the tank is filled up before attempting to move it!

- Do not reuse soda, juice or milk bottles for water storage. These containers are for one-time use and by reusing them you risk chemicals leaching into your water. Water with suspected chemicals should not be given to humans or animals!

- Don't just plan for your family, but plan for your pets and livestock. Water from rainwater tanks, ponds, storm water runoff, wells, and rivers can be used for animals in time of need. Animals can handle pathogens far better than humans.

- Common sense dictates that all of the mentioned methods have to be scrutinized for drinking safety. Use your purifiers and filters to clean your water first. Boil if you suspect pathogens. When in doubt, use your test kit and test the water before drinking.

- They come with a pump that can be used to empty the pod.

If you do not have a liner prepared, you can still fill every bath and basin in the house with water. This water can be used for many chores like washing clothes, bathing, flushing toilets and general sanitation.

Dehydration

Be prepared and therefore, be aware of the symptoms of dehydration. If you find yourself in a hot environment with a shortage of water, then you should already be on the lookout for signs of dehydration. Sweating, urination and breathing all cause the body to loose fluids. Everybody is susceptible to dehydration, but children, the elderly and sick people are especially at risk. Dehydration can also lead to heat stroke and cramps which can be absolutely life threatening.

When the body overheats, we get muscle cramps. This is not a serious problem and getting out of the sun and hydrating with water will take care of it.

When the body starts to overheat, and show signs of confusion, nausea and vomiting, headache, or high temperature, then we are dealing with heat exhaustion. This is still not a life-threatening condition, but if no action is taken to lower the

Typical Symptoms Of Mild Dehydration

- Dizziness and weakness
- Dry, sticky mouth; swollen tongue
- Thirst
- Few or no tears when crying
- Dry skin; inability to sweat
- Headache
- Little or no urine, especially from infants
- Muscle cramps
- Constipation

Typical Symptoms Of Severe Dehydration

- Fever and confusion
- Darkened urine color; little or no urination
- No tears when crying
- Rapid breathing and heartbeat
- Extreme thirst
- Low blood pressure
- Unconsciousness
- Sunken eyes
- Dry skin that doesn't bounce back when pinched into a fold; it's lost its elasticity

person's core body temperature, then we have the potential for heat stroke.

Typical symptoms of heat stroke:

- Headache (Don't administer any headache medicine)
- Rapid pulse
- Confusion
- Temperature elevation of up to 105 F (40.5 °C) or more
- Nausea and vomiting
- The skin can turn red due to the blood vessels trying to expel the heat
- Rapid breathing
- Possible loss of consciousness
- Seizures
- The person's skin might be cool, but the core temperature will be high

When a person suffers from heat stroke, immediately take action:

1. Bring the person indoors, out of the sun.
2. Remove the person's clothing and use ice or cold water to lower their body temperature. If water is scarce and an issue, wet a towel and use it to wipe the victim's body down.
3. Use a fan (hand or electric) to help with cooling the body of the victim.
4. Replace lost fluids through hydration. Get electrolytes back into the system.
5. Elevate the person's feet. It

should be elevated higher than the victim's heart.
6. Place icepacks wrapped in towels in the groin area, under armpits and behind the neck. Speaking from personal experience, this is a very effective method of lowering the body's temperature.

The key message concerning heat stroke: **If you stay hydrated, you will most likely not get heat stroke.**

This is where I have to reiterate: **You must store more water than what you think you will need in an emergency. Running out of water should not be an option.**

Conclusion

These days when it comes to water storage, there are plenty of options to choose from. There is of course no perfect solution that will fit everybody's needs. When selecting the right option for you and your family take into account that you might be mobile in a "get out of Dodge" scenario, or that you might be stuck in a "sit and wait" scenario. It is definitely best to be prepared for both types of scenarios. In addition, you will do well to buy a portable, emergency water filter for every family member. A convenient, lightweight option is the straw-type, that filters everything you drink from. They weigh about 50 grams and should be perfect for a group on the move. Keep it within reach if a disaster strikes, for if you have to flee your home this will be a life-saver.

Make sure that all your storage vessels are easily accessible. Your water should be located where you can quickly reach it in an emergency. Find a good balance between using large and small storage containers and arrange them in and around the house.

Make sure that you have access to all container valves and openings and rotate your water frequently. The latest research is pointing at some contradicting evidence concerning the safety of plastics. BPA (Bisphenol A) is a chemical agent added as a hardener to plastics. A common replacement for BPA is to use BPS (Bisphenol S). Once thought to be safe, it now seems that BPS also causes health problems of its own. If you are one of the growing amount of people who are totally against the use of plastics, then you have to consider stainless steel, ceramic, clay or glass. Unfortunately, these materials are all heavy and they are not always considered convenient for emergency storage. It will come down to creating a system that works for you and your family. Make sure that every family member is taken care of and also that they are informed about water conservation and the basics thereof.

Hidden Water Sources

Unknown to some people, you are already storing water in various appliances in your house.

In an absolute emergency and as a very last option, pay attention to the following:

- Your water heater. It has a hot water tank that stores water. Turn off the gas or electricity feed beforehand.

- Some homes rely on pumps with pressure tanks. These tanks are reservoirs filled with clean water.

- Melt ice cubes in your freezer and take advantage of the ice buildup at the back.

- The house pipelines are filled with water. Drain them by opening the highest faucet first and then go to the lowest faucet to drain the pipes. You might have to close off the main valve first.

- Toilet top tanks also store water. Boil and filter first.

- Swimming pools usually have chlorinated water. In an emergency, act quickly and fill every container you can find with water from your pool. Filter and boil it before drinking. The water in your pool will become contaminated very quickly if it's not treated and after a few days it will be like drinking swamp water.

- Aquariums have water. Filter and boil the water before you think about consuming it. This is contaminated water, so only consider this option as a last resort.

Take care to still boil or filter water obtained from these locations.

If you suspect any form of chemical contamination then discard the water immediately.

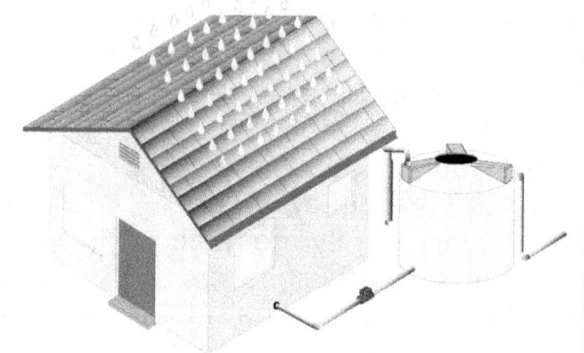

Chapter 4

Collection And Storage Of Rainwater

Collecting rainwater makes sense on so many levels and it is an obvious choice whether you live in the city or out in the country. A big benefit of rainwater is the fact that it's free of minerals and salts that cause corrosion in appliances. This is the very reason we sometimes refer to it as "soft" water. Moving outdoors, we see that gardens thrive under rainwater for it contains nitrates and has a nearly neutral (slightly acidic) pH. To benefit from this method and to join the long line of rain harvesters before us, we simply have to look up to the sky, to our roofs. Most roofs are already fitted with gutters and spouts which make them perfect for water harvesting. Take a walk around your house and other structures on your property. Examine the roof and gutters, checking for type of material used, the condition they are in and making notes of where your gutter downspouts are. If you are not familiar with the workings of pipes, fittings, pumps and water flow, consult a local plumber in your area to assist or, at the very least, to discuss your options with. Learning about plumbing is fun and it is also a very rewarding skill that you will use many times over. Inspect the ground around your house for stability and make sure that areas under downspouts are clear and level. Locate your storm-water drain and determine how overflow from tanks will reach the drain. Make a list of all the areas on your land that you want to supply with rainwater.

Be creative, but also be realistic. A simple system is less likely to break down and requires less maintenance.

Rainwater Collection

Rainwater is free, easy to collect and useful for various purposes. Under the right conditions it can provide enough water for a small household or garden. It is however, dependent on weather cycles and therefore it is generally not enough for large-scale irrigation.

Let's look at the numbers required to put an effective rainwater system into action:

Rainfall. The rainfall figures in the geographic region the property is located will simply tell you if it rains enough.

The roof collection area. The roof functions as the collection area and basically the larger the roof area, the more water you can harvest. Know how to size a catchment area.

Tank size. Appropriately sized tanks/cisterns can be used to store the harvested rainwater.

Household size. Does it rain enough to provide for every person (and pet) in the household? How much water do you need? (Numbers can be **estimated** or **calculated**.)

Rainfall

Local weather statistics will provide you with a relative accurate portrayal of monthly- and yearly rainfall, droughts, floods, as well as expected extreme weather patterns. It is best to use rainfall data from the nearest station with comparable conditions. These numbers can be obtained from your local municipality, government offices, or the local Meteorological Department, but a **quick search on a web browser** will also provide you with a realistic picture. When looking at rainfall statistics consider the following:

- ◇ intensity
- ◇ duration
- ◇ distribution
- ◇ visible patterns

These characteristics must be investigated to get an accurate picture and to decide whether your location is suitable. Rainfall is the most unpredictable of variables and to determine the potential rainwater supply for a given catchment, reliable rainfall data is required, preferably for a period of at least 10 years. The number of annual rainy days influences the need and design for rainwater harvesting. For example, the fewer the annual rainy days in a region, the higher the need for a rainwater collection system. In areas prone to droughts big storage tanks will be needed to store rainwater; it can even be feasible to recharge the rainwater into the groundwater for storage.

The Roof Collection Area

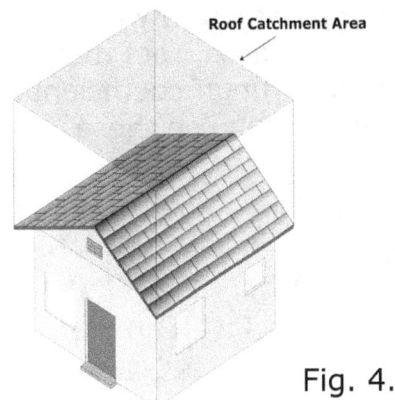

Fig. 4.1

First, let's look at the **catchment area** of the structure for this is the first point of contact for rainfall. Sizing your roof catchment area (fig. 4.1) will allow you to determine how much water you can harvest. Your roof surface area is the size of your roof, calculated in square meters or feet. This number is available on the building plans for the house. Alternatively, you can also use **Google maps** to determine the area of the roof. As a last resort you can measure it by hand with a measuring tape. If you look at figure 4.1 you will see that the angle of the roof's slope does not affect the calculation of the catchment area.

Roof Material – The material of the roof is not a crucial element of the process. Metal is best, but some modern asphalt shingles can also work depending on quality. The main issue with roof material is to make sure that contaminants don't enter your water tanks (see p.67).

Slope – During a rain event the slope of the roof will affect the runoff rate. A smooth and steep roof will shed runoff quickly and it will also ensure the removal of most contaminants. A flatter roof made of a rough surface (like shingles) will cause the water to flow at a slower rate and this raises the potential for contaminants to enter the gutter system. The runoff coefficient for any roof or catchment can be defined as the ratio of the volume of water that runs off a surface to the volume of rainfall that falls on the surface. By using the runoff coefficient of the roof, the water harvesting potential of a site can be estimated (see p.54).

Let's take a look at typical runoff coefficient values for the different roof types.

Type of roof	Runoff coefficient
Concrete roof	0.70
Tiled roof	0.75
Asbestos sheet	0.80
Galvanized iron sheet	0.90

Manual on Construction and Maintenance of Household Based Rooftop Water Harvesting Systems, Report prepared by AFPRO (Action for Food Production) for UNICEF

Now that we know what the catchment area, roof material, slope and runoff coefficient entail, we can use these variables to calculate *(below)* the harvesting potential of your dwelling. *(Easy alternative, use an internet online calculator.)*

Calculations - To accurately calculate the rainwater harvesting potential of a dwelling, we use the following formula:
$V = A \times R \times C \times 0.623^*$ **(gallons)**
$V = A \times R \times C$ **(liters)**

(* The value of 0.623 is a conversion factor that converts inches of rainfall over a square foot area, to total volume of gallons rainwater.)

V = Volume of available water, measured in gallons (gal) or liters (L).

A = Roof catchment area measured in square feet (ft^2) or square meters (m^2).

Length x Width = Catchment area

(No matter if roof is flat or pitched)

R = Yearly or Monthly rainfall, measured in inches or millimeters (mm).

C = Collection surface efficiency: 0.8 is an average. (Meaning the percentage of precipitation that appears as runoff. Smooth material like metal has a high efficiency, clay has a low efficiency. See the table on the previous page.)

Collection calculation for gallons (per month):
V = Catchment area (ft^2) x Rain (inches per month) x **0.8** x 0.623
Collection calculation for liters (per month):
V = Catchment area (ft^2) x Rain (inches per month) x **0.8**

Example (Imperial system):
V (gallons) = A (ft^2) x R (inches) x C x 0.623
1000 ft^2 x 2 inches (month) x **0.8** x 0.623 = 996.8 gallons (month)

Example (Metric system):
V (Liters) = A (m^2) x R (mm) x C
V = 90 m^2 x 50 mm x 0.8 = 3 600 Liters (month)
There is no conversion factor necessary for the metric system.

Note that you can make a **yearly or monthly** calculation with the same formula.

Tank Size

When determining the appropriate size for a water tank, it is crucial to consider the rainwater yield (supply) and/or water consumption (demand). These factors hold significant importance for two key reasons.

Firstly, if the tank is undersized, it means that you won't be able to collect the necessary amount of water. Conversely, if the tank is oversized, you will end up collecting more water than you actually need, resulting in water stagnation, which is far from ideal.

Take note that, for a rough estimation and without delving into precise calculations, you can simply consider an average of 270 gallons/1000 liters per person. Do this if you have a main water line and where rainwater will mostly be used to offset your monthly water bill.

> 270gal (1000 L) per person = Recommended Tank Size

Household Size

Knowing your water usage, helps you understand how much rainwater you need to be able to harvest for day-to-day use.

It is important to realize that water-use throughout the year often varies with the season. Most people use more water in the warmer months for gardening, washing cars, and other outdoor uses. If you conduct your water audit in the winter or fall, you should still consider the additional water you use in the summer months. The American Water Works Association (AWWA) **estimates that the average indoor water-use per person is 94 gallons of water per day**; this does not take into account outdoor water-use (watering lawns, washing cars, etc.).

> 94gal (355 L) per person = Avg. indoor use per person

If you live in an area with mains water, your historical water bills may be a handy way to gain some insight into your family's water consumption habits. Many water utilities provide customers with bills that contain information regarding the amount of water consumed and average daily consumption during the billing period. If the average daily consumption is not provided, you can calculate it by dividing the total amount of water used by the number of days in the billing period. Determine whether your water is measured in cubic meters (m^3), cubic feet (ft^3), gallons (gal), or liters (L) and convert to gallons. For converting into gallons, use the following conversion factors:

m^3 x 264 = gal

ft^3 x 7.48 = gal

L x 0.264 = gal

Calculating Water Use With A Meter

If your water bill does not provide water consumption data, then you can read your water meter to obtain this information. Water meters measure the total amount of water used in your home and are usually located at the property line or in the house. The meter may measure in cubic meters, cubic feet, gallons, or liters. To obtain your water-use over the course of a 24-hour day, read your meter at the same time on two consecutive days. You may want to measure water-use for several days and then calculate a daily average. A hose meter can also help calculate the volume of water you typically use for outdoor tasks.

Estimating Water Use Without A Meter

If you do not have a water meter you can estimate your water consumption. It will be important to measure all water-use, indoor and outdoor, to accurately estimate the quantity of water used. To determine how much water you consume in your home, it is necessary to measure water flow from each fixture in your house:

To calculate flow for faucets (indoor and outdoor) and shower heads, turn the faucet to the normal flow rate that you use, and hold a container under the tap for 10 seconds and measure

Water meters measure the total amount of water used in your home and are usually located at the property line or in the house.

the quantity of water in the container. Multiply the measured quantity of water by 6 to calculate the gallons per minutes (gpm).

To calculate flow for toilets, turn off the water supply to the toilet, mark the water line on the inside of the tank, flush, and then fill tank with water from tap. Measure the volume of water that is required to fill water back up to the water line mark on the tank and record this number. Turn water on to the toilet to resume normal use.

If your appliances or fixtures are relatively new, you may be able to obtain the flow rate from the manufacturer's specifications. Otherwise, use the following averages:

- Washing machine - 41 gallons per load
- Dishwashing machine - 9 gallons per load

Next, measure how many times per day or how many minutes each day you use each fixture or appliance. Multiply the water flow per fixture by the minutes per day the fixture is used.

Multiply the flow average for each appliance by the number of times the appliance is used each week. Don't forget to include the amount of time you use outdoor faucets each day.

According to the American Water Works Association (AWWA) the following table provides the average water-use for conventional and low-flow appliances.

Appliance	Water Use in Gallon
Vintage Toilet*	4-6/flush
Conventional Toilet**	3.5/flush
Low Consumption Toilet***	1.6/flush
Conventional Showerhead*	3-10/min.
Low-Flow Showerhead	2-2.5/min.
Top-load Washer	40-55/load
Front-load Washer	22-25/load
Dishwasher	8-12/load

*Manufactured before 1978
**Manufactured from 1978-1993
***Manufactured after January 1, 1994

Determine The End-Use Of Your System

When it comes to Rain Harvesting, there are two operative words we use when determining the end-use of our system – **Quantity** and **Quality**. How much water you need, and how you plan to use the water are the two main questions that will determine the effectiveness of your rainwater harvesting system.

Collecting rainwater means that you are taking a step closer towards self-sufficiency. An obvious incentive for you to collect rainwater is that it's free and that it reduces your water bill. Dry areas with depleted groundwater sources are perfect for rainwater harvesting, but even if you live in a water-rich environment, you can still conserve water through rain-harvesting. By not collecting rainwater, you are allowing this precious resource to go to waste and to flow into the ground or down the road into the storm-water drains. The cost of a rainwater harvesting system can vary depending on several factors, including the size of the system, the complexity of the installation, and the specific components used. However, rainwater harvesting systems can be affordable, especially if they are designed and installed with cost-effectiveness in mind.

Value For Money

A cost estimation is crucial if you are working on a budget and if you plan on using new system components. As with anything, the cost will vary on how big and how effective you want your system to be. It's necessary to point out that the operation and maintenance costs remain low after the initial investment. Let's look at some setup examples and costs:

(Remember that in most cases installing a water meter on your property is free.)

1. **Rain barrel systems**, which collect rainwater from a single downspout and store it in a barrel, can be an affordable option for those interested in rainwater harvesting. A basic rain barrel typically costs between $50 (used) and $200, making it a relatively inexpensive investment. Additionally, the installation and maintenance of a simple rain barrel system are also quite easy and cost-effective. One of the cheapest options is to simply place a rain barrel outside during a rain event to collect the water. (See p.112)

2. IBC Totes or Intermediate **Bulk Containers (IBCs) are large containers origin**ally used for shipping liquids. They have a higher storage capacity than rain barrels, typically ranging from 275 to 330 gallons (1,040 to 1,250 liters). The cost of IBC totes can vary based on factors such as condition (new or used) and material. Prices generally range from $100 to $500. (See p.112)

3. **Polyethylene (Poly) tanks** are above-ground tanks specifically designed for rainwater storage. They come in various sizes and shapes, ranging from a few hundred to several thousand gallons (1,000 to 15,000 liters). Prices for poly tanks depend on the capacity, quality, and features, and can range from a few hundred to several thousand dollars. (See p.107)

4. **Underground rainwater storage tanks** are installed beneath the ground, saving space and providing a more discreet option. These tanks can be made of various materials such as plastic or **fiberglass**, and their capacity can range from a few thousand to tens of thousands of gallons (10,000 to 100,000 liters). Costs for underground tanks can be higher due to excavation and installation requirements. Prices generally start from a few thousand dollars and can go up significantly depending on size and material. (See p.108)

5. **Complete installation.** Collecting and storing rainwater from an entire roof and utilizing it for multiple purposes throughout a building, like toilet flushing, laundry, and irrigation, can lead to substantial long-term savings on water bills. However, such a system may require a larger investment.

   a. A **dry system** tank setup, which is a typical setup that collects rainwater from a roof, can cost between $1,000 and $5,000. (See p.96)

a. For a more complex **wet system** setup, which collects rainwater through a network of underground pipes, the cost can range from $8,000 to $15,000, including installation costs. (See p.97)

Overall, the cost-effectiveness of a rainwater harvesting system will depend on the specific circumstances and end-goals of the user.

Potability

A discussion around potability of your rainwater is essential. Back in the day we were able to use rainwater for drinking, but these days it is advisable to test a few samples first to determine the quality. Depending on where you live, the two main issues with rainwater are contaminants and water purity.

The Centers for Disease Control and Prevention (CDC) recommends:

Rainwater might not be safe for household use without additional treatment.

Before using collected rainwater for drinking, bathing, or cooking, consider whether treatment is needed to make it safe. Testing the water can determine if there are harmful germs, chemicals, or toxins in it.

Water treatment options include:
- filtration
- chemical disinfection
- boiling

Filtration can remove some

> Before using collected rainwater for drinking, bathing, or cooking, consider whether treatment is needed to make it safe.

germs and chemicals.

Treating water with chlorine or iodine kills some germs but does not remove chemicals or toxins. Boiling the water will kill germs but will not remove chemicals. Using a simple device called a first flush diverter to remove the first water that comes into the system may help avoid some of these contaminants. The amount of water that should be removed by a first flush diverter depends on the size of the roof feeding into the collection system.

Consider adding a screen to the water inlet or emptying the rain barrel at least every 10 days to prevent mosquitoes from using the rain barrel as a breeding site.

Some people add purchased, treated water to the rainwater they collect in their cistern. This may make the treated water less safe.

Issues With Contaminants

When we track the journey of a raindrop (fig. 4.2) we see that the rain droplets move with the clouds through our atmosphere and that they cover great distances. On this journey, they come into contact

The Water Cycle

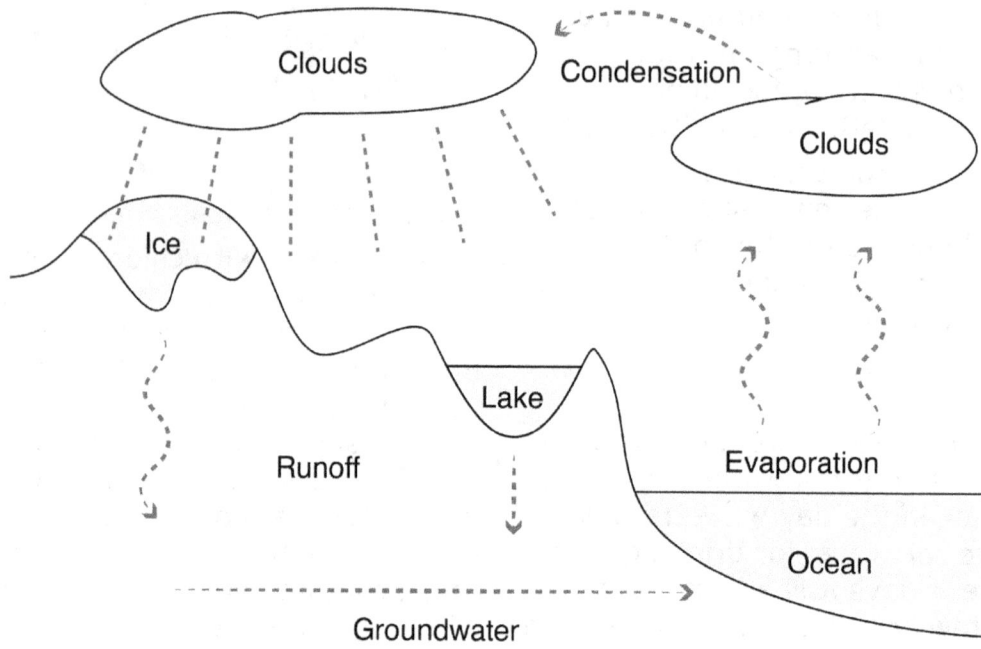

Fig. 4.2

with various chemicals and pollutants. Chemicals can be simple particles coming from chemical plants, paper mills, incinerators, power plants or even more seriously, radioactive particles from nuclear plants. Sadly, the day has come that acid rain, a scenario typical of science fiction movies, has become a reality of daily life for some.

Besides our atmosphere, the rain also has to fall and collect on various surfaces. These surfaces (normally rooftops) can be coated in chemicals or might simply just be dirty from bird droppings, chimney runoff, mold, etc.

These contaminants pose a real health risk and caution should be taken in this regard.

Issues With Quality

Water quality has become more of an issue of late. Whenever we discuss rainwater, the argument comes up that, in a perfect world, potable water should contain natural minerals, which are beneficial to human health. The main concern with rainwater, also called soft water, is that it's basically too clean and void of minerals since it accumulated through evaporation as part of the earth's water cycle. If you happen to live in an area with an abundance of clean rainwater, some modern-day experts advocate that ideally it should not be consumed for long periods of time.

However, many farmers will disagree and they swear by rainwater and have been drinking it for 20-30 years without any issues. Some people recommend supplementing your diet with minerals to compensate for the lack thereof in rainwater. Others recommend leaving rainwater in a clay pot that naturally adds minerals from the clay to the water.

These are theories that have not been tested properly and I cannot recommend or reject them. When it comes to this discussion, every landowner can make the decision for him- or herself.

Living in a remote area, far away from industrial activities and pollutants, can be a fortunate circumstance when it comes to accessing clean rainwater. If the water in your area is suitable for consumption without the need for extensive cleaning processes, then you have a reason to celebrate. Consider yourself lucky to have access to naturally clean rainwater in such circumstances.

Indoor Use

Once your tank is filled with rainwater, you will find that using the harvested water is just an absolute pleasure. It is a great relief knowing that you have water stored for a time of need and that nature provided it for free.

When it comes to indoor use, we see that a big benefit of rainwater is the fact that it's free of minerals and salts that cause corrosion in appliances. This, as mentioned earlier, is the very reason we sometimes refer to it as soft water.

Rainwater can be used indoors in a number of ways, including:

- ◊ **Flushing toilets:** Rainwater can be collected and used to flush toilets, which can help reduce water consumption and save money on water bills.
- ◊ **Laundry:** Rainwater can be used for washing clothes, especially if you use a non-toxic detergent.
- ◊ This can save money and reduce your environmental impact.
- ◊ **Showering and bathing:** Rainwater can be used to shower and bathe, but it may need to be filtered and treated before use to ensure it is safe and clean.
- ◊ **Watering plants:** Rainwater can be used to water indoor plants, which can help reduce the amount of tap water used.
- ◊ **Cleaning:** Rainwater can be used for cleaning floors, windows, and other surfaces.

Outdoor Use

Moving outdoors, we see that gardens thrive under rainwater for it contains nitrates and has a nearly neutral (slightly acidic) pH.

This is where you can start to get creative. Visualize your dwelling as your water harvesting system feeds different parts of the house and land. Use your senses to determine what will work and what not. Start small and make sure that you can build, repair and maintain the whole system. Set small goals at first and become master of your creation. Be conscious of waste and try to conserve as much water possible.

For gardening and irrigation, allow water to flow gently and to infiltrate the soil. Avoid erosion by having an overflow route for water in heavy downpours and try to utilize the overflow as a resource.

Use the water as a means to create a living area of growth; this is a resource from which you can harvest food, feed livestock and also produce other resources.

Here are your options:

Landscape irrigation: One of the most common uses of harvested rainwater is for landscape irrigation. It is suitable for irrigation since the water does not have to be treated first and because all minerals have been removed which is good for root health. Remember that the soil adds minerals back for the roots to absorb. Drip irrigation is an efficient way to water plants without wasting water. Don't flood your garden with water. Calculate how much water your plants need and what is the best delivery system.

Attract wildlife: Rainwater is also a good way to attract wildlife. Lush areas create an abundance of birds and insects, which in turn increases pollination and regeneration of the land. Man-made ponds attract wildlife and can be a supplemental source of water for these animals.

Livestock: Farm animals can also benefit from rainwater.

Fire protection: Having water at hand means that you can create a system for fire protection. Water quality does not play a role when fighting fires.

Building: When building, rainwater can be used to mix cement or concrete, and also as a dust suppressor and equipment cleaner.

Heaters and pools: Not only can we use rainwater for

showering, washing, cleaning and watering the garden, but it can also be used for your water heater or swimming pool.

Car washing: In regions where the local water supply comes from a desalination plant, it is important to be aware of the potential for corrosion. In such areas, rainwater can be a preferable alternative that should be considered.

Recharge: Rainwater can be used to recharge groundwater. The basic idea behind groundwater recharging is to collect and store rainwater during the rainy season (abundance), and then to use this water to recharge the groundwater aquifers.

Recharge

Rainwater can be effectively used to recharge groundwater through various methods.

Here are some of the more common approaches:

- **Rainwater Harvesting Pits or Recharge Wells:** Excavating pits or drilling wells in suitable locations allows rainwater to directly infiltrate into the ground. These structures are designed to capture and store rainwater, allowing it to percolate into the soil and replenish the groundwater table.
- **Recharge wells:** A recharge well is a narrow borehole dug

Having water at hand means that you can create a system for fire protection. Water quality does not play a role when fighting fires.

in the ground and lined with concrete rings.

The rainwater is collected from the rooftop or a recharge pit and allowed to percolate into the recharge well.

The water then flows down into the aquifer, recharging the groundwater.

- **Infiltration Basins or Swales:** A shallow trench is dug in the ground and filled with coarse sand and gravel. The rainwater is collected from the rooftop or a recharge pit and allowed to percolate into the trench. The water then percolates through the sand and gravel, recharging the groundwater.
- **Permeable Pavement and Pervious Surfaces:** Using permeable materials for pavements, driveways, and other surfaces allows rainwater to seep through the surface and infiltrate into the ground. This promotes groundwater recharge by preventing surface runoff and allowing water to reach the underlying soil.
- **Rooftop Rainwater Harvesting:** Collecting

rainwater from rooftops and directing it to storage tanks or directly into the ground can contribute to groundwater recharge. Gutters, downspouts, and pipes are used to channel rainwater to the desired location, such as a recharge pit or well.

- **Constructed Wetlands:** Wetlands can act as natural filtration systems and provide opportunities for groundwater recharge. By directing rainwater into constructed wetland areas, the vegetation and soil help filter the water and facilitate its percolation into the groundwater.
- **Check Dams and Contour Bunds:** Constructing small check dams or contour bunds along slopes or drainage lines can help slow down and capture rainwater runoff. These structures facilitate water infiltration, allowing it to percolate into the ground and recharge the aquifer.

It's important to note that the suitability and effectiveness of these methods may vary depending on factors such as soil characteristics, hydro-geological conditions, rainfall patterns, and local regulations.

Prior to implementing any groundwater recharge measures, it's advisable to consult with local water authorities, hydro-geologists, or relevant experts who can provide guidance specific to your region's conditions and regulations.

Overall, rainwater harvesting is an effective way to recharge groundwater, and it can be implemented in a variety of settings, including urban, suburban, and rural areas.

Conclusion

In areas where the local government or municipality does not provide main, municipal water or where water is just scarce, rainwater can provide an alternative option. Through careful planning and maintenance you can access remote areas on your land by utilizing pumps and cisterns.

Overflow and storm-water management is very important. When the precipitation rate is higher than the absorption rate (into the ground) it creates water runoff that eventually ends up in the rivers, lakes and oceans. Avoid erosion of your soil by creating an intelligent overflow system that allows the excess water to flow where it can be absorbed, channeled or stored (ponds).

The yearly mean average of rainfall will give you an estimate of how reliable rain-harvesting in your area will be.

Determine if the rainwater in your geographical area is affected by contaminants and if it needs purification or not.

The Conveyance System

This system consists first and foremost of the **roof** which acts as the catchment area and which allows rainwater to flow down into **gutters** and **downspouts** from where it's funneled into storage **tanks**. Houses are perfect for rain harvesting. Depending on the location, a house with a large roof of 2000 ft² (190m²) can provide about 50,000 gallons (190 000 liters) of water per year.

Common sense dictates that we can collect water by simply placing containers outside; more containers will equal more water. Although this can be quite laborious and time consuming, it is doable when the situation demands it. Makeshift structures with tarps can also be used to collect rainwater and this is often done in tropical zones. A better option is to harvest the water straight from the top of your house with the help of a conveyance system. This chapter will delve into the functioning of the conveyance system, exploring how a well-designed system can effectively transport water to your storage tank. Let's examine the catchment area, with a particular focus on rooftops, gutters, and downspouts.

Rooftops

The roof of a building or house is the obvious first choice for rainwater catchment. Roofs are usually the largest surface area of a building, which makes them ideal for catching rainwater. The larger the surface area, the more water you can collect.

Some other benefits include:
- Roofs are typically sloped, which means that rainwater runs off easily and quickly. This allows for a more efficient collection system, as water can be directed into gutters and downspouts.

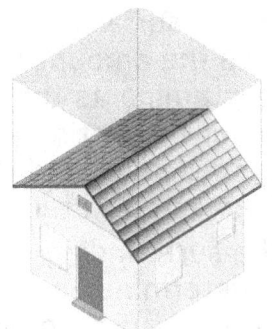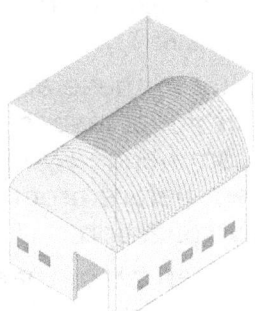

Roof Catchment Area

Fig. 4.3

Collection And Storage Of Rainwater

- Roofs are usually made of materials that are clean and free of contaminants, which means that the rainwater collected from them is typically cleaner and of higher quality than water collected from other surfaces.
- Roofs are easy to access, which makes it simple to install and maintain rainwater harvesting systems.

Overall, roofs are an excellent choice for rainwater harvesting. If you need additional capacity consider barns, garages or sheds to add volume to your system. Do not fret over the shape of the roof; round, flat or A-frame roofs all work well and will provide you with the footprint to harvest your water (see fig. 4.3).

The roof of each dwelling however, needs to be inspected before it can be deemed a proper catchment area.

Look for the following during the initial inspection:

◊ Lead paint and flashings
◊ Flaking or corroded material
◊ Cracked tiles
◊ Inferior roofing material (asbestos, treated timber, tar, gravel)

If any of these are found during your initial inspection, it is recommended you replace or fix them before progressing with any collection.

Another factor that we have to consider is roof design, and especially **roof pitch**. The roof

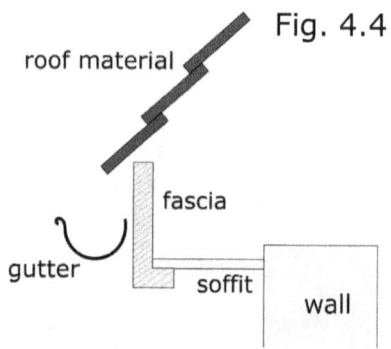

Fig. 4.4

pitch (fig.4.4), which is the angle or slope of a roof, can affect the rainwater harvest process in several ways.

Water Collection Efficiency: Roofs with steeper pitches tend to collect rainwater more efficiently. The angle of the roof allows water to flow more quickly and effectively into gutters and downspouts, minimizing runoff and increasing the amount of water that can be harvested. A steeply pitched roof also helps to reduce the chances of water pooling or stagnating, which can lead to leaks or damage.

Self-Cleaning: Roofs with steeper pitches are more likely to be self-cleaning. As rainwater flows down the roof, it helps to wash away debris, leaves, and dirt that may have accumulated on the surface. This can minimize the amount of sediment and contaminants that enter the rainwater harvesting system, leading to cleaner and higher quality harvested water.

Next, inspect the surrounding environment to determine external factors that can have an effect on water quality:

- Overhanging trees
- Animal nests
- Moss or heavy organic build up anywhere on the roof surface

If any are found, it is recommended to get the roof professionally cleaned. Keep in mind that neighboring industry or agriculture activities can cause contaminants to be carried across to your land and these particles can stick to your roof. If this is the case you will need to investigate the seriousness of the situation. Once the roof collection area is deemed suitable to capture rainwater, you can move forward to the next step, which is gutter selection.

Roof Surfaces For Harvesting Rain

1. Metal is the best texture because it is smooth and facilitates the efficient flow of water. Metal roofs are known for their durability and longevity. They can withstand harsh weather conditions, including heavy rain, strong winds, and UV exposure, without significant degradation. This durability ensures that the roof remains intact and prevents potential contamination of harvested rainwater.

Metal roofs have a smooth surface that allows rainwater to run off easily, effectively self-cleaning the roof during rainfall. This helps to minimize

Overall, roofs are an excellent choice for rainwater harvesting. If you need additional capacity consider barns, garages, or sheds.

---

the buildup of debris, leaves, or other contaminants that could affect the quality of harvested water.

Metal roofs are often lightweight, making them easier to install compared to heavier roofing materials. Their flexibility and compatibility with different roof designs and pitches allow for efficient rainwater collection system installation. Consider steel (galvanized or stainless) and similarly tin, aluminum or copper. Galvanized roofs are coated in zinc and you'll have to check the manufacturer's specs to see if the levels are within acceptable limits. Similarly, old tin roofs consisted of a sheet of metal that is covered in lead and tin. You don't want lead in your water. Modern tin roofs use a combination of tin and zinc. Make sure the material you are using is safe to use.

2. Clay or Concrete Tiles are porous which can cause as much as 10% of loss, but they are still a very good option for harvesting potable water. Clay tiles are known for their durability and longevity. They are resistant to weathering, UV exposure, and extreme temperature fluctuations. A well-maintained clay tile roof can provide a reliable and long-lasting surface for rainwater harvesting.

Similar to metal roofs, clay tiles have a smooth surface that promotes water runoff and self-cleaning during rainfall. This helps to minimize the accumulation of debris, leaves, or other contaminants on the roof surface. Clay tiles are commonly used on pitched roofs, which are ideal for rainwater harvesting due to their efficient water collection capabilities. Make sure the clay tiles are glazed and well-fired to improve the runoff coefficient. Clay tiles are prone to moss and algae growth so be sure to clean them often (yearly).

3. Slate is expensive, but it is very suitable for potable water, provided it's not coated with toxins or chemicals.

Slate is a dense and non-porous material, which means it has a minimal capacity to absorb water. It is a natural stone material that does not contain harmful chemicals or additives. Slate roofs are known for their exceptional durability and longevity. They have a lifespan of 50 to 100 years or even longer when properly maintained.

Slate typically installed on pitched roofs because it has a smooth surface that allows rainwater to run off easily, which promotes self-cleaning of the roof during rainfall, reducing the accumulation of debris, leaves, or other contaminants.

4. Asphalt shingles are very common and some of the well-known brands are generally considered fine for harvesting water. The older shingles were made of asbestos and should be avoided; check with the manufacturer first. Nowadays many shingles are imported from China where they are using a faster and cheaper manufacturing process. Look at the manufacturer's specs to see which materials are used and if they can potentially leach contaminants, and also if they release grit. The adhesives that are used during the installation process present a potential hazard and it is recommended to wait two to three years before you use new shingles for rainwater harvesting; this provides time to off-gas.

Asphalt shingles have a porous surface that can absorb water. This can lead to increased water retention on the roof, creating an environment for the growth of algae, moss, or mold. The presence of these microorganisms can negatively affect the quality of harvested water. Cleaning asphalt shingles can be challenging due to their textured and porous surface. Accumulated dirt, debris, or organic matter can be more difficult to remove effectively, increasing the likelihood of contaminants entering the harvested water. In general, the more common shingles should just be used for irrigation. Only when the manufacturer specifically states that it's safe for potable use can you consider shingles.

5. Wood shingles, tar, and gravel

These roofing materials are rare, and the water harvested is usually suitable only for irrigation due to leaching of compounds. (Don't use treated timber, asbestos, or a surface with exposed lead flashings.)

Gutters

Gutters are designed to capture and channel rainwater from the roof into a downspout, which directs the water into a storage tank. Gutters and downspouts help to prevent water damage to the roof, walls, and foundation of a building by directing rainwater away from the building. This is important because water damage can be costly to repair and can lead to other problems, such as mold growth. Gutters are easy to install and can be customized to fit the specific needs of a building. They can be made from a variety of materials, and come in different shapes and sizes. Gutters require regular maintenance to ensure that they are functioning properly. This includes cleaning out leaves and debris, checking for leaks, and repairing any damage. Gutters should be fitted to the edge of the roof area of all structures on the dwelling.

The gutters need to be inspected for:
- Flaking material
- Lead paint
- Corrosion
- Animal droppings or nests
- Moss or organic build up
- Pooling of water in the gutter (mosquito hazard)
- Improper installation
- Damage

Most of these problems will be visible with the naked eye, but to check for pooling of water you will have to do the following:

1. Get a hose out to fill the gutter.
2. Once filled, turn off the hose and let it drain.
3. Pooling will be visible if present.

Pooling is often caused by improper installation meaning that one of the support brackets (hangers) were installed too high and this caused an irregular gutter slope. If you are satisfied with your gutter inspection we can move on to different gutter characteristics.

Gutter Material

Gutters are made of various materials, but vinyl and aluminum are the two most common ones. The material of the gutter can affect the quality and quantity of the harvested water, as well as the durability and maintenance requirements of the gutter system. Let's take a look at each individual material.

Vinyl Gutters

Material: PVC or plastic

Lifespan: 10-20 years

Installation: Light and easy to install and perfect for the DIY enthusiast.

Recommended: These gutters are not recommended for very wet conditions (steel or copper would be better).

This is the gutter that most hardware stores will recommend.

Galvanized Steel Gutters

Material: Galvanized steel

Lifespan: 20-30 years and very durable, but may rust if not maintained

Installation: Heavier than vinyl gutters and they require professional installation due to the need for soldering.

Recommended: These gutters are recommended for homes that experience heavy rainfall and wet weather.

Aluminum Gutters

Material: Aluminum

Lifespan: 20-30 years, rust-resistant

Installation: Light and easy to install and perfect for the DIY enthusiast.

Recommended: These gutters suit most conditions, but aluminum can be brittle and drastic temperature changes can really affect your aluminum gutters. This causes the aluminum to contract and expand which can cause the material to crack and fall into disrepair much more quickly than anticipated.

Zinc Gutters

Material: Pre-weathered zinc, a corrosion resistant metal which won't suffer from harsh weather conditions.

Lifespan: 80 years, sturdy, rust-resistant and use a self-sealing patina to avoid formation of any scratches or cracks.

Installation: Professional installation recommended because of its high contraction and expansion rate when temperatures change. Seams are soldered, but the process is more difficult than with copper.

Recommended: Zinc gutters are a low maintenance option, but they aren't the best choice if you live near the coast. That's because the high salt content in the air might stain zinc gutters.

Copper Gutters

Material: Made of copper sheet, which can be coated with zinc or acrylic

Lifespan: Never rusts or needs painting; should last 100 years in any climate. They do not warp, bend in extreme weather conditions.

Installation: Professional installation, seams are soldered.

Recommend: These are very durable gutters and they give your home a unique look with a beautiful shine and traditional style. This is the most expensive option and not recommended for DIYers.

Gutter Types

Gutters come in different sizes. Surprisingly, gutter sizing is not as simple or straightforward as you would think. If it's an option, consult an expert about local plumbing codes to determine the appropriate gutter size for your area and your specific roof. Plumbing codes usually use the vertically projected roof area when calculating for drainage design.

It also depends on:
- the rain patterns, and
- how often your area experiences rainstorms, and
- the intensity of rainstorms

This is where the slope and size of the roof will have to be factored in.

Gutters come in various shapes and sizes, but every area will have its go-to gutter type that works in that specific environment. Go to your local hardware store and discuss which gutters will best fulfill your needs. More often than not they will recommend either a Half-round gutter or a K-style gutter.

Let's look at all the gutter types.

Half-Round Gutters

These are very common gutters (fig. 4.5) used all over the world. They feature a semicircular design and a curved lip. Due to the rounded design, they feature round downspouts.

Half-round gutters come in 5-inch and 6-inch widths. This simple, minimalist style of gutter is preferred for old historic homes that used a similar gutter back in the day.

Good:
- They are found in aluminum, vinyl, copper, steel, galvalume, and zinc
- Fairly easy to clean (with chemicals)
- Easy to replace
- They limit debris buildup

Bad:
- They have seams which can cause leaks or pooling
- Can accumulate moss build up

Half-round gutter

Fig. 4.5

K-Style Gutters

K-style gutters (fig. 4.6) are a very common style of gutter and they are also found all over the world. They are fairly easy to install and thus suitable for self installation.

These gutters come in 5-inch to 6-inch widths, and they tend to feature rectangular downspouts. Due to their flat backs, you can nail K-style gutters directly to your fascia boards without brackets.

Good:

- They are found in copper, aluminum, galvanized steel, galvalume and vinyl
- They are square and they hold more water than rounded gutters
- They're also tougher, doing a better job of standing up to water pressure, winds, and other stresses without bending or protruding
- They don't need brackets to fasten to fascia boards

Bad:

- They are harder to clean and require more maintenance
- They gather more debris

Fascia Gutters

Fascia gutters (fig. 4.7) are installed directly onto a fascia board, or the siding panel that is between your gutter and the exterior wall of your home.

Fascia gutters are often larger than K-style or half-round gutters, making them perfect for extreme weather and homes and buildings with large roofs given their ability to handle large and sudden rushes of water. They are also extremely secure and will hold up to violent storms and adverse weather conditions.

Good:

- Can handle heavier water flow due to larger design
- Great-looking, seamless gutters
- Attach securely to your home's fascia board
- Can handle extreme weather

Bad:

- Can be difficult to install and you'll probably need to hire a

K-style gutter

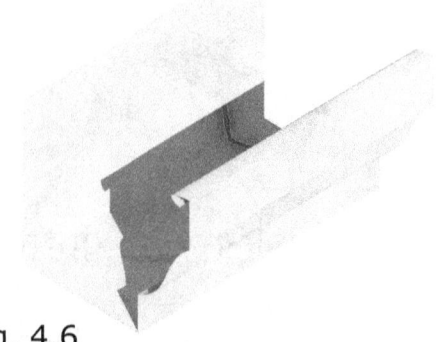

Fig. 4.6

Fascia gutter

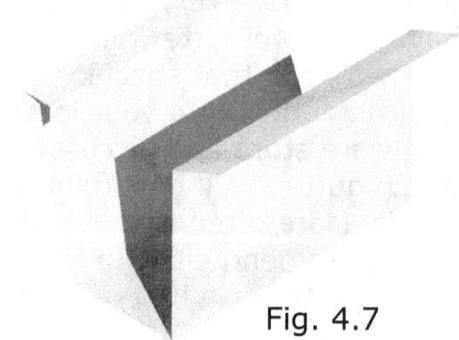

Fig. 4.7

professional
- The increased width means frequent cleaning (although gutter guards can help)
- Quite expensive

Box-Style Gutters

These are gutters (fig. 4.8) that are often seen on industrial or commercial buildings, but they can be tailored to fit a residential building too. These are very wide gutters and they can handle extremely heavy rain events. Due to their size, you will need a larger roof to install them.

Keep in mind that, unlike other gutters, box gutters are not hung on your roof's edge. Instead, they use a high back section that tucks under a roof's shingles. Due to this, box gutters must be installed while your home is being built.

Good:
- Can handle very heavy downpours
- Very durable

Bad:
- Quite expensive
- Lack in aesthetic appeal
- Must be installed when house is being built

Gutter Installation

Every gutter style will require a specific installation process. You should receive installation instructions when buying your gutters. When installing your gutters keep the following basics in mind:

- As a general rule, gutters should be at least 5 inches (13cm) wide. Depending on where you live and what kind of rain events you will encounter, your gutters can also be as wide as 10 inches (25cm).
- Gutter length should not exceed 50 ft. (15.2m).
- Gutter supports are brackets (also called hangers) that should be able to carry the weight of a full gutter with water or ice. Provide gutter supports (evenly spaced) for every 3 feet (90cm) of gutter length. In areas with heavy snow, you need a support for every 12 inches (30cm). Gutter supports should also be placed 4 inches (10cm) in from the ends of the gutters.
- The gutter outlet capacity should suit the downspout capacity. It cannot exceed the bottom width of the gutter. The size of the downspout

Box-style gutter

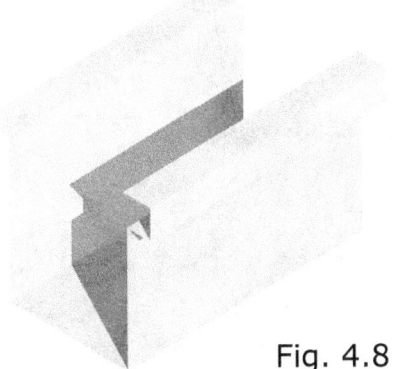

Fig. 4.8

should be constant throughout its length. As a general rule, provide one square inch of downspout area for every 100 square feet of roof area.

- To drain, gutters need to be sloped towards the downspout at an angle. Having a 1 to 2 inch (2½-5cm) slope over 40 feet (12m) is sufficient.
- If the gutter run measures longer than 40 feet (12.2 m), it should be positioned to pitch down from the middle, aimed toward a downspout at each end. If the gutter is shorter than this length, it will slope down to the left or right, toward a single downspout.
- Keep the front of the gutter ½ inch (1.25cm) lower than the back to prevent water from splashing against the building.
- The golden rule for placement height of your rain gutters is that they should always be placed about three inches (7½cm) below your roofline. If they are placed too high, they can allow runoff water to spill out over their backside.
- To prevent UV sunlight breakdown, you can paint PVC gutters and pipes.

Another factor that we have to consider is **roof pitch**. The roof pitch (fig. 4.9), which is the angle or slope of a roof, can affect the placement height of

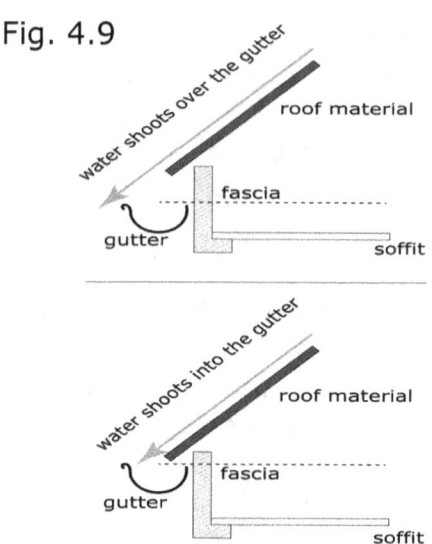

Fig. 4.9

gutters in several ways.

- Firstly, if the roof has a steep pitch, the gutters will need to be installed higher up on the roof to ensure that they can effectively catch and redirect rainwater. This is because a steeper roof pitch means that rainwater flows more quickly and forcefully down the roof, and the gutters need to be positioned higher to ensure

You Will Need

- Downspout/elbow sections
- Gutter sections
- End caps
- Bracket hangers with screws
- Ladder
- Measuring tape and pencil
- Hacksaw
- Spirit level
- Drill
- Screws (for metal gutter)
- Tin snip (for metal gutter)
- PVC cement and sealant

that they can capture the water before it runs off the edge of the roof. Take note that gutters that are too high can enable water runoff to drip down their backside, which can cause deteriorating fascia boards.

- Conversely, if the roof has a flatter pitch, the gutters can be installed lower down on the roof since the water flows more slowly and is less forceful. In this case, the gutters can be installed closer to the edge of the roof without risking overflow or other water-related issues.

How To

Installing a roof gutter requires some basic tools and knowledge of the process. Here are the general steps to install a basic roof gutter:

1. **Gather all the material:**
   Keep safety in mind and make sure that you have a sturdy ladder and a second pair of hands to assist. Use the manufacturer's instructions as a guide where available.

2. **Establish the positions of the main components:**
   Sketch an outline of the house on paper and measure the heights of the locations where you plan to place downspouts.
   On the sketch, also mark every spot where you need elbows, corner pieces,

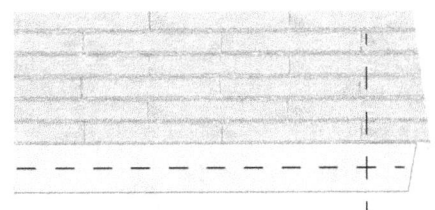

downspout outlets, and downspout extensions. Your outlet hole will connect to the downspout; mark this position on the fascia with a pencil. This position is from where the water will flow through the outlet, into the downspout, and into your tank.

3. **Length of roof:**
   Measure the length of the roof where the gutter will be installed. Make sure to account for any corners or turns that the gutter will need to make.

4. **Cut the gutter:**
   Move down to ground level, and use a saw to cut the sections of the gutter to the appropriate length.

5. **Cut and fit the gutter outlet hole:**
   While your gutter is at ground level, proceed to cut the hole for the outlet. For some **vinyl gutters** (round), you can simply snap the outlet into place, but for most you will need to cut a hole in the gutter at the marked spot using a hole saw or a drill with a hole saw attachment. Apply PVC cement and insert the outlet into the hole of the gutter.

The outlet should fit snugly into the hole. Hold the outlet in place for a few seconds  to allow the cement to set.

For **metal gutters** use a tin snip or drill, and cut out the shape on the gutter to make room for the outlet. Insert the outlet into the hole you created in the gutter, and use sheet metal screws to secure it in place. Be sure to tighten the screws firmly but not too tight to avoid damaging the gutter. The outlet running into the downspout will carry water from the gutter to the tank or drain. (See p.89 for downspout installation)

6. **Gutter end caps:**

   Install the gutter end caps on each end of the gutter. This will prevent water from flowing out of the sides of the gutter.

   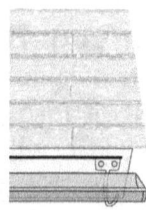

7. **Gutter brackets:**

   Secure you ladder and attach the gutter brackets to the roof or fascia depending on type of gutter used.

   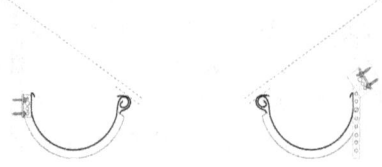

   These brackets will support the weight of the gutter. Remember to slope your gutter (see p.78).

   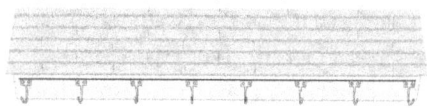

8. **Secure the gutter sections:**

   To ensure the gutter sections are securely attached to the brackets, it is important to follow the guidelines provided by the manufacturer. Depending on the specific installation, different methods may be required. Some installations may involve using screws, while for certain gutter types, they can be snapped into slots on the brackets for attachment.

   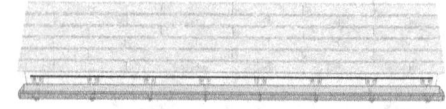

9. **Join gutters:**

   If necessary join two gutters with a gutter union bracket. Seal any gaps or joints in the gutter with sealant. This will prevent water from leaking through any small openings.

10. **Test:**

    Test the gutter system to make sure it is functioning properly. You can do this by running water through the gutter and observing how it flows through the gutter outlet. Make sure it is functioning properly. You can do this by running water through the gutter and observing how it flows through the gutter outlet.

11. **Inspect your system:**

    After a few weeks you should inspect your gutters to look for any indication of failure. From the ground, watch for the following signs:

- Loose attachment of the gutters.
- Lack of slope causing insufficient drainage toward the downspouts.
- Presence of water stains or dripping on the outside of the gutters, indicating water overflow.

Have a look on page 78 at an example of how to install a half-round gutter in a mild climate without any snowfall.

Gutters and Snow

Installing gutters in areas with heavy snowfall requires special attention and consideration to ensure that the gutters can handle the weight and potential damage caused by snow and ice buildup. Some tips for gutters in areas with heavy snowfall:

Right material: The material of the gutter is an important factor to consider in areas with heavy snowfall. Aluminum and steel gutters are strong and durable options that can handle the weight of snow and ice. However, it is important to ensure that the gutters are properly installed and reinforced to prevent damage.

Install a larger gutter: Installing a larger gutter can help to accommodate the increased volume of water and debris that may accumulate during heavy snowfall. A 6-inch gutter is recommended in areas with heavy snowfall, compared to a standard 5-inch gutter.

Install gutter guards: Gutter guards help to prevent snow and ice buildup in the gutters, reducing the risk of damage and clogging. Gutter guards, such as mesh screens and leaf filters, can be installed over the gutter.

Slope the gutter: It is important to slope the gutter properly to allow water to flow freely and prevent the buildup of snow and ice. A slope of at least ¼ inch per 10 feet of gutter is recommended.

Reinforce the gutter: In areas with heavy snowfall, it may be necessary to reinforce the gutter with brackets to prevent sagging or damage.

These reinforcements should be spaced no less than 12 inches (30cm) and no more than 24 inches (60cm) apart (depending on snowfall) and installed securely to the fascia board.

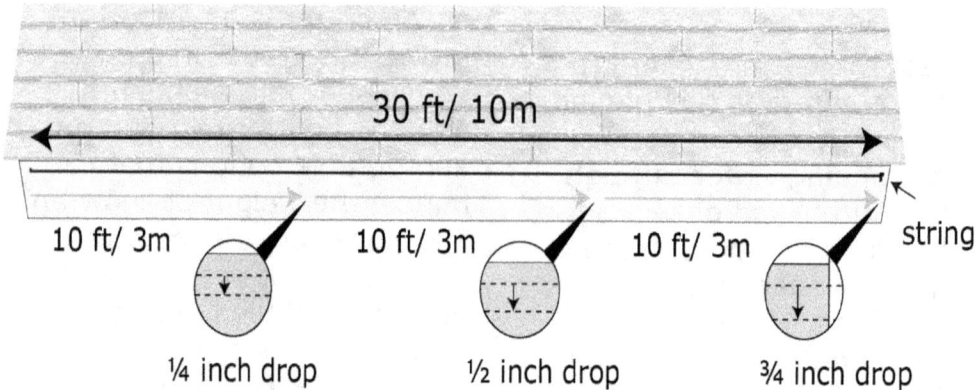

1. Stretch a string (or draw a line) across the fascia where the gutter should be placed. As a rule, a 1 to 2 inch slope over 40 feet of gutter is enough. Use a spirit level to check that there is indeed a fall towards the outlet!

In this example, for a 30ft gutter, a ¾ inch slope is sufficient.

2. Measure 10ft sections, with a ¼ inch drop per section.
3. Use a pencil and space the gutter hangers 3ft apart.

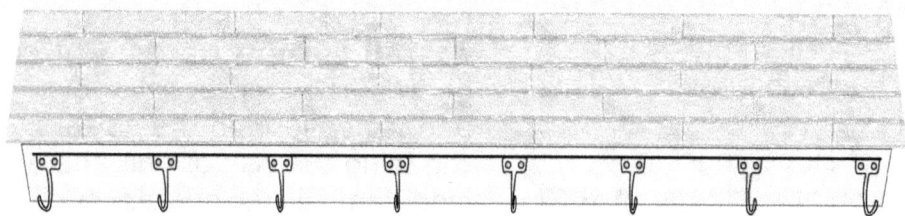

4. The height of your hangers should be around 3 inches below your roofline, but the pitch of your roof might necessitate adjusting this height.
5. Use your drill to make small pilot holes where your screws will go.
6. Secure each bracket into place using the manufacturer's screws.
7. With all the brackets in place, you can now start fitting the lengths of gutter, beginning at the outlet and working back.
9. Carefully lift the gutter length into place and rest it on the brackets.

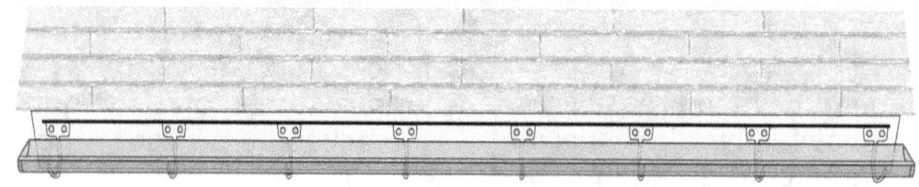

10. Most gutters "snap" into their slots. Follow the manufacturer's instructions.
11. Fit a union bracket to join up the next length of gutter. Use sealant if needed.

The gutter needs to extend out to the end of the roof, past the fascia board.

SIZING GUTTERS
(Imperial measurements)

Diameter of Gutter sloping 1/16 inch per foot	Maximum Rainfall Rate				
	2 inch/hr	3 inch/hr	4 inch/hr	5 inch/hr	6 inch/hr
4 inch	720 ft²	480 ft²	360 ft²	288 ft²	240 ft²
5 inch	1,250 ft²	834 ft²	625 ft²	500 ft²	416 ft²
6 inch	1,920 ft²	1,280 ft²	960 ft²	768 ft²	640 ft²

Adapted from Uniform Plumbing Code, 1997 Edition, International Association of Plumbing and Mechanical Officials, Table 11-3.

SIZING GUTTERS
(Metric measurements)

Diameter of Gutter sloping 5.2mm per meter	Maximum Rainfall Rate				
	50.8 mm/hr	76.2 mm/hr	101.6 mm/hr	127 mm/hr	152.4 mm/hr
101.6 mm	66.9 m²	44.6 m²	33.4 m²	26.8 m²	22.3 m²
127 mm	116.1 m²	77.5 m²	58.1 m²	46.5 m²	38.7 m²
152.4 mm	178.4 m²	119.1 m²	89.2 m²	71.4 m²	59.5 m²

Adapted from Uniform Plumbing Code, 1997 Edition, International Association of Plumbing and Mechanical Officials, Table 11-3.

Collection And Storage Of Rainwater

Gutter Protection

When your rain gutters clog with leaves, pests, and debris, water can back up and cause damage to your roof, walls, and foundation. Gutter guards can address this issue by blocking debris from entering your gutters while allowing water to flow freely. Choosing the best gutter guard from the numerous available styles can be challenging, though.

Below are the types of gutter guards you should be familiar with when assessing your options.

Screen Gutter Guards

Screen gutter guards (fig. 4.10) are perforated metal or plastic sheets that lay on top of the gutter. The material's holes block medium to large debris while still letting water flow through. Because the holes are fairly large, screen gutter guards are inefficient at blocking tiny debris like pine needles, pollen, and shingle grit. Both micro-mesh and mesh gutter guards are a type of screen gutter guard.

A screen gutter guard can be installed professionally or on your own. The screens fit right on top of your old gutters, so there is no retrofitting that needs to be done.

The only issue with screen gutter guards is that they need to be cleaned out at least yearly. A complete gutter cleaning job with screen leaf guard removal is necessary at least once per year to ensure that the screen gutter guards continue to work for years.

Good:
- They are affordable
- Good for DIY installation
- They block large debris from ending up in the gutter

Bad:
- Holes in the screen are fairly large and some debris can pass through
- Will need to be removed and cleaned once per year
- Very light and can be blown off by the wind (solved with professional installation)

Mesh Gutter Guards

Mesh gutter guards (metal or plastic) are quite durable and often perform better than screen guards. They provide a good balance between filtration and water flow. Mesh gutter guards (fig. 4.11) have small holes and will block out most debris and should rarely need to be removed for cleaning. They should be installed by a professional for the whole

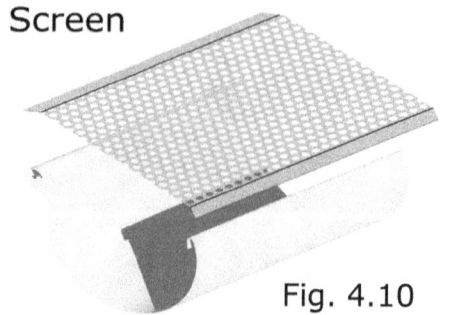

Screen

Fig. 4.10

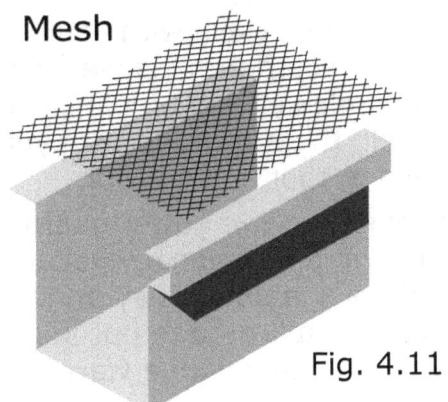

Mesh

Fig. 4.11

installation process can present quite a few challenges. It's also best to avoid products that slide under your shingles, which can jeopardize your roof warranty, and flimsier options, which might blow away in the wind.

With mesh gutter guards the way the debris collects on top typically allows it to flow over and off the side of the roof. Most homeowners would agree that a quick raking around the house's perimeter is better than a monthly gutter cleaning.

These are fairly expensive guards.

Good:

- Considered very efficient gutter guards
- Will block a large variety of debris from large to small
- It helps improve water flow
- Keeps homeowners off roofs

Bad:

- It should be professionally installed
- Can be expensive

Micro-Mesh Gutter Guards

Micro-mesh gutter guards (fig. 4.12) are very similar to mesh gutter guards with the only significant difference being that the holes in the mesh are even smaller. The finely woven mesh screens keep even the smallest debris at bay. If you live in an area where pollen, pine needles, and dirt are common, then this is the gutter guard for you. Experts generally agree that this is the most effective design on the market today. However, their superior filtration comes at the cost of flow rate; it's important to be careful when installing them in high-flow areas, such as roof valleys.

Micro-mesh gutter guards need to be professionally installed, and you will have to be careful when you look at the overall quality of the option you choose; however, for the most part, micro-mesh gutter guards will not disappoint.

Good:

- A very effective type of gutter guards
- Easy to clean and to remove debris
- Allows for decent water flow

Micro-mesh

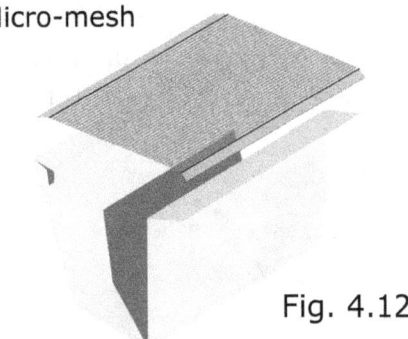

Fig. 4.12

away from the foundation

Bad:
- Higher pricing of installation
- Lower general flow rate

Reverse-Curve Gutter Guards

Reverse-curve gutter guards (also surface-tension guards) is an old but effective design (fig. 4.13) for blocking debris from entering your gutters. They feature a plastic or aluminum hood with a curved edge that leaves a slot between the guard and the gutter lip. Water flows down the hood, around the curve, and back into the gutter while debris slides off to the ground below. This is a great design, but it is not perfect. Surface tension diminishes when water gains too much momentum, making it possible for water to shoot over the edge of your gutter during heavy rain. It also inserts under your shingles (meaning you have to lift them up) to match your roof's pitch, which may conflict with your roof warranty.

Surface-tension gutter guards are visible from the street, which could clash with your home's aesthetics, as well. Most companies offer multiple colors to help them blend in with your existing roof and gutters.

A reverse curve gutter guard must be professionally installed most of the time. They are a bit more complicated to put in but typically come with a warranty.

Good:
- Typically come with a warranty
- Do an excellent job of allowing water to flow properly
- Can handle blocking of the large debris

Bad:
- Must be professionally installed
- Can cause damage to the roof if the installation is not done properly
- During heavy rain you can lose a lot of water as it shoots off the edge
- Visible from the ground
- Clogs can still be a problem even with the high pricing of reverse curve gutter guards

Brush Gutter Guards

Brush gutter guards (fig. 4.14) look like large bristle brushes that sit inside the gutter. The brush gutter guards stand up inside the gutter and allow water to pass through; however, the other debris must remain outside the gutter. Because

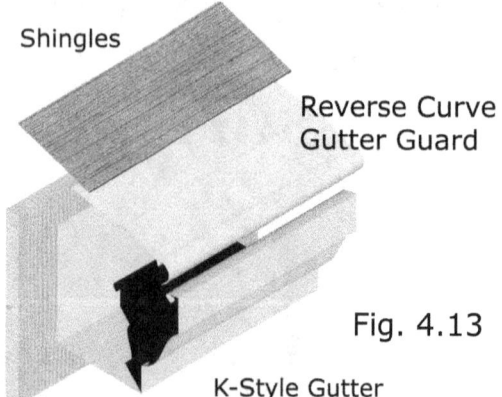

Fig. 4.13

Fig. 4.14

Brush

gutter brushes trap debris instead of shedding it, they have to be removed for cleaning more frequently than other designs. If you have a roof that collects a lot of pine needles and other smaller tree debris, expect to be in for a lot of work with the brush gutter guards. They're also not durable under harsh sunlight, meaning you may have to replace them every few years.

Good:
- Easy to DIY install
- Will blocks large debris from entering the gutter system
- Easy to remove for cleaning

Bad:
- Increased maintenance
- Constant issues with debris blockage
- It can get easily clogged
- Not durable

Foam Gutter Guards

The foam gutter guard (fig. 4.15) is essentially a sequence of triangular foam that fits into the gutter while allowing water to flow through the porous material and that causes debris to stay on top. Foam gutter guards are inexpensive and easy to install but should be a temporary solution, if used at all. When the foam is wet, it presents favorable conditions for seedlings and trees to grow inside the guard, damaging your gutters or possibly catching fire. Additionally, many customers have reported that the foam deteriorates quickly due to sun and other weather conditions

Fig. 4.15

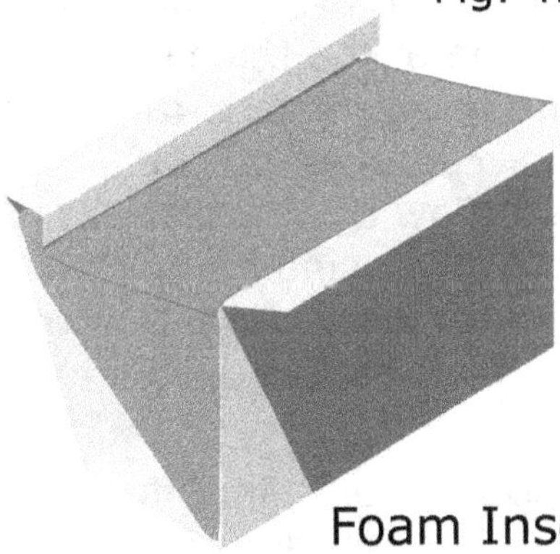

Foam Insert

and requires frequent cleaning.

Good:
- Very inexpensive
- Very light and easy to handle
- It can be installed without the help of a professional
- Easy to get to fit in a variety of gutter styles

Bad:
- Debris will collect on top of the foam (frequent cleaning)
- Removing and cleaning will be a significant project
- Not durable at all

Downspouts

A downspout, also known as a downpipe or a drainpipe, is a vertical pipe that runs from the roof gutters down to the ground or a drainage system. Downspouts are typically made of metal or plastic and are attached to the side of a building, often at the corners.

The primary function of a downspout is to collect rainwater from the gutters on a roof and channel it away from the foundation of the building. Without a downspout, rainwater would accumulate on the roof and flow down the walls, potentially causing damage to the building's foundation or leading to moisture problems inside.

Downspouts come in different sizes and styles, depending on the amount of rainwater that needs to be handled and the architectural style of the building.

They can be round or rectangular in shape and may be concealed within the walls of the building or attached to the exterior.

Overall, downspouts play an important role in managing rainwater runoff and protecting buildings from water damage. By capturing rainwater that would otherwise be lost or wasted, downspouts can also help conserve water and reduce water bills.

Downspout Material

Downspouts can be made from a variety of materials, including metal, plastic, and concrete. The material can affect the quality and quWantity of the harvested water, as well as the durability and maintenance requirements of the downspout system. The most common materials used for downspouts are:

Aluminum

Aluminum downspouts are lightweight, durable, and resistant to corrosion.

They are often used for residential and commercial buildings and can be painted to match the color of the building.

Copper

Copper downspouts are popular for their aesthetic appeal and durability.

They have a distinctive color that develops a natural patina over time. Copper downspouts are often used in historic or high-end buildings.

Galvanized Steel

Galvanized steel downspouts are strong and durable, making them a popular choice for industrial and commercial buildings.

They are coated with a layer of zinc to prevent corrosion.

PVC

PVC downspouts are lightweight, easy to install, and affordable.

They are often used for residential buildings and are available in a variety of colors to match the building's exterior.

Concrete

Concrete downspouts are typically used for large commercial or industrial buildings where strength and durability are important.

They are heavy and difficult to install, but they can withstand extreme weather conditions and are resistant to damage.

Overall, the choice of material for a downspout depends on factors such as the building's style, the climate, and the budget. When you purchase gutters of a specific material, you should also decide on matching outlets, downspouts and diverters, which are all normally made of the same material. This keeps things simple and makes the installation process easier.

Downspout Filters and Diverters

Downspout Diverters are a family of devices which divert water away from the natural flow of a downspout toward the rainwater storage tank.

Some **downspout diverters** provide a filtration function in addition to diverting the water.

Some diverters also eliminate the need for a tank overflow pipe.

A **Downspout Filter** is a device that traps debris to prevent clogs in a storage tank or drainage system.

A filter can be installed at the gutters, in the downspout pipe, or at the downspout outlet. Without a downspout filter, drainage systems can overflow, causing water to pool around a home's foundation and causing costly damage. All downspout filters require some form of maintenance, but some require more than others. There are four main types of products that filter or divert water before it enters your downspouts or while it flows through them:

Strainers

A strainer (fig.4.16) is a type of filter installed in the hole where the gutter and the downspout meet. The purpose of a downspout leaf strainer is to prevent clogs in the

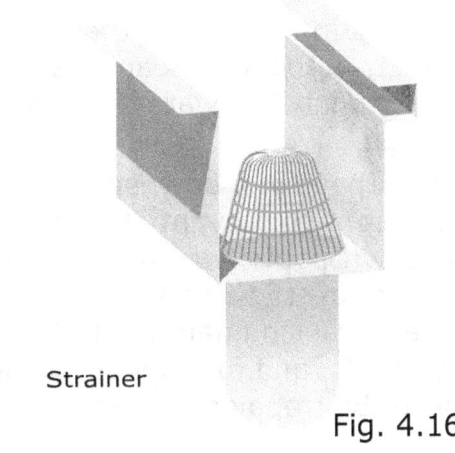

Strainer

Fig. 4.16

downspouts, which can lead to water overflowing from the gutters, causing damage to the roof, walls, and foundation of the house.

They are typically made from plastic, copper, or aluminum arranged in a net-like structure that strains debris from water as it passes through the gutter and into the downspout. Strainers normally are formed in a round, conical shape, allowing debris to pass around them.

Good:
- They are easy to install
- Inexpensive

Bad:
- Do not eliminate the need to clean gutters
- Can become clogged if gutters are not properly cleaned

Leaf Traps

Leaf traps (fig.4.17) function as a type of filter that can be placed anywhere from the middle, to the bottom outlet of the downspout.

As the name suggests, these filters trap the leaves and debris allowing only the water to flow through.

The purpose of a downspout leaf trap is to prevent clogs in the downspouts, which can lead to water overflowing from the gutters, causing damage to the roof, walls, and foundation of the house. By trapping the leaves

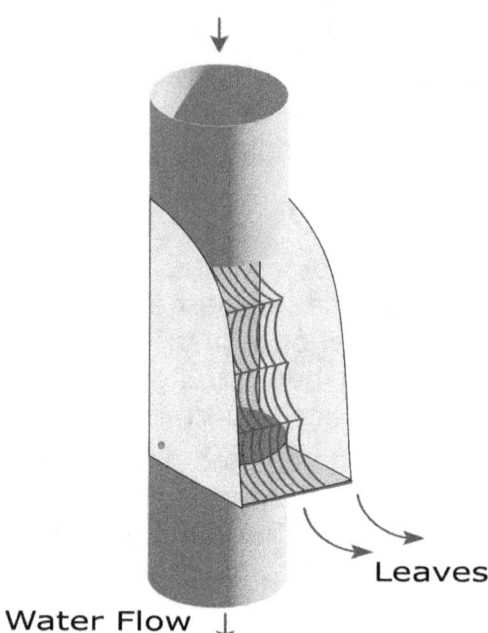

Fig. 4.17

and debris before they enter the downspout, the leaf trap helps ensure that water can flow freely through the gutter system and be directed away from the house as intended.

Open traps are easy to clean (debris is visible), but they hold less debris than closed traps.

Closed traps have a splash lid and are harder to clean, because the debris is not visible and this makes it hard to tell when maintenance in needed.

Good:
- They can handle a lot of debris matter
- They catch debris while allowing the flow of water

Bad:
- Quite hard to install
- They require frequent

maintenance if you have a yard with large trees

Downspout Diverters

Downspout Diverters (fig. 4.18) are a family of devices which, unlike filters, have the option of directing the flow of water away from the natural flow of the downspout.

Downspout diverters are installed directly to the downspout to create a seamless flow of water from the downspout, and in most cases, into the storage tank. The downspout diverter has a damper (a device used to control the flow of fluids in a system) inside which you can manually change to open or close the water flow to one leg or the other. In other words, diverted water is routed to a barrel (or specific location) when the damper is switched one way, or to the downspout when the damper is switched in the other direction.

An interesting function of a diverter is that it automatically handles overflow—when the tank is filled, excess water backs up through the hose and back into the downspout. Some diverters contain their own filters, and others do not require an overflow pipe. The main purpose of a downspout diverter is to prevent water from pooling around the foundation of the house, which can lead to water damage and structural problems.

Good:
- Really good at preventing water waste
- Helps you save money and conserve water

Bad:
- Can only be used on a downspout that fully extend from the roof to the ground
- Does not handle heavy rainfall too well
- Must have appropriate overflow system in place
- Not effective in cold weather with possibility of ice

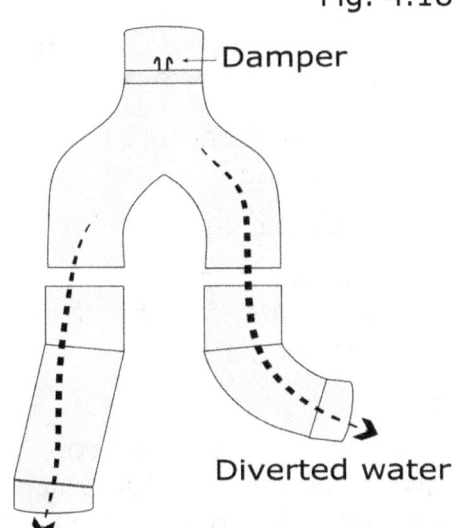

Fig. 4.18

First Flush Diverter

As it rains, water builds up in the gutters. Often gutters are filled with debris containing animal droppings, insects or dead animals like mice or lizards. This means that the first flush of water can be contaminated with

First Flush Diverter

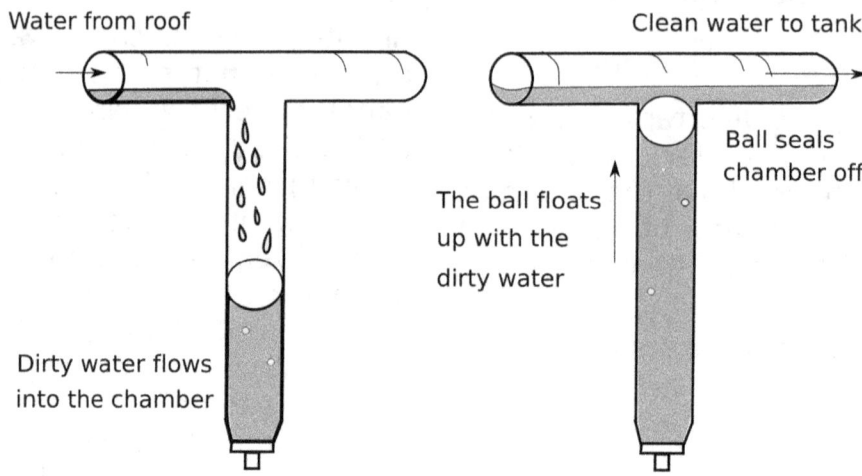

Fig. 4.19

unwanted microbes that you do not want in your tank.

The solution to this dilemma is the First Flush Diverter (fig. 4.19).

This system consists of a chamber (pipe) with a ball float on the inside. As the water flows from the gutter towards the tank, it is diverted into the chamber. Once the chamber is full, the ball float rests on a seat at the top of the chamber and thus it seals off the top of the chamber. This prevents the new water from entering the diverter. Now the rest of the water can flow towards the tank. It is advisable to fit the longest length chamber possible.

Fit a diverter at every pipe that supplies water to a tank. You can install the chamber below the rainhead filter.

Allow enough space underneath the chamber to allow opening of the release valve.

These diverters come in kits and are very easy to assemble (see p.129).

Good:
- Prevents sediment, bird droppings, spiders, insects, mosquito eggs and debris from entering the rainwater tank/cistern
- Improves water quality, protects pumps and internal appliances
- Easy installation, just add pipe and glue
- No mechanical parts
- Low maintenance requirements, but should be cleaned regularly

Bad:
- A damaged First Flush can divert too much water which will result in slow fill rates

Downspout Installation

Installing a downspout on a gutter is a straightforward process that can be completed in a few simple steps. It's always best to follow the manufacturer's instructions.

Take note that if your downspout connects to a fixed location like an underground drainage pipe at ground level, then it's easier to start the process from the ground up, in order for the gutter outlet and the drainage pipe to line up. The reason is that the gutter, where the downspout connects at the top, can be moved into position, but a fixed drainage pipe at the bottom cannot.

Let's look at a standard assembly:

1. **Gather all the material:** Keep safety in mind and make sure that you have a sturdy ladder and a second pair of hands to assist. Use the manufacturer's instructions as a guide where available.

2. **Determine the location of the downspout:** Decide where you want to place the downspout. It's typically best to place the downspout at a corner of the house or at a point where the slope of the gutter changes.

   Mark a 90° vertical line down the wall where the center of the downpipe will be to help guide your installation. You can use a plum line or spirit level. This line for the downspout should be parallel to the building corner.

You Will Need

- Downspout/elbow sections
- Gutter outlets
- Bracket hangers
- Ladder
- Measuring tape and pencil
- Framing square
- Saw
- Spirit level
- Drill
- Screws
- Tin snip
- PVC cement and sealant

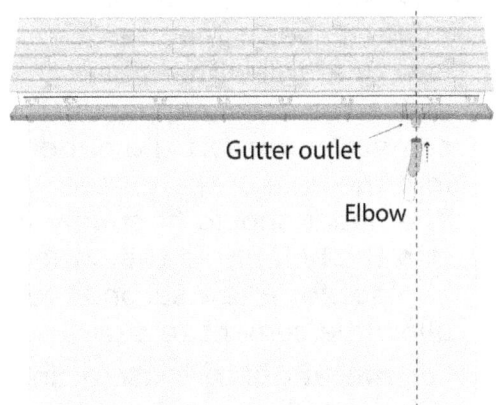

3. **Use a framing square to transfer the location of the downspout (line drawn) onto the gutter.** Mark this

spot clearly on the gutter and then remove the gutter to do the next step on the ground.

4. **Cut the hole for the gutter outlet and downspout:** Modern gutters come with a separate gutter outlet section and you should follow the instructions of the manufacturer.

Most vinyl gutter outlets snap into place. If you are dealing with an older type of gutter, then you will have to cut the hole yourself. Use a drill or a saw to cut a hole in the gutter at the location you've chosen for the downspout. Make sure the hole is the same size as the downspout outlet. Do a quick dry assembly (test fit) to check the position of the outlet relative to your downspout line.

For **vinyl gutters** cut a hole in the gutter at the marked spot using a hole saw or a drill with a hole saw attachment. Apply PVC cement and insert the outlet into the hole of the gutter. The outlet should fit snugly into the hole. Hold the outlet in place for a few seconds to allow the cement to set.

For **metal gutters** use a tin snip or drill, and cut out the shape on the gutter to make room for the outlet. Insert the outlet into the hole you created in the gutter, and use sheet metal screws to secure it in place. Be sure to tighten the screws firmly but not too tight to avoid damaging the gutter.

5. **Next you need to mark the positions of your downspout bracket hangers** that will secure the pipe to the wall. These bracket hangers should be spaced evenly (no more than 6 feet/1.8m apart) along the length of the downspout to provide adequate support. Hold a downspout bracket hanger horizontally over the line you've drawn. Position it just below where the elbow joint for the downpipe will be. Mark the positions for the fixing holes with a pencil. When 100 percent sure of the hole positions, you can go ahead and with a masonry bit and drill the holes for the bracket hangers.

Fit them with appropriate wall plugs ready for attaching your downspout.

6. **Connect the downspout to the outlet:** Insert the downspout into the outlet and secure it using gutter cement, screws or rivets. You may need to trim the length of the downspout to fit it properly.

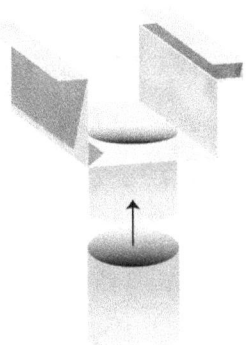

7. **Install the elbows and extensions:** If necessary, attach elbows and extensions to the downspout to direct water away from the house. Secure these components according to the manufacturer's instructions (cement, screws or rivets).

8. **Secure the downspout to the wall:** Use your hanger brackets to secure the downspout to the wall of the house. Remember to follow the manufacturer's instructions.

9. **Test the system:** Once the downspout is installed, test the gutter system to make sure it's working properly. Run water through the gutter and downspout to ensure that water is flowing freely

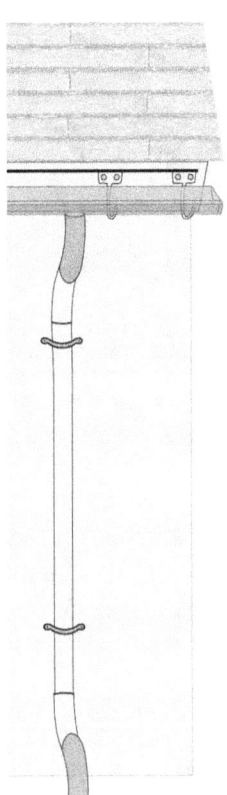

and being directed away from the house.

10. **Apply sealant to all seals:** If you live in a hot climate leave tiny gaps between joints for expansion of the plastic in hot weather.

Overall, installing a downspout on a gutter is a relatively simple process (see p.92) that can be completed with basic tools and materials.

If you're not comfortable doing the work yourself, it's best to hire a professional to ensure that the job is done correctly.

Collection And Storage Of Rainwater

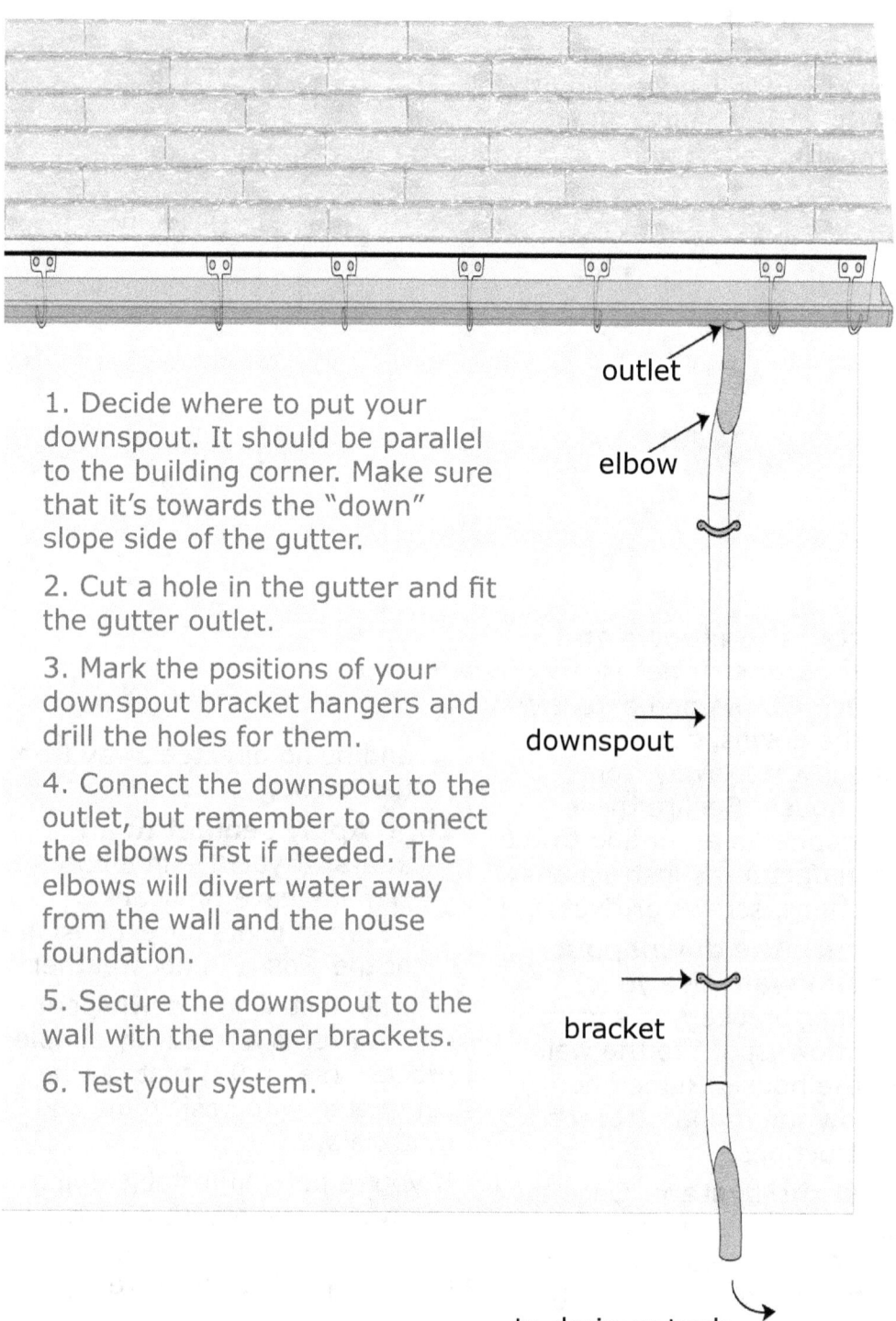

1. Decide where to put your downspout. It should be parallel to the building corner. Make sure that it's towards the "down" slope side of the gutter.

2. Cut a hole in the gutter and fit the gutter outlet.

3. Mark the positions of your downspout bracket hangers and drill the holes for them.

4. Connect the downspout to the outlet, but remember to connect the elbows first if needed. The elbows will divert water away from the wall and the house foundation.

5. Secure the downspout to the wall with the hanger brackets.

6. Test your system.

SIZING VERTICAL DOWNSPOUT PIPES

(Imperial measurements)

Diameter of downspout pipe	Maximum Potential Rainfall Rate				
	1 inch/hr	2 inch/hr	3 inch/hr	4 inch/hr	5 inch/hr
3 inch	6,440 ft²	3,220 ft²	2,147 ft²	1,610 ft²	1,288 ft²
4 inch	13,840 ft²	6,920 ft²	4,613 ft²	3,460 ft²	2,768 ft²
5 inch	25,120 ft²	12,560 ft²	8,373 ft²	6,280 ft²	5,024 ft²

Adapted from Uniform Plumbing Code, 1997 Edition, International Association of Plumbing and Mechanical Officials, Table 11-1.

SIZING VERTICAL DOWNSPOUT PIPES

(Metric measurements)

Diameter of downspout pipe	Maximum Potential Rainfall Rate				
	25 mm/hr	50 mm/hr	75 mm/hr	100 mm/hr	125 mm/hr
75mm	600 m²	300 m²	200 m²	150 m²	120 m²
100mm	1,286 m²	643 m²	429 m²	321 m²	257 m²
125mm	2,334 m²	1,117 m²	778 m²	583 m²	467 m²

Adapted from Uniform Plumbing Code, 1997 Edition, International Association of Plumbing and Mechanical Officials, Table 11-1.

SIZING HORIZONTAL DOWNSPOUT PIPES
(Imperial measurements)

Diameter of pipe sloping ⅛-inch per foot	Maximum Potential Rainfall Rate				
	1 inch/hr	2 inch/hr	3 inch/hr	4 inch/hr	5 inch/hr
3 inch	3,288 ft²	1,644 ft²	1,096 ft²	822 ft²	657 ft²
4 inch	7,520 ft²	3,760 ft²	2,506 ft²	1,880 ft²	1,504 ft²
5 inch	13,360 ft²	6,680 ft²	4,453 ft²	3,340 ft²	2,672 ft²
Diameter of pipe sloping ½-inch per foot	1 inch/hour	2 inch/hour	3 inch/hour	4 inch/hour	5 inch/hour
3 inch	6,576 ft²	3,288 ft²	2,192 ft²	1,644 ft²	1,310 ft²
4 inch	15,040 ft²	7,520 ft²	5,010 ft²	3,760 ft²	3,101 ft²
5 inch	26,720 ft²	13,360 ft²	8,900 ft²	6,680 ft²	5,320 ft²

Adapted from Uniform Plumbing Code, 1997 Edition, International Association of Plumbing and Mechanical Officials, Table 11-2.

SIZING HORIZONTAL DOWNSPOUT PIPES
(Metric measurements)

Diameter of pipe sloping 10mm/m	Maximum Potential Rainfall Rate				
	25 mm/hr	50 mm/hr	75 mm/hr	100 mm/hr	125 mm/hr
75 mm	305 m²	153 m²	102 m²	76 m²	61 m²
100 mm	700 m²	350 m²	233 m²	175 m²	140 m²
125 mm	1,241 m²	621 m²	414 m²	310 m²	248 m²
Diameter of pipe sloping 40mm/m	25 mm/hr	50 mm/hr	75 mm/hr	100 mm/hr	125 mm/hr
75 mm	611 m²	305 m²	204 m²	153 m²	122 m²
100 mm	1,400 m²	700 m²	465 m²	350 m²	280 m²
125 mm	2,482 m²	1,241 m²	827 m²	621 m²	494 m²

Adapted from Uniform Plumbing Code, 1997 Edition, International Association of Plumbing and Mechanical Officials, Table 11-2.

Conduits

We are already familiar with gutters and downspouts, and the next step is to look at the pipes that will transport the water to your storage tank. Rainwater conduits (or pipes) have the function of transporting rainwater from rooftops to the ground-level harvesting system.

These conduits are essential to the conveyance system for various reasons:

- Roof conduits are great for harvesting rainwater because they provide a direct and safe pathway for rainwater to be collected from the roof and directed into a storage tank or other collection system.

- Roof conduits are connected to gutters, which help to channel rainwater into the conduit and prevent it from flowing off the roof and being lost.

- Roof conduits are also efficient because they take advantage of the natural slope of the roof to move the water towards the collection point. This means that very little energy is required to move the water, which helps to keep the system simple and low-cost.

Overall, roof conduits are a great way to harvest rainwater because they are easy to install and maintain, efficient, and can help to reduce reliance on municipal water supplies.

Conduits can be made of material like polyvinyl chloride (PVC) or galvanized iron (GI), and these are materials that are commonly available.

Typically, there are two setups that will need our attention:

◊ **Dry System.** In a standard setup the pipes have to be connected and drawn to the required location by providing sufficient slope so that the water flows with gravity and no water is stored in the pipes after the rain stops.

◊ **Wet System.** In a more complicated setup the catchment area is connected directly to the storage tank, which means that the water is stored in a closed system and is not exposed to the elements.

When considering the sizing of a conduit system, it's important to consider that the size of the water pipes will vary based on the rainfall rate in the area.

Properly sized conduits are essential to handle the intensity of precipitation and effectively transport rainwater to a storage container for end-use.

It is crucial to have an efficient conduit system that can withstand challenging environments and harsh weather conditions, as it plays a vital role in feeding your tanks.

Tanks serve as a key component of a homestead, not only for rainwater harvesting but also for storing water from springs and wells. If the tank is located in a remote area, it is necessary to ensure that there is clear access and a reliable conduit system for water delivery, especially during emergencies when it may be the sole source of water.

In our discussion on tanks and conduits, we will explore different pipe setups that facilitate the connection between the conveyance system and the storage tank.

We will learn that tanks provide water storage for both dry and wet systems.

The Dry System

The dry system is a simple form of conveyance system and it is also the cheaper option. Pay attention to the following components:

1. **Roof collection**: In a dry system, the rainwater is typically collected from rooftops or other impervious surfaces and directed into a gutter or downspout system.

2. **Filtration:** From there, it is conveyed to a filtration system, usually consisting of screens or filters, which removes any larger debris or

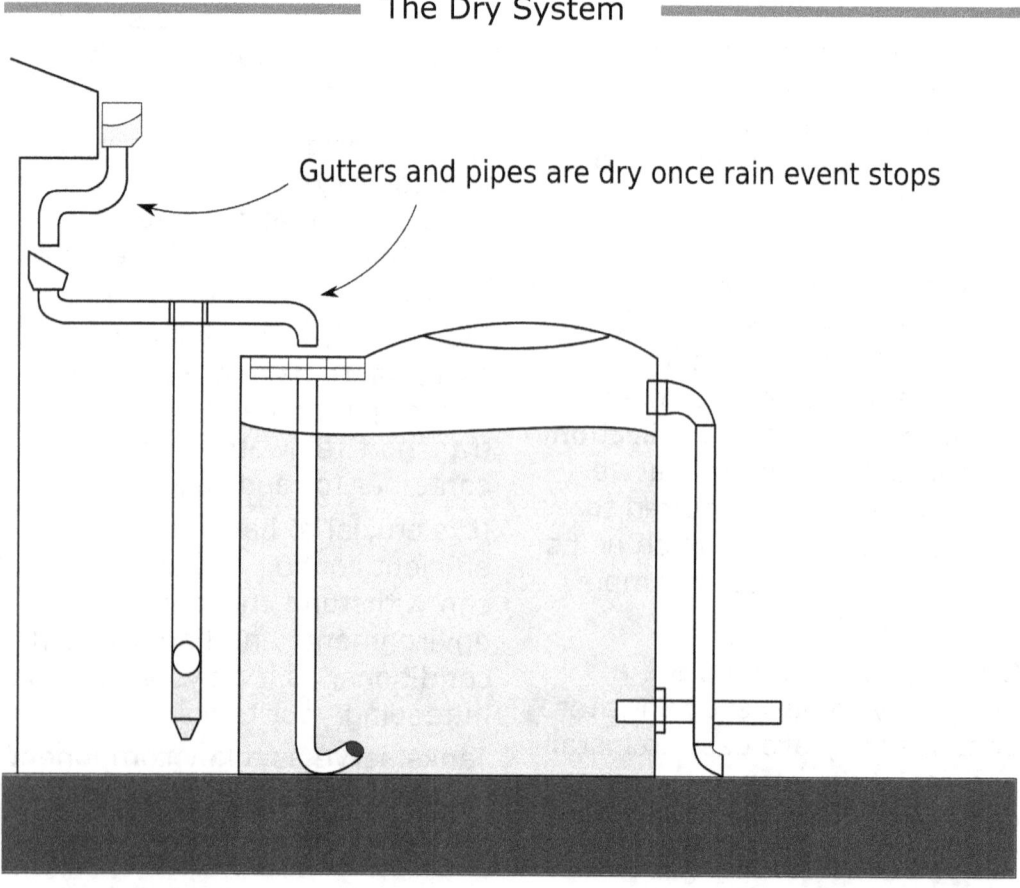

The Dry System

Gutters and pipes are dry once rain event stops

particles.

3. **Conduits:** The filtered rainwater then continues to flow through the conduits and into a storage tank or cistern for storage.

4. **Storage:** The rainwater is stored in tanks or reservoirs designed specifically for rainwater harvesting. These tanks can be above ground or underground, depending on available space and local regulations.

Let's look at the main characteristics of this system:

- When looking at the pipes, we quickly notice that this system's main feature is the fact that all pipes are "dry", until it rains. The pipes do not hold water after the rain stops. Water flows from the highest point to the lowest and when the rain stops, the pipes are dry due to gravity which assists in pulling every drop of water down into your tank.

- All pipes are above ground which means that they are visible, reachable and thus easy to maintain. The collection tank is usually located right next to the catchment area, so there's no need to run piping underground.

- The pipe system runs directly from the gutter of the roof, down into the tank.

- This inexpensive system allows you to store a large amount of rainwater and is great for climates with infrequent, larger storm events.

- This system is generally found on structures with only a couple of downspouts.

Good:
- Very simple and fairly cheap
- Pipes are easily accessible and easy to maintain
- Can store a large amount of water

Bad:
- Pipes are exposed to sunlight and weather
- The storage tank must be located next to your house

The Wet System

The wet system is a bit more complicated and expensive to implement. Pay attention to the following components:

1. **Roof Collection:** The first step involves collecting rainwater from rooftops. Rainwater runs off the roof surface and is channeled into gutters or downspouts. This inflow of water creates the pressure necessary for rain to travel through your pipes underground.

2. **Filtration:** To remove debris and contaminants, rainwater is usually filtered before entering the storage tanks.

This can be done through various filtration methods such as mesh screens, leaf diverters, or leaf traps. The filtration system prevents larger particles like leaves, twigs, and debris from entering the storage tanks, ensuring cleaner water.

3. **Conduits:** Once filtered, the rainwater is directed through pipes or conduits towards the storage tanks. The water then exists the ground via a riser pipe until it finds its way into the tank for storage. The pipes can be made of materials such as PVC or HDPE (high-density polyethylene).

4. **Storage:** The rainwater is stored in tanks or reservoirs designed specifically for rainwater harvesting. These tanks can be above ground or underground, depending on available space and local regulations.

The main characteristic of this system:
- The water runs from the roof into pipes that continue down into the ground and then over to your tank. Once it reaches the tank, the pipe

The Wet System

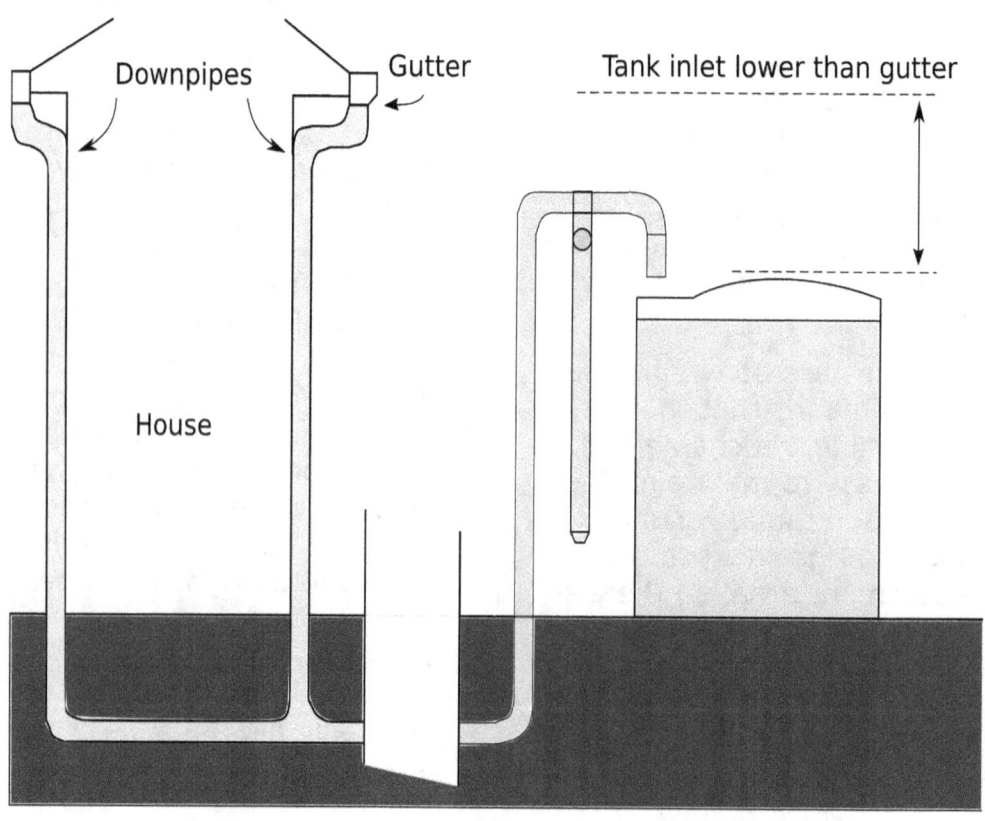

Special care should be taken to protect pipes when installed under driveways and roads. Pipes are typically placed deep enough to provide sufficient cover and protection.

---

rises up and empties out inside the tank.

- The water rests in the pipes, until the next rain event. Due to the nature of gravity, there will always be some water left in the pipes and that is why we call it a "wet" or "charged" system.

- If we look closer, we see that underneath the gutters there are multiple downspouts with pipes. They continue down along the wall and along (or under) the ground and then run up against the tank and into the tank inlet. The elevation of the tank inlet must be below the lowest gutter on the house to ensure the uninterrupted flow of water.

- A benefit of this system is that it gives you the freedom to locate your tank as far away from the house as is needed.

When implementing this system, it is important to be aware of the following considerations:

- In areas with mosquitoes, special care should be taken to insect-proof the buried pipes and all connections and fittings must be absolutely watertight. Use non-corrosive mosquito screens and ensure holes are smaller than 1mm in diameter.

Stagnant water in pipes can create a potential breeding habitat for mosquitoes to lay their eggs.

- It can also create an environment with low levels of oxygen which will cause leaves and organic material to decompose. Once your water has been contaminated through anaerobic fermentation, it poses a health risk.(You will be aware of a rotten egg smell.)

- Special care should be taken to protect pipes when installed under driveways and roads. As a general guideline, pipes installed under driveways and roads for rainwater harvesting using the wet system are typically placed deep enough to provide sufficient cover and protection. This depth can range from 18 to 36 inches (45 to 90 centimeters) below the surface.

Pipes in the ground are susceptible to cracking in freezing conditions and should be emptied in winter and buried beneath the freeze line.

- Make sure you drain your pipes between rainfall events!

Collection And Storage Of Rainwater

The wet system requires a significant amount of space to install the pipes underground, which can be a limitation for some properties.

This system should be installed with the help of a professional who can recommend underground first flush diverters, filters and screens.

Good:
- Aesthetically pleasing
- Practical solution when tanks are far away from roof

Bad:
- Vulnerable to mosquito infestations
- Prone to anaerobic fermentation
- More expensive, more work
- Pipes should be drained between rainfall events

Space requirements:
- The wet system requires a significant amount of space to install the pipes underground, which can be a limitation for some properties
- Pipes are prone to cracking in freezing conditions
- Requires professional installation

Conclusion

No matter which conveyance system you choose, special care should be taken to protect your pipes from the elements and physical damage:

- **Proper Pipe Placement:** Ensure that the rainwater harvest pipes are installed in locations where they are less prone to damage.

- **Pipe Covering:** Use protective covers, enclosures, or paint to shield the pipes from the sun.

- **Adequate Clearance:** Maintain sufficient clearance around the pipes to prevent any obstruction or accidental damage.

- **Burial Depth:** Ensure they are buried at a sufficient depth to protect them from external factors such as freezing temperatures, root growth, or accidental digging.

- **Proper Backfilling:** If the pipes are buried, make sure they are surrounded by a suitable backfill material, such as gravel or sand, that provides support and helps distribute external loads evenly.

Once you have a conveyance system in place, you can move on to the next phase, which is *where* and *how* to store the harvested water.

Storage Tanks

Rainwater tanks come in various sizes and shapes to suit different needs and spaces. They can be small enough to fit under a sink or large enough to store thousands of gallons of water. They can be installed above ground or below ground to fit different spaces and preferences.

These days rainwater tanks are available in different materials, such as plastic, fiberglass, concrete, wood or steel. Each material has its own benefits, such as durability, flexibility, or insulation. Most tanks require minimal maintenance, such as cleaning and checking for leaks.

In a **basic setup**, the tank typically consists of a durable, watertight container designed to store the harvested rainwater. It is equipped with a simple inlet pipe that directs the water into the tank and an outlet pipe for accessing the stored water when needed. While this setup fulfills the fundamental purpose of storing water, it may lack certain features that contribute to water quality maintenance.

On the other hand, an **advanced setup** incorporates additional features to ensure optimal water quality. These may include a pre-filtration system, such as a mesh screen or sediment trap, that removes debris and contaminants before the water enters the tank.

An overflow mechanism helps divert excess water during heavy rainfall to prevent overfilling and potential damage.

Access points, such as inspection ports, facilitate inspection, cleaning, and maintenance of the tank.

Lastly, a secure lid or cover prevents unauthorized access and keeps out sunlight, insects, and other contaminants.

Understanding the distinction between a basic and advanced setup provides insights into how to maintain the cleanliness and suitability of the water stored in your tank.

Basic Tank Setup

With the basic tank setup we can see from the diagram on page 102 that water travels from the roof into the gutter and from there it travels down the downspout and into the tank.

Not all rainwater systems have a first flush diverter for contaminated water. If you inherited a system without a diverter, just make sure that you drain the sludge from your tank more often.

Invest in a good **filter screen** at the inlet area and also add a filter at the outlet.

The **tank inlet** is where the water enters and it can be modified to suit your needs. It can be attached to the tank or you can leave an open space between tank and downspout. Just remember that all tank

Basic Tank Setup

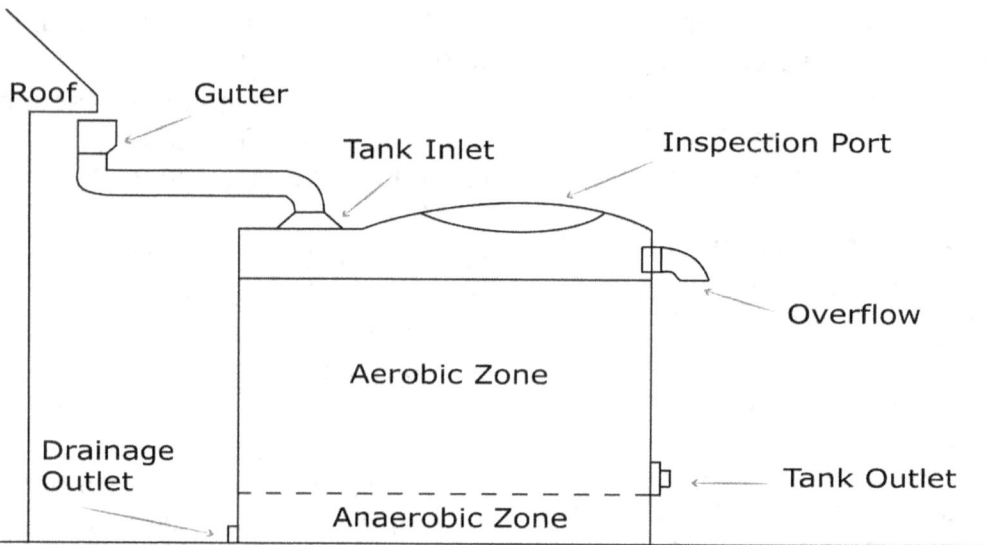

openings should be closed or fitted with screens or strainer baskets to keep bugs and critters out.

The **anaerobic zone** is where the unwanted water rich in sediment settles.

The **tank outlet** is raised to avoid the sediment from the anaerobic zone. It should be the same size as the tank inlet.

The **aerobic zone** is where the clear water void of heavy particles is found. If a rain event delivered too much rain, the overflow will allow excess water to escape. This pipe can be connected to the storm-water runoff or even to other tanks or pools.

The **inspection port** must be closed at all times. Remember to do inspections often to ensure longevity of your system.

Let's compare this system with a more advanced setup.

Advanced Tank Setup

From the diagram on page 103 we can see that these days most advanced setups have a First flush diverter system. This takes care of the initial flow of contaminated water.

Inlet areas. Some inlets are connected to the tank to prevent mosquitoes and debris from entering. Others have a space between the downpipe and tank. These tanks use tank screens and all water must pass through this screen before entering the tank.

Calming inlets are specially designed to prevent the stirring up of sediment in the tank.

Floating Pick-up. Inside the tank, the heavier sediment will settle at the bottom or anaerobic zone of the tank. The lighter particles float at the top. As

Advanced Tank Setup

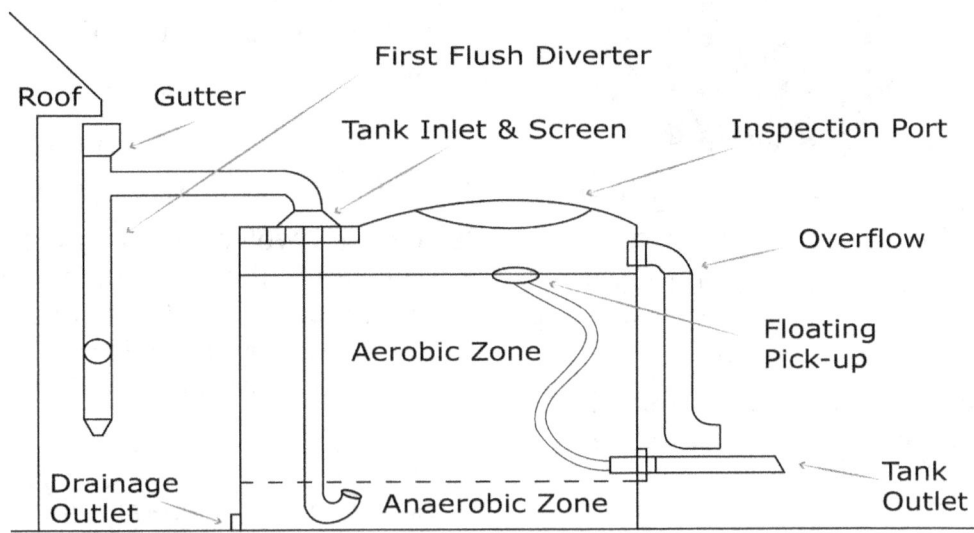

the name suggests, a floating pick-up floats in the calm, clear water in the middle of the tank. This intake avoids the heavy sediment at the bottom of the tank and lets the best quality water in and also filters floating debris.

You'll find the best quality water in your tank around 6-8 inches from the surface. Ensure that your pick-up is collecting water within this range.

The **floating pick-up** connects to the pipe outlet, typically near the very bottom, where water is drawn for household and irrigation use. Your overflow outlet can have a flap or cover to keep critters out.

When looking at both the basic setup as well as the advanced setup, we see that you'll need to secure the entry and exit points, as well as any other openings to keep potential contaminants out.

Protecting Your Tank Inlet

At your inlet, the first port of call is to keep leaves, debris, mosquitoes, pests, and sunlight from entering your tank. Tank screens and solar shields are effective tools that help to preserve your rainwater quality by reducing nutrient loads, light and algae growth – and preventing mosquitoes from using your rainwater to breed in.

Tank Screen

The tank screen (fig. 4.20) consists of fine mesh that keeps most debris and mosquitoes out, but it allows large volumes of water to pass through. When it comes to mosquitoes make sure that the mesh openings are smaller than 1mm. If you want,

Fig. 4.20

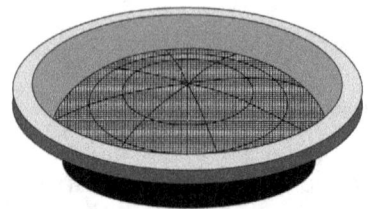

you can even go smaller to keep dust and fine sediment out.

The tank screen is suitable for any tank type and it is very easy to remove and clean.

Most tank screens come with a solar shield.

Solar Shield

The solar shield (fig. 4.21) is a plastic flap that keeps sunlight out which will limit the chance of algae growing in your tank inlet.

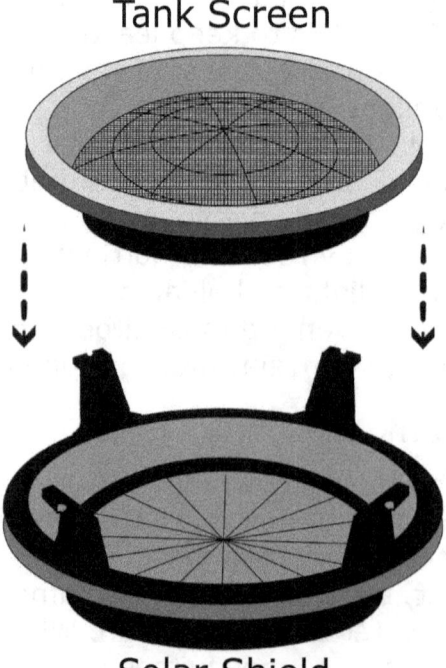

Tank Screen

Solar Shield

Fig. 4.21

There are a variety of options that can either mount underneath the tank screen or can be placed on top of the cover.

Securing Your Tank Overflow And Outlets

Now that we've taken care of the tank inlet, we can follow the same principles to ensure that the tank overflow is secure from contamination.

Overflow

Overflow systems are simple devices that manage the overflow of water while also improving ventilation. The overflow directs any additional water that can't fit into the tank to the attached storm-water drain.

Most overflow devices will have mosquito screens that have

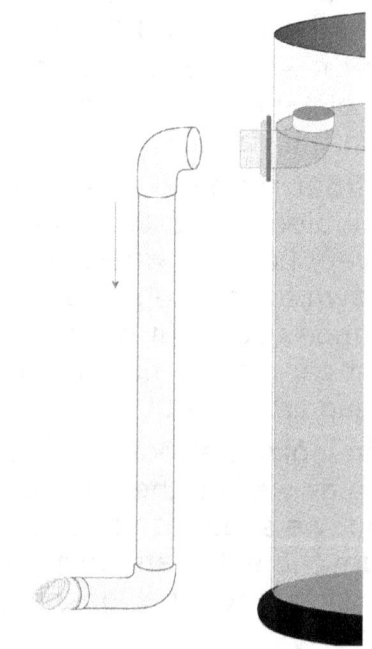

been specifically designed to keep mosquitoes and other insects out of your tank. You can install a mosquito screen at the overflow fitting on the tank, or further down the line where it meets the ground. These screens are very convenient and are easy to maintain. One option is to install the mosquito screen in the overflow and another is to put it on the outside with a flap that simply lifts up to let the water pass.

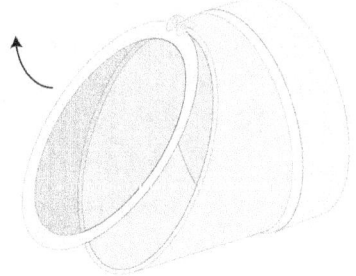

When sizing your overflow system, keep in mind that you want your overflow rate to match the flow rate of the pipes that enter the tank. This means the overflow has to be the same size or greater than the inlet, so you're not in a situation where there is more water coming into your tank than can get out. This means that the overflow will always flow effectively during a heavy rain event. An overflow system is very easy to install and perfect for the DIY enthusiast.

Backflow (Air gap)

The backflow device prevents storm-water backflow from entering a rainwater tank, by creating a physical air break on the outlet of the tank overflow and before the storm-water drainage system. The Backflow also acts as a visual inspection point to detect trickle top-up system failure.

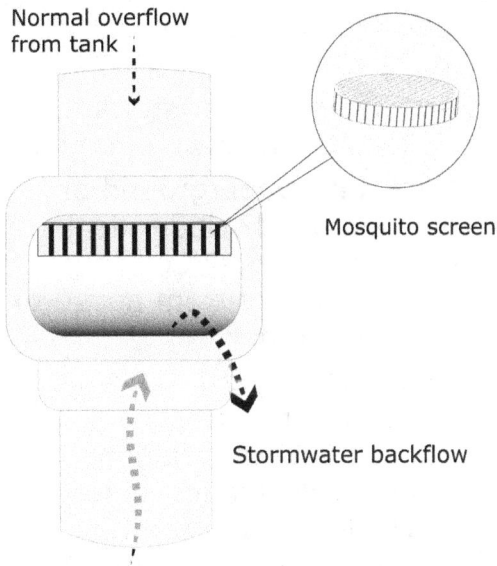

The action of backflow can occur when the storm-water system gets flooded due to heavy rain. This causes storm-water drains to get backed up and to push water back towards your tank. You do not want this contaminated water to enter your tank, and to prevent this you can install a Backflow (Air gap) that will make it impossible for storm-water to flow back into your tank. Most Backflows come with one or two removable mosquito screens made from mosquito proof stainless steel mesh for easy cleaning and low maintenance.

Tank types

Next let's take a look at the actual storage tanks available at hardware stores all over the world. Whatever your needs, rest assured that you have plenty of options to choose from. Tanks come in all shapes and sizes and every year we see new designs with better design characteristics.

When buying a water storage tank, keep the following in mind:

◊ **Location.** Underground or above ground? Under decks or in basements? Consider piping distance from house to tank required.

◊ **Size and design.** Consider what size and design will suit your installation location. How much space is available to place the tank? If a tank does not fit into the designated space provided, you can look for a "slim tank" or go smaller and use two or three appropriately sized tanks. There are also modular tanks on the market that form a linkable "wall" where you can store your water. Larger tanks suit colder climates with freezing conditions. Fewer tanks generally means less maintenance. For most households one large tank will suffice. If a tank does not fit into the designated space provided, you can look for a "slim tank" or go smaller and use two or three

Above Ground Tanks

Above-ground rainwater tanks are more commonly chosen for rainwater harvesting:

- Cost-effective: They are generally less expensive to install compared to underground tanks.
- Ease of installation and maintenance: They are easier to install since they don't require extensive excavation. They can be placed on a suitable foundation or stand, and connections for gutters and pumps are easy to find.
- Flexibility and scalability: They offer flexibility in terms of size and placement. They come in a variety of shapes and sizes to fit different spaces and water storage requirements.
- Visual inspection and monitoring: They allow for easy visual inspection of water levels, quality, and potential issues such as leaks/ contamination.
- Ease of customization: They can be customized with additional features such as filters, overflow systems, or diverters.
- Portability: They can be moved or relocated more easily than underground tanks.

Underground Tanks

Underground water tanks can be useful for rainwater harvesting in several situations:

- Limited space: If you have limited space on your property, an underground water tank can be a practical solution.
- Aesthetics: Some people prefer to maintain the visual appeal of their outdoor space without the presence of above-ground tanks or structures.
- Property regulations: In certain areas, local regulations or homeowner association rules may restrict the use of above-ground rainwater tanks.
- Temperature regulation: Underground water tanks tend to have more stable temperatures compared to above-ground tanks.
- Security and protection: Underground tanks are generally more protected against vandalism, theft, and potential damage from extreme weather events like storms or high winds.
- Preservation of views: If you have scenic views on your property that you don't want to obstruct, an underground water tank can be an ideal solution.

appropriately sized tanks. A small roof with a small tank is good for toilet flushing and washing. A large roof with a large tank allows for a variety of options, both inside and outside the house.

- **Tank material.** Will it survive the environment you intend to use it in? Consider climate, temperatures (especially freezing temperatures), and conditions like the occurrence of wildfires. How long is your tank guaranteed to last? Do not use a translucent plastic tank for water storage! These tanks allow sunlight into the tank and this causes the water to grow algae.
- **Water quality.** Does the tank material affect the taste and quality of the water?
- **Cost.** Consider that delivery, installation and setup charges will all affect the price.

Talk to the tank distributor and make sure that you get the right tank for your needs. Make sure that your tank meets or exceeds NSF 61 standards. You should gather as much information as possible before buying a tank.

Let's take a look some rainwater storage options.

- Polyethylene

Material: Plastic (polyethylene)
Size: Wide range
Price: Affordable
Convenience: They are light

and come with plumber's fittings that are easy to fit and assemble. Available at most hardware stores.

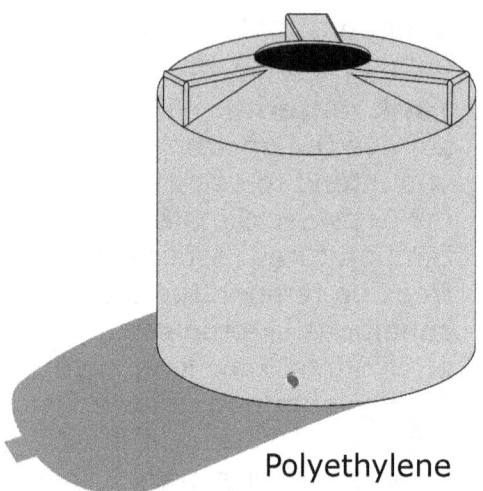
Polyethylene

Characteristics: These are very popular tanks. Polyethylene tanks are tough and lightweight. They are UV and rust resistant. Painting these tanks is not advised since most paints don't stick to plastic and will flake in the elements. Buy pre-painted and remember that dark colored tanks absorb heat. Place them in the shade where possible. When placed outside, they will last for a long time. Placing them inside, in the shade, can increase their lifespan to much longer. They are normally placed above ground.

For buried installation, you need a specially reinforced tank. Buried tanks are more expensive because the tank has to be reinforced and the excavation required is costly. Don't put a tank in ground with high clay content, for the clay expands and contracts and this can potentially damage the tank.

These tanks are definitely not to be used in areas prone to wildfires.

Get product specs when buying from your distributor.

Make sure that your tank is government approved and that it meets all industry standards.

On a side note, it's worth mentioning that we are still dealing with a plastic container that you are using for water storage. Plastic brings an element of toxicity to the table and this can be a turn-off for some buyers. Make sure you do thorough research on this subject, to give you the peace of mind that you require for you and your family's wellbeing.

- Fiberglass

Material: Layered fiberglass and resin

Size: Wide range

Price: Affordable

Convenience: Light, tough and easy to transport

Characteristics: Fiberglass tanks are also tough and lightweight. Years ago they were mostly used above ground, but these days you can find excellent **underground** fiberglass tanks. Tanks come in a wide range of colors and sizes and they are resistant to heat, rust and chemical erosion and won't degrade. The tank fittings are built into the tank and

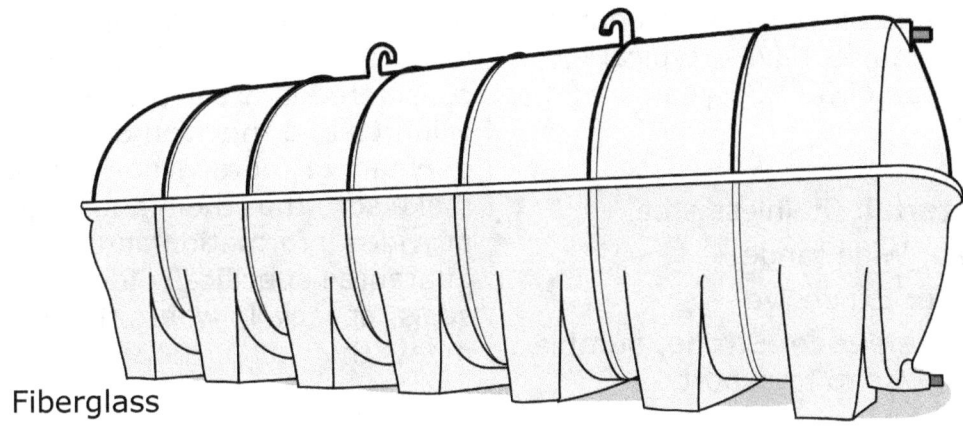
Fiberglass

therefore you are guaranteed few issues with leaks. These are long-lasting tanks and they are actually reasonably priced for what you get. They should be opaque to prevent algal growth and there should be a liner on the inside of the tank that's approved for storage of drinking water for human consumption.

Get product specs when buying from a distributor. Make sure that your tank is government approved and that it meets all industry standards.

Worthy of consideration is the fact that most fiberglass tanks are not recyclable and will most likely end up as landfill. They contain polyester resin and other toxic substances, which are released into the environment when they degrade.

- Metal

Material: Galvanized sheet metal or corrugated steel
Size: Wide range
Price: Affordable
Convenience: Light, easy to transport and can be custom made on site
Characteristics: Metal tanks are tough, strong and lightweight. They provide a great solution for household uses, irrigation and fire protection. Depending on brand and manufacturer, they can be used above or below ground. These tanks should be considered in areas prone to wildfires. They also do not crack in sub-zero conditions. There should be a food-grade liner or epoxy coating on the inside of the tank that's approved for storage of drinking water for human consumption. Get product specs when buying from a distributor. Make sure that your tank is government approved and that

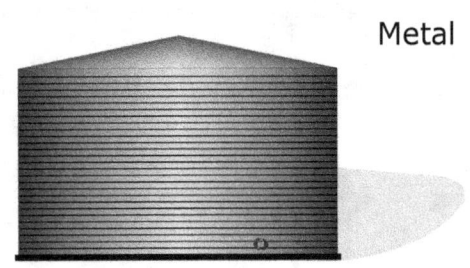
Metal

it meets all industry standards. These tanks have a typical lifespan of 30 - 35 years.

• Stainless Steel

Material: Stainless steel
Size: Wide range
Price: Expensive
Convenience: Strong, durable and easy to transport

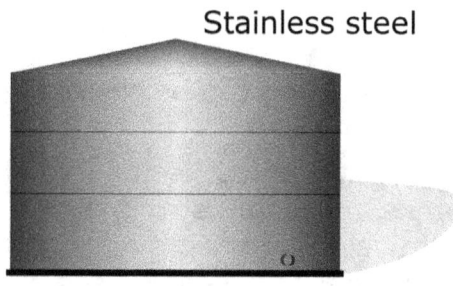
Stainless steel

Characteristics: These are very good tanks. Stainless steel tanks by definition are superior in the categories of strength, durability, and hygiene. They are mostly used above ground, except where the manufacturer specifically states that it is to be used below ground. Certain metals do add the possibility of rust, but stainless steel will not readily rust or discharge in water. These tanks should be considered in areas prone to wildfires and they can easily last 50 years or more. When buying, make sure that the manufacturer provides information and a guarantee specifically for the seals, especially where the floor meets the wall of the tank.

• Concrete/ Ferro-Cement

Material: Concrete with or without rebar
Size: Large sizes
Price: Fairly expensive
Convenience: Durable and strong and usually manufactured on site

Characteristics: Concrete tanks are tough, permanent structures and can be custom made. These tanks can be poured in place or they can be prefabricated off-site. Some manufacturers also provide underground holding tanks. These are structurally strong tanks that last a long time. The concrete keeps the water temperature low and stable, which also prevents algal

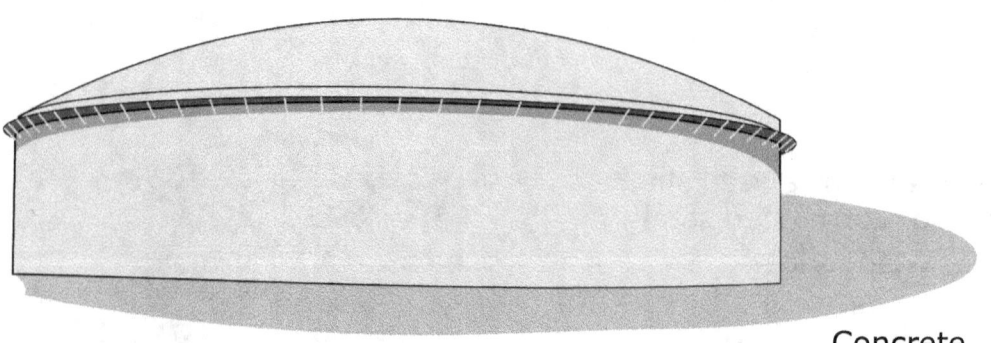
Concrete

growth. They are perfect for areas prone to wildfires. Make sure that the inside of the tank is plastered with material that is suitable for drinking water. These tanks have an advantage in that the concrete dissolves calcium into the water (due to the acidic nature of rainwater), which provides a pleasant taste. These days you can buy waterproof plaster to fix cracks and leaks.

- Wood

Material: Redwood, pine, cedar
Size: Wide range
Price: Expensive
Convenience: Looks great, all natural and no contaminants

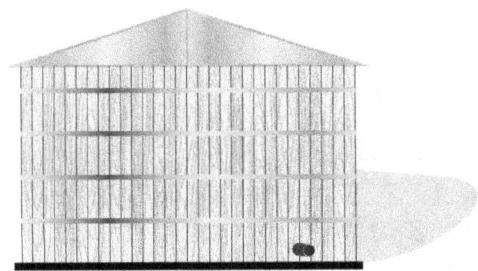

Wood

Characteristics: Wood tanks are still available, but they have become more expensive and therefore less popular. They used to be made of redwood, but you can also find tanks made of pine, cypress or cedar. They are custom built by skilled technicians who wrap the wood with steel cables under tension. The natural qualities of redwood is very appealing since it contains no resin and the tannins in the wood function as a natural preservative which keeps boring insects at bay. They are not to be used in very dry climates for they tend to shrink and to leak. Used above ground only. A food-grade liner is placed on the inside for potable use. Wood tanks will outlast any of the alternatives.

- Bladders/Pillows

Material: Typically flexible reinforced material (polymer alloy, PVC, polyurethane)
Size: Wide range
Price: Fairly expensive
Convenience: Fits under decks and between walls (out of sight)

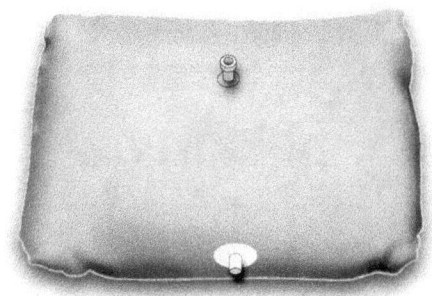

Bladder

Characteristics: Bladder or pillow tanks are manufactured from a flexible material. This type of tank rises and expands as you fill it with liquid. These bladders are very convenient in that they can fit into tight storage spaces and thus they are ideal for storage under decks, in basements or in between walls. For potable water storage make sure that the bladder comes with a food-grade liner. It connects easily to downspouts, pipes and pumps and is manufactured from a high strength fabric. These bladders come in sizes ranging from 25-5,000 gallons (94-10,000 liters).

- Rain Barrels/Butts

Material: Plastic, wood, or metal

Size: Small sizes (55 gallons or 210 liters)

Price: Plastic ($100-160) and wooden($100-500) barrels are fairly cheap, but the metal ones are expensive ($200-700).

Convenience: Durable and strong and easy to handle (light)

Rain Barrel

Characteristics: The biggest benefit of these barrels is that they are small and light. They are ideal for rain harvesting in urban areas and if you live in an area with limited water resources or where water is expensive, rain barrels can be an effective way to conserve water and to save money on your water bill.

Keep in mind that rain barrels require regular maintenance, including cleaning and winterizing. If you don't have the time or inclination to maintain a rain barrel, it may not be worth it for you.

When purchasing a rain barrel, look for ones made from food-grade materials such as high-density polyethylene (HDPE) or polypropylene (PP). These materials are safe for storing and using water.

- IBC (Intermediate Bulk Containers) Totes

Material: Plastic

Size: 275 gallons (1000 liters) and 330 gallons (1250 liters)

Price: New $100 to $300

Convenience: Durable, strong and easy to handle (can stack them and they are square) and they are cheap

Characteristics: For rainwater storage, it's recommended to use IBC totes made of food-grade materials, such as high-density polyethylene (HDPE) plastic, which are safe for storing potable water. If you don't want algae in your water make sure you paint the tank

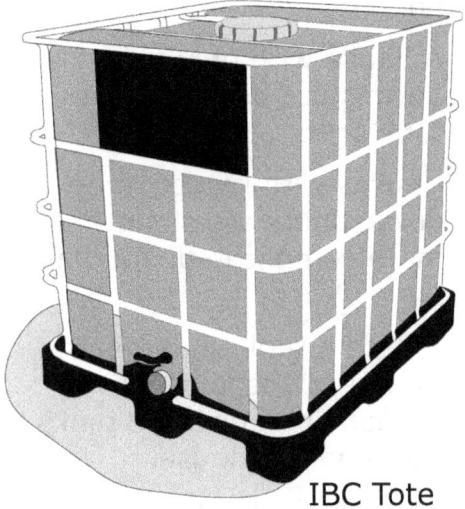
IBC Tote

black or to get a cover for it. IBC totes are designed to be sturdy and long-lasting, but it's important to select a tote that is well-made and suitable for outdoor use. Look for totes that are UV-resistant and have a sturdy frame (caged) to prevent cracking or leaking. The standard poly composite tote tank (275gal) dimensions are a 40" x 48" base, with an overall height of 46".

Caged and empty poly IBC totes in 275 – 330 gallon capacities, are acceptable for stacking 4 high.

If full, caged totes are recommended for two-high stacking only (maximum cargo weight).

IBC totes should have a valve at the bottom to allow for easy drainage of collected rainwater. Totes used for rainwater storage should have a filtration system to remove debris and contaminants from the collected water. This can include a mesh screen or sediment filter.

They should be placed on a sturdy, level surface to prevent tipping or damage.

Conclusion

These are the most popular and widely available rainwater storage options, but keep in mind there are still other kinds of tanks available out there. For best results consult a professional in your area. No matter which water tank you select, always make sure that your water does not come into contact with any unwanted chemicals or pollutants and that where necessary, a food-grade liner is placed on the inside of the tank. Make sure that the fabric used is NSF 61 rated.

When it comes to storing water for human consumption, you might want to consider the following information:

Most of these tanks use plastics, liners made of plastic, resins or coatings, that will come into contact with your water. Although they are all, for the most part, government approved, you will do well to do some of your own research to decide if plastics (made out of oil) or similar, is something you want your water to be stored in.

This might the deciding factor when purchasing a new water tank for your home.

Preparations For Tank Installation

Whether you are using a rain barrel, an IBC tote or an actual water tank, you still have to prepare the area around the roof and the ground underneath the tank, install tank components and distribute the water for end-use.

Pay attention to the following:

1. First and foremost you will have to determine the size and type of tank(s) you will use for rainwater harvesting.

2. Determine where your tank will be located. The position of the tank will be determined by its size, shape and type.

3. What kind of foundation will provide you with a sturdy, level platform to place your tank?

4. Locate and assess the quality of your downspouts and gutters. Make sure that the gutters are of quality material and that the hangers are still securely attached to the fascia. Look for cracks in downspouts. Replace old and faulty components.

5. Add gutter guards or screens.

6. Decide on the kind of tank you need.

7. Next, you need to decide if you need one, two or more tanks.

- A house with only one downspout deserves one tank or barrel.

- Houses with two or three downspouts are slightly more complicated. You have two options:

  a. Place a tank underneath each downspout. This means two or three tanks around the house.

  b. Alternatively, you can connect gutter spouts from one side of the house, to connect with spouts from another side, to all flow into one tank. This involves a bit more work, but the end result is worth it.

You can hire an expert to take care of the assembly and construction components, or you can just do it yourself.

Doing it yourself brings you closer to the project and means that you can maintain, repair and replace everything yourself when necessary.

Go to your local hardware store and make sure you get local advice about rainfall, gutter sizes, tank types and sizes, pumps, etc.

To begin, let's explore the various options available for selecting a suitable tank foundation.

The Foundation Under Your Tank

When it comes to positioning and preparing a foundation for a water tank most people underestimate the weight of these vessels. In fact, a water tank can weigh more than a small car when it's filled with water.

Make sure of the following:

- Placing the tank on a sturdy, flat (level) surface in a suitable location is crucial to its longevity. A thin slab (foundation) will definitely break apart due to moisture infiltration. Aim for a minimum of 4 inches (10cm) of reinforced concrete for a water tank.
- Your foundation (fig. 4.22) must be at least 8 inches (20cm) greater than the diameter of the tank. The foundation diameter should never be smaller than the tank diameter.
- Your foundation must be level and clean. No hollow or protruding areas.
- Your tank must be placed at least 3 feet (1m) away from the house and on solid, compacted ground or on a hard surface. The tank overflow should flow away from both the tank and house foundations.

There are six common positions to place a water storage tank, each with its own advantages and disadvantages:

On The Floor

Placing the tank directly on the floor of your dwelling is a cheap option, but the weight of the tank can damage the ground, and there is a risk of overflowing.

Patios, decks, floors and basements are not recommended to place a storage tank.

Pros
- Cheap
- Convenient

Cons
- Can damage the ground
- Can drain into the foundation of your dwelling
- Can flood your home

Soft, Sandy Areas

This is not an ideal foundation for a water tank, but if it's all

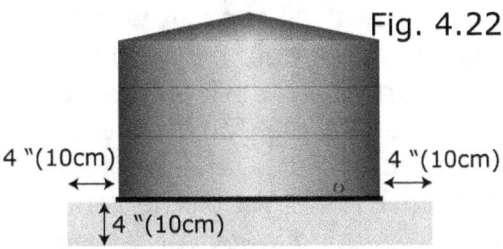

Fig. 4.22

Aim for 4 inches (10cm) of reinforced concrete for a water tank.
Your foundation must be 8 inches (20cm) greater than the diameter of the tank.

you've got then you will need to reinforce the terrain.

Flatten the area and use a support structure with wide wooden beams or stone slabs to make a flat, sturdy level for the tank to rest on. Use wood that can resist flooding and does not become waterlogged.

There should be ample drainage underneath and if not, consider creating a layer of pea gravel to remedy the situation.

Dig about a hand's width into the ground and create a clean, sunken rectangular area.

Fill it with gravel and make sure it drains away from the house. Prepare a level layer of bricks, concrete blocks or thick wooden beams on top.

The tank can rest on top of this structure.

Pros

- Easy to work with as it is a fine-grained material that can be easily shaped and compacted.

Cons

- Needs lots of reinforcement
- It can shift and settle over time, which can cause the tank to become unstable and potentially damage it
- Sand is prone to erosion

Concrete Slab

A concrete slab is a recommended option for tanks of all sizes. A thick, flat slab of concrete is strong and as such it makes the tank easily accessible for maintenance which will extend the life of the tank.

A concrete slab may require a pump, and due to its footprint size, space may be an issue. For a concrete slab, you have to make sure that the area is level. Pour more cement if necessary, but make sure it drains away from the house into the stormwater drain.

Pros

- Best option, very stable
- Not that expensive
- Will extend the life of the tank
- Can handle tanks of all sizes
- Makes the tank easily accessible

Cons

- May require a pump
- May require more space

Steel/Wood Platform

If the ground is not suitable, you can raise the tank onto a structure made of metal or wood. This must be a very solid structure capable of holding the weight of a full tank. A wooden platform can only be used for rain barrels. A fabricated steel stand is suitable for heavier tanks under 1300 gallons (5000 liters). The height of the tank provides good water pressure and is a space saver. However, it can be expensive to erect, and the tank may be exposed to harsh weather conditions.

Pros
- Does not require a water pump. For every 10 feet (3m) of elevation, you can expect 4.3 psi of water pressure.
- Can save space

Cons
- Expensive
- Only good for tanks under 1300 gallons (5000L)
- Difficult to access
- The steel platform has bars which might cause the tank to crack
- Expose the tank to the wind and sun

On The Roof

Placing the tank on the roof of a building is another option that can provide excellent pressure flow. Having a tank on the roof of your own home provides protection from vandalism, but it can be difficult to access and may cause damage if not planned well.

Pros
- Great for water pressure
- No need for a pump
- Save from vandalism
- Can save space

Cons
- Not easy to access
- A leak in the tank can flood or damage the house
- Hard to replace a roof tank

Underground

Finally, an underground enclosure is a suitable option if space is limited, but it can be challenging to access for maintenance.

To protect the water tank, factors such as the cost of erecting a platform, available space, vandalism, use of pumps, and access for maintenance should be considered.

Underground installations require the help of a professional.

Pros
- Aesthetically pleasing
- Protected against sunlight
- Can save space

Cons
- Very hard to access in case of breakdown
- Hard to maintain
- Expensive

Possible Problems

If a water storage tank leaks water into a house's foundation, it can lead to several potential problems:

- **Structural Damage:** Water intrusion into the foundation can compromise its structural integrity over time. The continuous presence of water can weaken the foundation walls or footings, leading to cracks, shifting, or settlement. This can affect the stability and overall

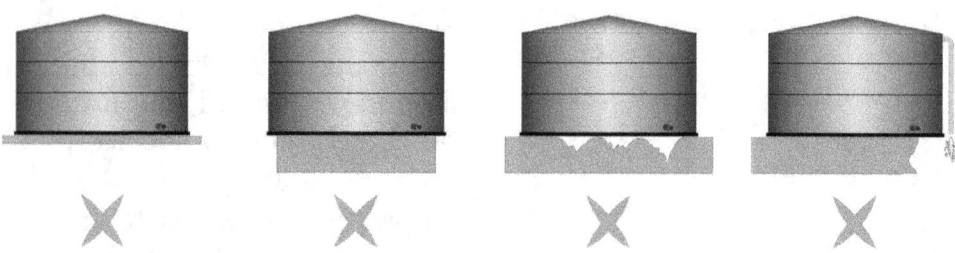

structural performance of the house.

- **Mold and Mildew Growth:** Excess moisture from the leaking water can create a favorable environment for mold and mildew growth. Mold spores thrive in damp conditions and can spread throughout the house, leading to health issues and damage to building materials.
- **Damage to Interior Finishes and Belongings:** Water leaking into the foundation can seep into the interior spaces, causing damage to walls, flooring, carpets, furniture, and other belongings. Stains, warping, rotting, or deterioration of materials may occur, necessitating repairs or replacements.
- **Increased Humidity and Poor Indoor Air Quality:** Water intrusion can result in increased humidity levels inside the house. Elevated humidity can lead to discomfort, promote the growth of mold and mildew, and contribute to poor indoor air quality, potentially causing respiratory issues and allergies.
- **Higher Utility Bills:** A leaking water storage tank can result in continuous water loss, leading to increased water consumption and higher water bills. The financial impact can be substantial if the leak goes unnoticed or unaddressed for an extended period.
- **Pest Infestation:** Persistent moisture from a leaking tank can attract pests, such as termites, ants, or rodents, that are drawn to damp environments. These pests can further damage the house's structure and pose additional challenges for homeowners.

Solutions

It is essential to plan ahead and to address any water storage tank leaks promptly to mitigate the aforementioned risks:

◊ Regular inspection, maintenance, and repair of the tank, as well as proper waterproofing measures for the foundation, can help prevent water intrusion and protect the integrity of the house.

◇ If you suspect a leak or notice signs of water damage, it is advisable to consult with professionals, such as plumbers, waterproofing specialists, or structural engineers, to assess the situation and determine the necessary remedial actions.

◇ In addition to placing the tank on a flat surface, another tip to protect the tank includes ensuring it's not placed directly on steel bars and always placing a flat board under the tank for protection against damage. Plan ahead, for tanks are heavy, really heavy, when full. A full tank cannot be moved without draining it first.

◇ You do not want a leaking tank to drain into any structure and therefore you have to provide a barrier between your tank and the building. Concrete is best for this.

◇ It is generally recommended to follow local building codes and guidelines for the placement of storage tanks. The specific requirements may vary depending on factors such as the type and size of the tank, the soil conditions, and the local regulations. Consulting with a

Plan ahead for tanks are very heavy when full. A full tank cannot be moved without draining it first.

structural engineer, builder, or local authorities can provide guidance on appropriate distances and considerations specific to your location and circumstances.

Overall, it's essential to choose the best position for the water storage tank based on your specific needs and to take the necessary steps to protect the tank to ensure its longevity. The most common tank foundations, that have been tried and tested for decades, are concrete slabs and steel platforms. You cannot go wrong with these two options.

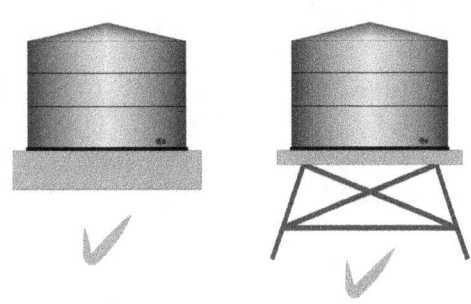

Having gained some insight into tank foundations, let's now shift our focus to the water tank itself and delve into the process of installing fittings onto it.

Collection And Storage Of Rainwater

How To Pour A Concrete Slab

You will need:
- Necessary permits
- Suitable concrete mix
- Reinforcement: Depending on the size and load of your water tank, you may need to reinforce the concrete slab with rebar/wire mesh.
- Formwork: Build or purchase formwork to create the shape and size of the slab.
- Gravel: Use a layer of compacted gravel or crushed stone as a base for the slab.
- Shovel, rake, wheelbarrow, bull float, screed board: These tools will be helpful for moving, spreading and smoothing the concrete.
- Level, measuring tape, and string: Use these tools to ensure the slab is level and aligned.

How To:
1. Prepare the site: Clear the area and excavate the area to the required depth.
2. Install formwork: Set up the formwork according to the dimensions of your desired slab. Apply a release agent to the inside of the formwork to prevent sticking.
3. Add a gravel base: Spread a layer of compacted gravel or crushed stone evenly over the excavated area. This is for drainage.
4. Reinforce the slab (if necessary): If your water tank is large or heavy, reinforcing the concrete with rebar or wire mesh is recommended.
5. Mix and pour the concrete: Prepare the concrete mix according to the manufacturer's instructions. Start pouring the concrete into the formwork, working from one end to the other. Use a shovel or rake to spread and level the concrete.
6. Consolidate and smooth the concrete: Use a screed board or straightedge to level the surface of the concrete, moving it back and forth in a sawing motion. Float the surface with a bull float to smooth out any imperfections.
7. Cure and protect the concrete: Cover the freshly poured concrete slab with plastic sheeting or damp burlap to retain moisture and promote proper curing. Leave it undisturbed for at least 48 hours or follow the curing time recommended by the concrete manufacturer. After curing, keep the slab moist by regularly spraying it with water for several days.

Remove the formwork: Once the concrete has cured sufficiently, carefully remove the formwork. Take caution not to damage the edges of the slab.

How To Install Fittings On A Water Tank

Polyethylene (PE) tanks are equipped with water tank fittings that are intentionally designed to be user-friendly, requiring no specialized plumbing knowledge to work with. This makes it an easy DIY project for the average enthusiast, who will only need a few basic tools. Moreover, the knowledge gained in this section is transferable and can be applied to other types of tanks and fittings. First, let's have a look at the fittings:

Water tank fittings are also known as bulkheads, bulkhead fittings, and through-wall fittings. By definition, a bulkhead is a plumbing connection that passes through the wall of a sealed vessel or object to allow the flow or continued route of a liquid.

When we look at plastic tanks we see that bulkhead fittings come in an assortment of materials, styles, sizes and threads.

- The most common materials include PVC, PE, PP, CPVC, ABS and stainless steel.
- The styles include threaded, bolted, self aligning, and garden hose adapter.
- Frequent sizes include ½", ¾", 1", 1-¼", 1-½", 2", 3", and 4", (12mm, 20mm, 25mm, 32mm, 40mm, 50mm, 75mm, 100mm).
- National Pipe Thread (NPT) style for union to other NPT plumbing is the most common thread.

With all these options at your disposal, it's easy to see why no special knowledge is required for the DIY enthusiast to succeed at installing fittings on a water tank. Let's take a look at the 2" threaded PP (polypropylene) bulkhead fitting which is considered one of the most common.

In essence, a bulkhead fitting (fig.4.23) comprises of three key components: the body, the washer or gasket, and the locknut.

The **body** serves as the primary element of the through-wall fitting, while the **washer or**

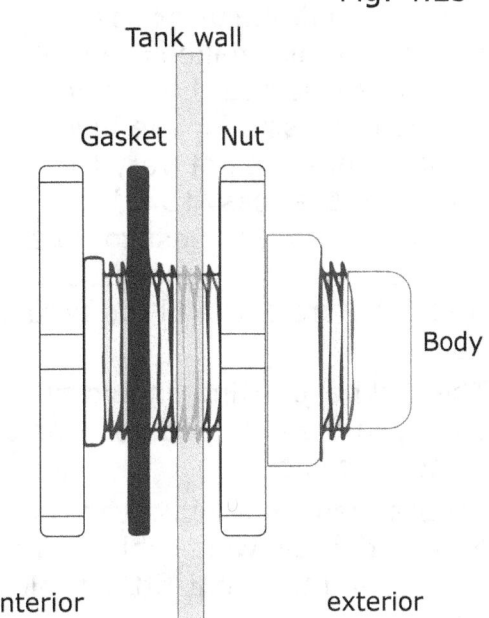

Fig. 4.23

Collection And Storage Of Rainwater

gasket ensures a watertight seal, and the **locknut** secures the entire fitting in place.

How To Install A Bulkhead Fitting

1. Prepare Tank and Location

Drain and clean the tank. If you are only working at the top of the tank with overflows and inlets, then drain the tank to a level that is beneath the location for the new fitting. Use a permanent marker to mark the location where the fitting will be installed.

Look for flat sections on the tank wall. Water tanks often have deliberately created flat surfaces called fitting flats, which serve as designated areas for attaching bulkhead fittings.

2. Cut the Tank Wall

Before you drill, cut or saw into your tank, make sure that you have selected the correct location and size for the hole. Plastic tanks are not very thick and it's rather easy to drill through them. It's best to do a practice run on a piece of spare plastic before you drill the actual hole.

The bulkhead fitting should fit snugly in the hole and you don't want a lose fit.

For standard type bulkheads, look at the following table to aid in selecting the right fitting hole size:

Imperial

Bulkhead Size	Min Hole Size	Max Hole Size
½"	1.25"	1.38"
¾"	1.58"	1.75"
1"	1.9"	2"
1-¼"	2.16"	2.375"
1-½"	2.375"	2.5"
2"	3"	3.15"

Metric

Bulkhead Size	Min Size	Max Size
12mm	32mm	35mm
20mm	40mm	45mm
25mm	47mm	52mm
32mm	55mm	60mm
40mm	59mm	65mm
50mm	75mm	80mm

When drilling the hole for a bulkhead fitting in a plastic rainwater tank, it's important to use the right type of drill bit and hole saw to ensure a clean and accurate cut. A hole saw is typically recommended for this purpose, as it can create a clean circular cut without damaging the surrounding material.

It's better to use a hole saw that is specifically rated for cutting plastic, as this will ensure that the blade is designed to handle the unique properties of the material. You can typically find plastic-rated hole saws at hardware stores or online retailers.

Another option is to use an

auger type bit, which can also create a clean and accurate cut. However, it's key to make sure that the bit is sharp and in good condition, as a dull or damaged bit can cause the plastic to crack or break. Your bulkhead should fit snugly. Cleaning up around the edges of the cut and smoothing out with sandpaper is recommended.

Bolted Bulkhead Fittings

To install bolted bulkhead fittings on a poly tank, a hole of a specific size must be cut, which will differ from a standard bulkhead fitting. Use a hole saw for this cut. Additionally, pilot holes for the bolts used to secure the fitting should be drilled, using a power drill and drill bit. It's critical to note that the required hole size for bolted bulkheads will vary between polypropylene and stainless steel fittings. Look at the following table to aid in selecting the right fitting hole size:

Imperial Polypropylene

Bulkhead Size	Hole Saw Size	Bolt Qty	Bolt Drill Size
¾"	1 ⅜"	4	11/32
1"	1-⅝"	4	11/32
1-¼"	2"	4	13/32
1-½"	2-¼"	4	13/32
2"	2-¾"	4	13/32
3"	3-⅞"	6	13/32
4"	5"	8	13/32

Metric Polypropylene

Bulkhead Size	Hole Saw Size	Bolt Qty	Bolt Drill Size
20mm	25mm	4	13mm
25mm	35mm	4	13mm
40mm	44mm	4	17 mm
50mm	57mm	4	17 mm
75mm	83mm	6	17 mm

Stainless Steel Bolted Bulkhead Fittings

Bulkhead Size	Hole Saw Size	Bolt Qty	Bolt Drill Size
20mm	35mm	4	13mm
25mm	41mm	4	13mm
32mm	51mm	4	17 mm
40mm	57mm	4	17 mm
50mm	70mm	4	17 mm
75mm	98mm	6	17 mm
100mm	127mm	8	17 mm

Stainless Steel Bolted Bulkhead Fittings

Bulkhead Size	Hole Saw Size	Bolt Qty	Bolt Drill Size
¾"	1"	4	11/32
1"	1-⅜"	4	11/32
1-½"	1-¾"	4	13/32
2"	2-¼"	4	13/32
3"	3-¼"	6	13/32

3. How to insert the Fitting

It is noteworthy to ensure that the bulkhead fitting is inserted properly through the hole in the tank, with the gasket and fitting body on the inside of the tank and the locknut and thread side on the outside. This will allow for proper plumbing connections to be made.

If tank access makes it difficult to insert the fitting from the tank interior, a weighted string can be used to guide the fitting into position. This can be done by feeding the string through the tank access port (or lid) and threading it through the newly cut hole, then guiding the fitting along the string and into position. Having another person assist with this process can make it easier, as they can help guide the string and fitting into position. Pay special attention to this part, since it can be the most difficult part of the fitting installation process, especially if access to the tank is limited. As an alternative, a gadget called a **Fish Tape** can also be used. A fish tape is a flexible and slender tool that is used to feed wires, cables, and other objects through small or narrow openings, such as holes in tanks, ceilings, or floors. It is commonly used by electricians, network technicians, and other professionals who need to run wires and cables through confined spaces.

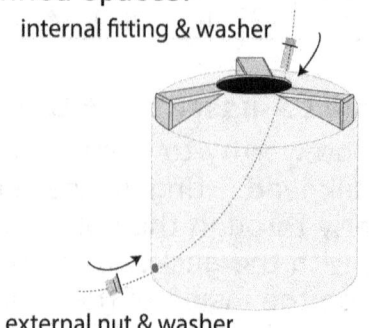

Fish tapes are typically made of a long, narrow strip of steel or fiberglass, which is coiled up for storage. The tape is fed through the hole or opening until it reaches the other side, and then the wire or cable is attached to the end of the tape and pulled back through the opening.

4. Fasten the Locknut or Bolts

Insert the gasket into the fitting body and place it on the inside of the tank, with the flat, solid end facing inward.

Thread the locknut onto the fitting body from the outside of the tank and tighten it snugly by hand.

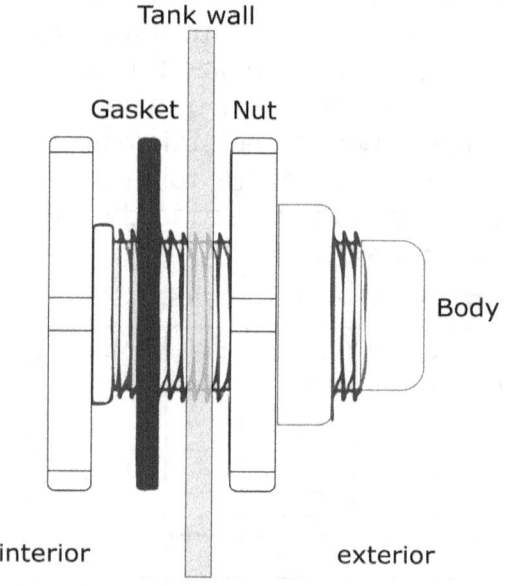

Use your hand to tighten the locknut until it is snug and the gasket is compressed. Do not over-tighten, as this can damage the gasket and cause leaks. Once the locknut feels secure and it resists your hand movement, then you can perform another quarter turn to one-half turn with a wrench to ensure the fitting is tightened securely. Over-tightening can damage the gasket and cause leaks.

It is also important to note that Plumber's (Teflon) tape should not be used on the locknut during fitting installation. This is because the tape can interfere with the proper compression of the gasket and lead to leaks. However, Plumber's tape can be used on the threads of the fitting when connecting valves or plumbing to the bulkhead fitting after it has been installed.

On bolted style fittings thread the bolts through the fitting body from the outside of the tank, making sure the bolts pass through the holes in the tank and line up with the holes in the fitting body.

Place the washers and nuts onto the threaded end of the bolts and tighten them by hand until they are snug.

Use a wrench to tighten the nuts in a crisscross pattern, starting with one nut and moving diagonally to the opposite nut, until they are tight and the gasket is compressed. Do not over-tighten, as this can damage the gasket and cause leaks.

5. Check Seal

It is crucial to perform a water test after installing a bulkhead fitting in a water tank to ensure that the seal is tight and there are no leaks. Filling the tank up to the level of the fitting and checking for any water coming from around the seal is an effective way to perform this test.

If a leak is detected, it is important to take corrective action immediately. For compression-style bulkhead fittings, this may involve loosening the locknut slightly and rotating the entire fitting in a clockwise turn to reposition it and create a tighter seal. For bolted-style bulkhead fittings, checking the tightness of the bolts may be necessary.

Flexible Plumbing vs. Rigid Plumbing

What is flexible plumbing?

- Flexible plumbing is any hose that complies with the chemical that is in the tank.
- A flexible hose will not put any stress on the fitting when it moves because it will flex when needed.
- Flexible plumbing is perfect for interconnecting tanks. Because of the material, when a poly tank is filled with liquid, it may flex a small amount. This causes the bottom fitting to tilt, putting extra pressure on the plumbing connection.

What is rigid plumbing?
- Plastic, steel or any other rigid pipe.
- Rigid pipes are generally considered pipes that cannot deflect 2% of their diameter before failing.
- Rigid pipes are most commonly installed for low-pressure or gravity-flow water and wastewater applications.

The choice between flexible hose and rigid PVC piping depends on various factors.

Flexible hose:
- Flexibility: Flexible hoses are more adaptable and can accommodate slight movements or shifts between the tanks.
- Installation: They are relatively easier to install, requiring fewer fittings and connectors.
- Cost: Flexible hoses tend to be more affordable compared to rigid PVC piping.
- Maintenance: They are generally easier to maintain and repair, as they can be easily replaced if necessary.
- Material: Select a flexible hose specifically designed for water applications to ensure durability and compatibility.

Rigid PVC piping:
◊ Stability: Rigid PVC piping provides greater stability and structural support due to its rigid nature.
◊ Longevity: PVC pipes are known for their durability and can withstand long-term exposure to water and weather conditions.
◊ Pressure and flow: Rigid piping can handle higher water pressure and maintain consistent flow rates.
◊ Material: Ensure that the PVC piping used is suitable for conveying potable water.
◊ Consider your specific requirements, such as the distance between the tanks, potential movements or shifts, desired flow rate, and budget when deciding which piping option is best for your situation.

In short, rigid plumbing, such as PVC pipes, is suitable when you can ensure that the tank will maintain its rigid structure (no flex) regardless of being full or empty. This is commonly seen in tanks made of concrete, steel, fiberglass, or wood.

On the other hand, flexible plumbing, like using hoses, is a better choice when you anticipate movement or flexing in the tank's walls. This can be caused by factors such as seismic activity, unstable ground surfaces, or the material of the tank itself. Poly tanks are an example of tanks that may benefit from flexible plumbing solutions.

Flexible Fittings for Connecting Tanks

Let's take a look at a typical flexible plumbing kit (fig.4.24) for connecting two tanks.

- 2 x 2" Threaded Male nipples (goes into each bulkhead fitting)
- 2 x 2" ¼ turn Ball Valves (connects to nipples)
- 2 x 2 " Male hose camlock couplers (connects to Ball Valves)
- 2 x 2 " Female hose camlock couplers or 2 x hose clamps
- 1 x 3ft (100cm) length of 2" flexible reinforced hose (connects to camlock couplers)

How To Install Flexible Hosing

A typical use of flexible hosing is to connect the two outlet ports of two tanks. Follow these steps:

a. First, you will need to measure and cut your high-quality, flexible hose to the correct length.
b. Ensure the tanks are level and positioned correctly, allowing for easy connection.
c. Clean the tank outlets to remove any debris or obstructions that could interfere with the hose connection.
d. Thread one male nipple into the outlet port of the first tank, and also another male nipple into the outlet port of the second tank. Attach a ball valve to each nipple.

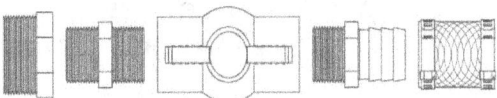

e. Attach one end of the flexible hose to the ball valve of the first rainwater tank. This can typically be done using hose clamps. Alternatively you can use a combination of male hose camlock couplers to female hose camlock couplers.
f. Repeat this process when you connect the other end

Flexible Fittings

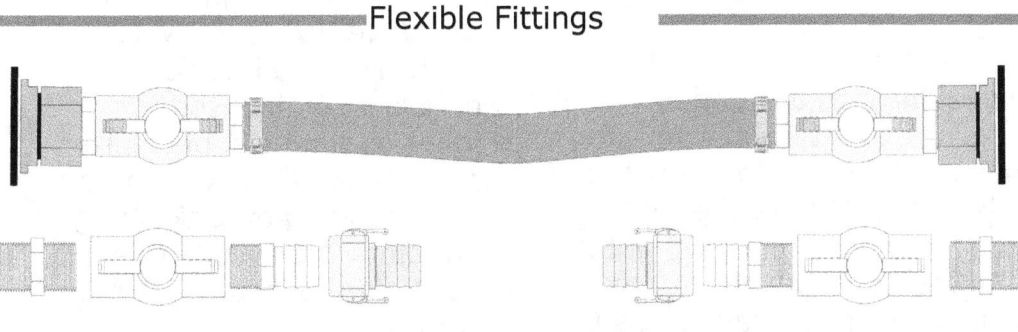

Fig. 4.24

of the flexible hose to the inlet of the second rainwater tank, again using appropriate fittings and ensuring a secure connection.

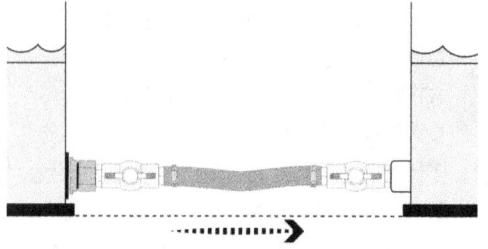

g. Double-check all connections to make sure they are tight and leak-free. Use Plumber's (Teflon) tape where necessary.
h. If required, support the flexible hose at intervals along its length using clamps or brackets to prevent sagging or undue strain on the connections.
i. Test the system by turning on the water supply and checking for any leaks or issues. Adjust the connections if needed. Make sure that the connections are water-tight.

Fittings Under Pressure

Not all fittings are under pressure in a typical rainwater harvest system. Normally the inlets at the top of the tank act as funnels, directing water into the tank, but they do not experience significant pressure. On the other hand, fittings such as outlet ports and drain valves, which are located at the bottom of the tank, do experience moderate pressure due to the weight of the water volume pushing through them under the influence of gravity and atmospheric pressure. When purchasing fittings for your rainwater harvesting system, you will find different options at the hardware store. **Schedule 40 PVC** is generally recommended for water tank fittings because it is designed to handle pressurized uses. However, **DWV** (Drain, Waste, and Vent) **PVC** parts may look very similar to standard white Schedule 40 PVC pipe and fittings, but they are not intended for pressurized applications.

DWV fittings can be suitable for inlet pipes, first flush diverters, and the overflow outlet of the tank, since these parts do not operate under pressure.

The most common schedule numbers for PVC pipes used in rainwater tank installation are Schedule 40 and Schedule 80. **Schedule 40** PVC pipes are thinner and lighter than **Schedule 80** pipes, which are thicker and heavier and are typically used for higher commercial pressure applications. Now, with a better understanding of tank fittings and joints, let's apply this knowledge to install a First Flush diverter.

Installing A First Flush Diverter

Installing a First Flush diverter is a relatively simple process. First Flush diverters are used in rainwater harvesting systems to divert the initial runoff from a roof during rain events, which often contains debris and contaminants, away from the storage tank. This helps to improve the quality of harvested rainwater. Let's look at a step-by-step guide on how to install a four inch First Flush diverter.

You Will Need

- First Flush diverter kit (includes diverter unit, fittings, and pipe)
- Tape measure & marker/pencil
- PVC cement
- Pipe cutter or hacksaw
- Pipe wrench or pliers
- Plumber's (Teflon) tape

Typical 4" First Flush diverter kit:
- First Flush Tee
- Diversion Chamber
- Ball Valve
- Ball Seat
- End Cap
- 2 x wall brackets (4"/100mm pipe)
- Flow control washers

Step 1: Choose the location

Select a suitable location for the First Flush diverter along the downspout or pipe that collects rainwater from the roof. Choose a spot that is easily accessible and allows for proper installation and maintenance of the diverter.

Step 2: Measure and cut the pipe

Measure the length of pipe (Diversion Chamber) required for the First Flush diverter installation. As a rough guide, 3 feet of 4 inch (1m of 100mm) pipe holds approximately 2.3 gallons (8.8L) of water. Cut the pipe to the appropriate length using a pipe cutter or hacksaw. Make sure the cut is clean and square.

Step 3: Decide on the orientation

Decide on the orientation of the First Flush Tee that is most suitable for your installation. The Tee can be installed horizontally or vertically, and the outlet must be easily accessible for maintenance and inspection. Refer to figure 4.25 on page 131 for suggested installation orientations.

Step 4: Select which socket you will use

Select the appropriate socket outlet, on the First Flush Tee, that will attach to your diversion chamber. Follow the manufacturer's instructions to guide you.

Step 5: Orient the Ball Seat

Orient the Ball Seat to fit inside the First Flush Tee socket (that will later attach to your diversion chamber) then apply PVC cement and press it hard up inside the socket. If you are attaching the First Flush to existing downpipes, then pipe reducers might be required depending on your downspout pipe width. For similar size downspouts (4"/10cm), a reducer will not be required.

Step 6: Connect First Flush Tee to the downspout

Measure the existing downspout and cut it to create space for the First Flush Tee. Ensure that the End Cap (outlet) of your diverter sits at least 5.9 inches (15cm) from the ground when fully assembled. Apply PVC cement to the First Flush Tee sockets and pipe to position your First Flush Tee in place.

Step 7: Install the End Cap

Take your Diversion Chamber pipe and apply PVC cement to one end. Insert the pipe firmly into the End Cap.

Make sure that the ball can fit through the End Cap, otherwise you will have to install it after Step 10.

Step 8: Install the Diversion

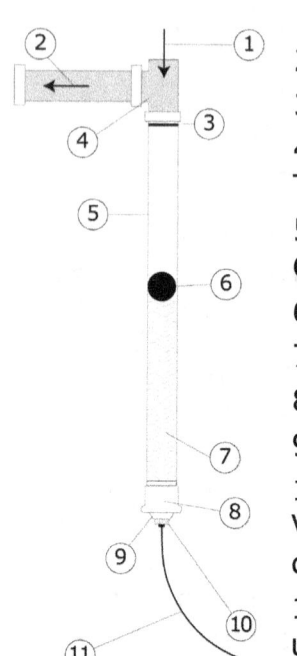

1- From roof
2- To tank
3- Ball Seat
4- First Flush Tee
5- Diversion Chamber
6- Ball Valve
7- Dirty water
8- End Cap
9- Exit Funnel
10- Release Valve screw cap
11- Irrigation use

Chamber pipe

Earlier in Step 5 you fit the Ball Seat inside the First Flush Tee. Take your Diversion Chamber pipe and apply PVC cement to the other end. Insert the pipe firmly into the socket of the Ball Seat.

Step 9: Mount it to the wall

Attach the unit to the wall using the supplied brackets, ensuring that it is fully secured. The upper bracket should sit directly under the First Flush Tee to support the weight of the unit.

Step 10: Insert the Ball Valve

Place the ball inside your First Flush diversion chamber.

Step 11: Install fittings

Attach the rest of the fittings like the exit funnel and release

Installation Orientations

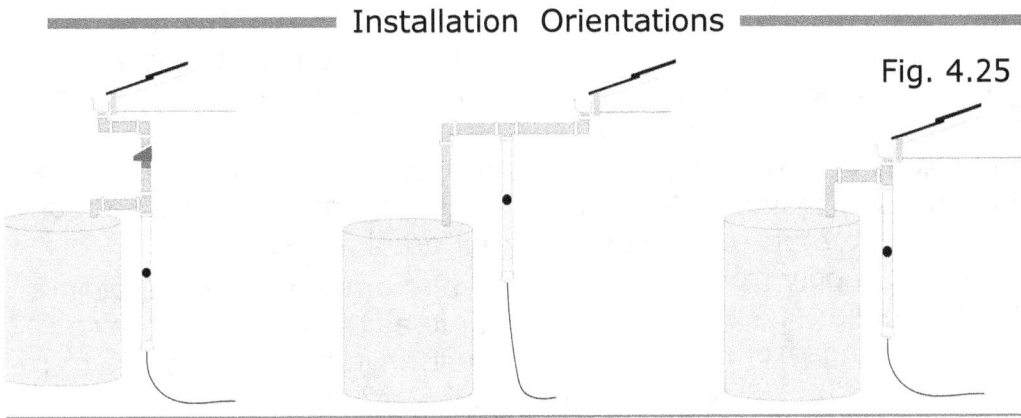

Fig. 4.25

valve screw cap according to the manufacturer's instructions. These parts normally screw on to the end cap.

Step 12: Select a flow control washer

Remember to select the appropriate flow control washer and fit it into its designated spot. Start with the flow control washer with the smallest gauge hole (lowest number), and use a larger gauge washer if blockages occur. Save the remaining washers for possible future use.

If your diverter came with a filter screen, then follow the manufacturer's instructions to install it.

Step 13: Connect the overflow pipe

If your First Flush diverter kit includes an overflow pipe, connect it to the designated outlet on the diverter unit.

The overflow pipe is used to divert excess water away from the storage tank once the diverter unit is full.

Step 14: Test the installation

Turn on your rainwater harvesting system and check for any leaks around the First Flush diverter unit and fittings. Adjust and tighten connections as needed.

Step 15: Maintenance

Regularly inspect and clean the First Flush diverter unit to ensure it is functioning properly. It is important to ensure that your First Flush diverter outlet remains clear of any debris. If your outlet becomes blocked, the chamber will not empty and the first flush of water will not be diverted when it rains. Follow the manufacturer's instructions for cleaning and maintenance.

Note: It's important to follow the manufacturer's instructions provided, as installation steps and requirements may vary depending on the brand and model. If you're not confident in your DIY skills, it's recommended to seek the help of a professional installer.

Vents

Storage tanks must have dedicated vents, overflow and discharge pipes, and drain lines to operate. Keep in mind though that, to prevent the water supply from being contaminated, you must protect these openings from birds, bats, other animals, insects, rain and windborne contaminants.

Vents Explained

Vents play a crucial role in maintaining the air pressure balance inside and outside of a tank. As the water level in the tank rises and falls, the air pressure inside the tank tries to equalize with the external air pressure. When the tank is being filled with water, the air inside the tank gets compressed as it has less space. This compression creates pressure on the tank, and the excess air is forced out of the tank through the vent and overflow port to maintain equilibrium. Conversely, when water is drawn out of the tank, the volume of air inside the tank increases, creating a vacuum. This vacuum causes outside air to be drawn into the tank through the vent and overflow, again maintaining the balance of air pressure. A vacuum is the absence of pressure. If there is no pressure internally, there is no force to combat atmospheric pressure. There is and always will be a constant force (atmospheric pressure) acting on the exterior of the tank walls. Atmospheric pressure is 14.7 pounds per square inch. When a full vacuum is created there is no internal pressure in the tank to combat atmospheric pressure (external force).

Also, storage tank vents cannot serve as the overflow; tanks must have a vent separate from the overflow.

Vent Design

Vents must have screens to keep organisms out of the reservoir. We recommend using durable 24-mesh non-corrodible screen backed with 4-mesh screen. Vent openings must face downward or have shielding to minimize the entrance of insects, rainwater, rain splash, and excessive dust. Modern vent designs protect against icing, vacuum conditions, and tampering. Vent size must be adequate to relieve vacuum during peak-flow conditions.

1. Simple Screen Patch

A simple option is to drill a hole in the top of the tank, ⅝" or 15mm will suffice, and then to glue a small piece of 24-mesh screen over the hole with silicone sealant.

A plumber's fitting (like a PVC male adapter with thread) will also work. Just glue the mesh screen in the fitting and screw it into your tank.

These are makeshift solutions and not recommended as long-term options. Due to the fact that the mesh screen is facing upwards, dust, droppings, and other particles will gather on the screen surface and clog the openings and render it less effective.

2. Gooseneck Or J-Type Vents

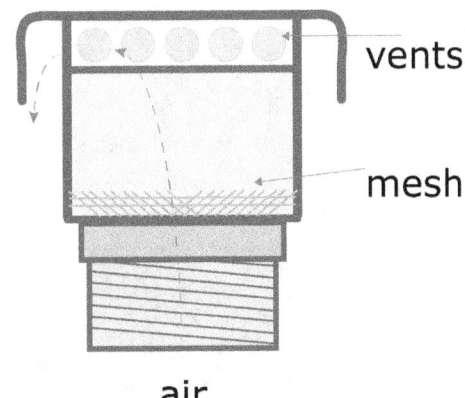

A better option is a gooseneck vent. These vents are generally considered to provide adequate sanitary protection. They may be constructed out of PVC, steel, ductile iron, or other material. All vents must be sealed and secured to the top of the tank to keep out contaminated water and to deter vandalism. The screened open end faces downwards, and should be the recommended minimum height above the top of the tank.

3. Mushroom-Style Vents

These vents, made of durable non-corrodible steel, have an internal downward or vertical screen with hood shield and are secured or sealed to the tank

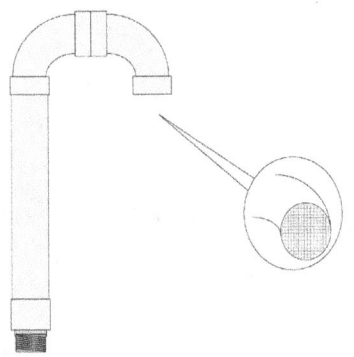

top to keep out contaminants. If constructed of sturdy material they provide added security from tampering.

Water tank vents may be more expensive compared to normal PVC fittings for several reasons:

- **Specialized Design:** Water tank vents are designed to perform specific functions such as allowing air to escape during tank filling, preventing vacuum formation during emptying, and maintaining proper pressure inside the tank. These vents require specialized designs and features to ensure proper functionality.
- **Materials:** Water tank vents are often made from high-quality materials that are durable, resistant to corrosion, and capable of withstanding varying environmental conditions.
- **Regulations and Standards:** Water tank vents need to meet specific regulatory standards.

How To Make A Vent

Many manufacturers sell their vents in hardware stores or online. If buying a vent is not an option, then you'll have to make your own.

To make a simple gooseneck vent you will need:
- ¾" PVC pipe of around 4-6" (10-15cm)
- 2 x ¾" PVC 90° Elbow
- A base fitting to attach to your tank (i.e. ¾" PVC Male Adapter)
- PVC primer and cement
- 24-mesh non-corrodible screen
- Tin snips

How to:
1. Cut the non-corrodible screen to the appropriate size to match the size of the opening in the downward-facing gooseneck.
2. Fit the screen into the opening of the downward facing gooseneck (90° PVC Elbow). Use the tin snips to trim the screen if necessary.
3. Apply silicone sealant to seal the sides of the screen, ensuring it is securely in place.
4. Assemble the PVC pipes and fittings as shown, using the 90° PVC Elbows and the base fitting (PVC Male Adapter).
5. Use PVC primer and cement according to the manufacturer's instructions to join the PVC components securely. Allow sufficient time for the cement to cure.

A less efficient, secondary option is to get flexible mesh (will not last as long) and to secure it with a hose clamp to the downward facing gooseneck (90° PVC Elbow).

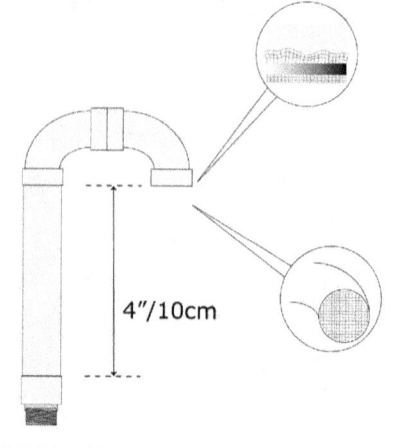

Overflow Installation

We already know the basics of a tank overflow system (see p.104), but it's also necessary that we look at the finer details and mechanics of the whole process:

a. A water tank overflow is a mechanism that is designed to prevent a water tank from overfilling and spilling excess water onto the surrounding area. The overflow works by allowing excess water to escape from the tank when the water level reaches a certain point. Diverting the excess water that enters the tank is helpful in that it prevents overflow back up the downspout and towards the pre-tank filters. Creating an exit-path allows you to control the excess water and to guide it to a more suitable location like your garden (irrigation) or the stormwater drain.

b. Typically, a water tank overflow consists of a pipe that is connected to the tank and extends to the outside of the tank. The overflow pipe is positioned at a height above the normal water level in the tank, usually near the top of the tank. Due to the fact that the overflow outlet is situated at the very top of the tank it also allows you to remove unwanted floating particles which in turn improves the quality of your water.

c. The overflow prevents the water level from rising and overfilling the tank. This release of water helps the tank to equalize and to release internal pressure which can put unwanted strain on the tank walls and fittings. The overflow guarantees pressure equilibrium.

d. Many water tank overflows also include a screen or filter to prevent debris or insects from entering the overflow pipe or port.

e. It is important to regularly inspect and maintain water tank overflows to ensure that they are functioning properly and prevent water damage from occurring.

Having a water tank overflow provides several benefits, including:

a. **Preventing water damage:** The primary benefit of having a water tank overflow is that it prevents water damage by ensuring that the tank does not overfill and spill excess water onto the surrounding area.

b. **Conserving water:** Water tank overflows can also help to conserve water by ensuring that any excess water is not wasted. Instead, it can be directed to other uses, such as irrigation or

watering plants.
c. **Reducing strain on the tank:** Water tank overfilling can strain the tank and its components, increasing the risk of damage or failure. To mitigate this, a water tank overflow is crucial as it releases surplus water, relieving the strain on the tank. It is important to note that a water tank overflow does not serve as a vent for the tank. You must still install a separate vent to address the fluctuation of air pressure in the tank.
d. **Improving water quality:** A water tank overflow with a screen or filter can help to improve water quality by preventing debris or insects from entering the overflow pipe or channel.
e. **Meeting regulatory requirements:** In some areas, it may be a regulatory requirement to have a water tank overflow.

Regarding the drainage of water from the overflow, you have several options to consider:
1. You can choose to direct the overflow pipe to drain directly into the storm-water system.
2. Another option is to connect the overflow to a lower lying tank that can be used for secondary water requirements.
3. Alternatively, you can connect the overflow to flow into a pool, pond, or cistern, which can then be utilized for purposes such as livestock water supply or firefighting needs.

How to

1. Determine the height of the overflow:
The overflow should be positioned at a height above the normal water level in the tank. Check the manufacturer's instructions or consult a professional to determine the appropriate height.

2. Choose the location for the overflow:
The overflow outlet should be located in a position that allows excess water to flow away from the tank and prevent water damage to the surrounding area. We recommend the outlet of the overflow to dump water a

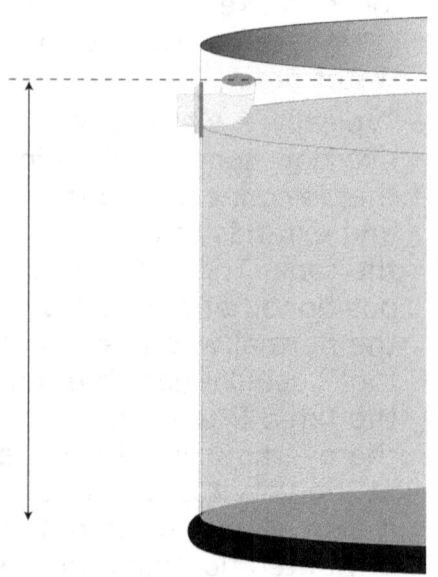

distance of no less than 20 feet (6m) away from the tank base. The location should be easily accessible for maintenance and inspection. Be wary of letting water flow directly onto the ground. Not only does it cause erosion and loss of soil, but it can potentially also damage structure foundations.

3. Install the overflow fitting:

Check the manufacturer's instructions that came with your overflow installation kit, otherwise follow these steps:

a. Drill a hole in the tank with a hole saw that is ¾" (2cm) larger than the fitting size.
b. Clean the hole by breaking away the small fragments of plastic (burrs) on the surface of the hole drilled. They can cause leaks.
c. Place a rubber gasket behind the fitting. This will stop water leaking out the fitting joint when the water level is high.
d. Insert your overflow fitting from the outside of the tank. The inside curve must be facing upwards.
e. Screw the overflow fitting into the tank wall while maintaining pressure on the screws. Tighten the screws evenly: Begin tightening the screws gradually and evenly, working your way around the port. Tighten each screw a little bit at a time until the port is snug against the tank. Don't let the gasket bulge, since it will cause leaks.
f. If needed, add an overflow pipe that matches your inlet pipe diameter.

4. Install a screen or filter:

The overflow should include a screen or filter to prevent debris or insects from entering the pipe or port. The screen or filter should be installed at the top of the overflow pipe or port.

5. Secure the overflow:

The overflow should be securely attached to the tank and the surrounding structure. Use appropriate fittings and brackets to ensure that the overflow is stable and can withstand water flow.

6. Test the overflow:

Once the overflow is installed, test it to ensure that it is

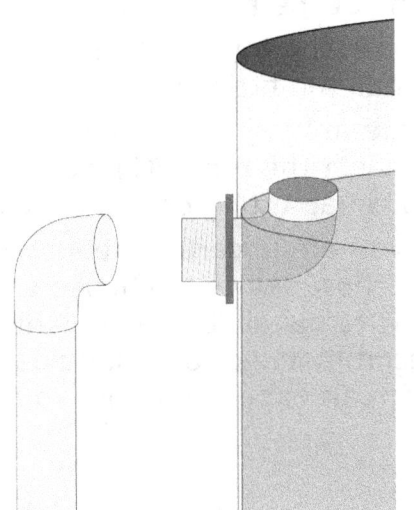

functioning properly. Fill the tank with water and monitor the overflow to ensure that excess water flows out of the tank and away from the surrounding area.

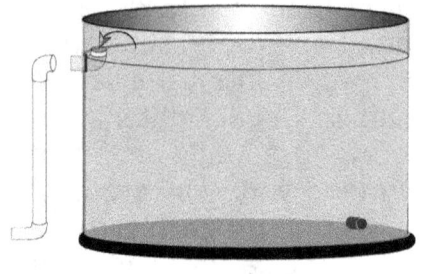

When it comes to managing water in a rainwater storage tank, there are several crucial factors to consider regarding overflow fittings.

1. Firstly, it is important to match the size of the tank's overflow pipe to the size of its inlet pipe. This ensures that the overflow can keep up with the incoming water when the tank is full. For instance, if the inlet pipe is 3½ inches in diameter, then the overflow pipe should also be 3½ inches in diameter.

2. Secondly, it is crucial that the overflow pipe is installed at a lower elevation than the inlet pipe. It is recommended that the overflow be a minimum of half the pipe diameter below the inlet pipe. This allows for overflow to occur before the water backs up into the pre-tank filter or inlet pipe.

3. Thirdly, it is advisable to have a lip or bevel on the overflow pipe. This creates a siphon effect, which helps to draw water into the pipe and siphon floating particles through the overflow. If PVC piping is used, an angle of 11 or 22 degrees cut across the pipe is recommended.

4. Finally, to prevent mosquitoes and other bugs from entering the rainwater storage tank, it is essential to create an airtight connection on the tank. This will keep the tank free from pests and ensure that the water remains clean and safe for use.

Nowadays, there is a wide range of very efficient overflow devices available on the market. You can also make your own overflow pipe by simply taking a predetermined length of PVC pipe and attaching a 90° PVC Elbow facing the tank outlet at the top end, and another directing water away from the tank at the bottom end.

Use PVC primer and cement to install the fittings.

Once assembled you can connect the top PVC Elbow to a bulkhead fitting on the tank wall.

If you are uncertain about which overflow device to choose or how to install it, then it is highly recommended to seek guidance from an expert to assist you.

Single Tank Installation

Once you've done your research and completed your calculations, you can buy your tank. New is best, but a good second hand tank will also do.

- Make sure that it is completely clean and remove all coats of paint, solvents or chemicals. If unsure, do not use it!
- New tanks should come with plumber's fittings and a warranty.
- Translucent tanks and pipes should be avoided because they will encourage algal growth.
- Ask about options concerning a water level indicator. This is a very useful accessory to have.
- It's recommended that the downspout and the gutters are of the same material and that the gutter outlet matches the diameter of the downspout.

In the following installation we will look at a traditional dry system installation of a single, standard, polyethylene tank. Every individual's vision is different, but the information provided here will get you on track to create your own, unique rainwater harvest system to meet your needs.

How To

Step 1: Suitable location

Make sure your tank is in a spot where it can collect rainwater and also where runoff or leaks

You Will Need

- 1 x Tank (clean/ food-grade, with an inspection port)
- Pre-prepared level surface or suitable platform for placement
- Gutters and downspouts as needed
- Gutter screens, guards and fasteners as needed
- Tank inlet kit
- Tank overflow kit
- Tank outlet kit
- Tank drainage outlet kit
- PVC pipe (as needed)
- PVC 90° elbow (as needed)
- PVC Tee (as needed)
- PVC threaded ball valve
- Tank vent
- First Flush diverter kit
- Plumber's (Teflon) tape
- PVC primer and cement
- Pipe wrench or pliers
- Drill or hole saw
- Tin snips (option)

won't cause any damage to your property or your neighbor's.

Step 2: Prepare the foundation

Make sure that you selected a suitable location away from your house's foundations and also that it's flat and sturdy enough to handle the weight of a full tank. A concrete foundation is generally considered the best option. Four to six inches (10-15cm) of reinforced concrete should be enough for a standard tank. The foundation diameter should be eight inches (20cm) greater that the diameter of the tank (see p.115).

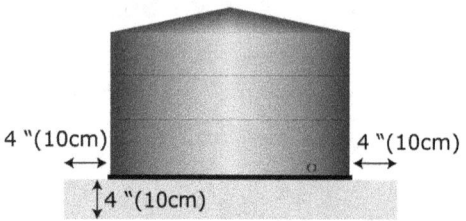

Aim for 4 inches (10cm) of reinforced concrete for a water tank.
Your foundation must be 8 inches (20cm) greater than the diameter of the tank.

Step 3: Place your water tank

Depending on the type of tank, use a crane or simple manpower to move the empty tank to the area that you have prepared. Visualize the full effect that it will have on your dwelling such as irrigation, household, water heating systems, general use, car washing, swimming pools, fire fighting) and prepare accordingly. Your tank must be placed at least 3 feet (1m) away from the house.

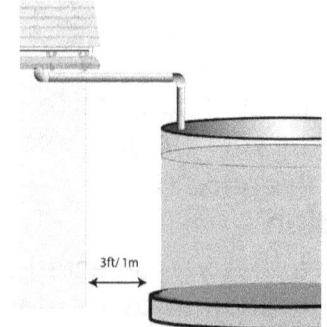

Step 4: Check your roof surface

Make sure there are no chemicals, paint, paint flakes or mold on the roof surface. Chimneys cause smoke residue that gets washed down into gutters. You want a clean roof, preferably metal, slate or tile. Clear overhanging branches from gutters.

Step 5: Place guards or screens over the gutter holes

These screens or strainers will keep debris and leaves from clogging your downspouts (see p.85).

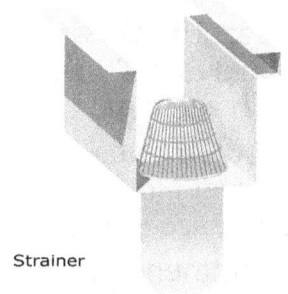

Strainer

Step 6: Install your gutter outlet

Determine the location on the gutter where you want to install the outlet. Mark the spot with a pencil or marker. Most modern gutters come with a separate gutter outlet section and you should follow the instructions of the manufacturer. Most vinyl gutter outlets snap into place. If you are dealing with an older type of gutter, then you will have to cut the hole yourself.

For older **vinyl gutters** cut a hole in the gutter at the marked spot using a hole saw or a drill with a hole saw attachment. Apply PVC cement and insert the outlet into the hole of the gutter. The outlet should fit snugly into the hole. Hold the outlet in place for a few seconds to allow the cement to set.

For **metal gutters** use a tin snip or drill, and cut out the shape on the gutter to make room for the outlet. Insert the outlet into the hole you created in the gutter, and use sheet metal screws to secure it in place. Be sure to tighten the screws firmly but not too tight to avoid damaging the gutter.

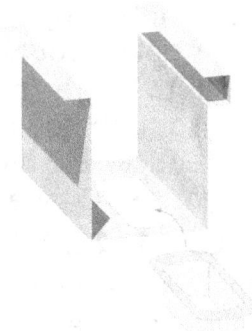

Step 7: Slide the downspout or elbow over the outlet

Make sure that the gutter outlet matches the diameter of your downspout. Apply PVC cement around the joint to ensure a watertight seal (vinyl) or use screws or rivets to secure the downspout to the outlet (metal). If you used PVC cement, hold the downspout in place for a few seconds to allow the cement to set.

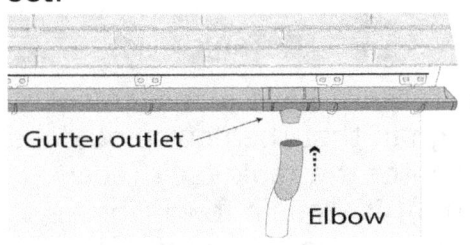

Depending on the layout of the structure, you have three options:

a. Install a downspout (without elbow) that goes straight down into a tank. Do this only when the distance between gutter and tank is very short.

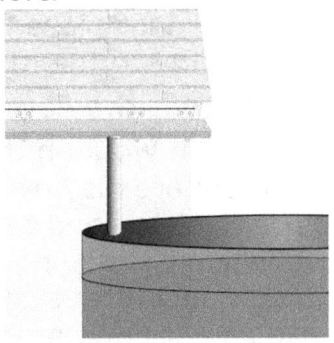

b. Install an elbow and then a downspout that is anchored to the structure. This is a normal vertically aligned setup.

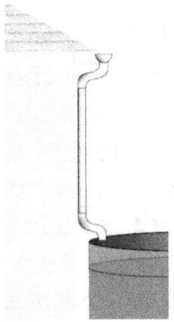

c. If your tank is farther away from the house, then you will first need to install an elbow to guide the water (horizontally) in the direction

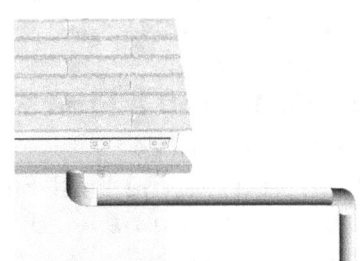

Collection And Storage Of Rainwater

of the tank.

Use PVC cement and primer to connect these joints. Make sure you have the correct size and fittings. All downpipes must be uniform in size, and secured to the building structure. Avoid random "flying" pipes that run from the roof to the tank.

Step 8: Test the connection

At this point work on the roof catchment area is finished, and you should stop and test your installation by running water through the gutter system to make sure it flows properly without any leaks. Make the necessary adjustments if needed.

Step 9: Decide on a pipe configuration

After determining the location of your water tank and the configuration of your downspouts, you may need to decide whether to install a first flush diverter system. If you choose not to install a first flush diverter, you can rely on the downspouts to carry the water directly into the tank. However, it is important to note that without a first flush diverter, debris and contaminants may enter the tank, potentially affecting the quality of the collected water.

Step 10: Install a First Flush diverter (see p.129)

If you opted for the First Flush diverter, then you have to install the diverter between the downspout and the tank. Every downpipe that flows into the tank should have a diverter. Leave enough space at the bottom of the diverter chamber to be able to open and close the release valve by hand. Try not to waste the water, since you can still use the contaminated water from the diverter for irrigation. If your First Flush diverter came in a larger diameter than your downspouts, then you will have to use a reducer bushing, i.e., something like a PVC 3" to 2" reducer bushing, to connect to your PVC pipes.

Step 11: Install the tank inlet

Install the tank by carefully following the manufacturer's instructions. Connect fittings provided for the inlet, outlet, overflow and drainage outlets. The fittings come with the tank and you should check if they have o-rings.

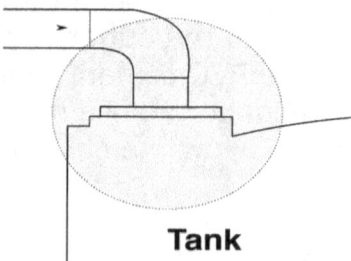

The connection is sealed. It keeps insects and debris out.

Tank

Some tanks have a convenient design in that the downpipe does not enter the tank itself (see below); it stops a few inches above the tank. The tank inlet is then covered with a screen guard that keeps debris and mosquitoes out.

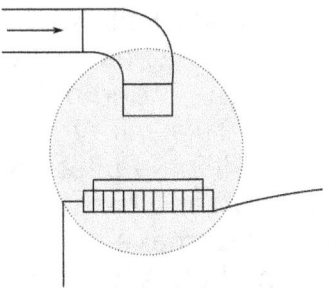

Tank with inletscreen, to keep insects and debris out.

Step 12: Install the overflow outlet (see p.135)

The tank must have an overflow outlet. Remember that your overflow outlet must be able to match the inflow of water. During a serious rain event the overflow will determine how much stress the fittings and joints will experience. When it comes to overflows you have the following options:

You have the option of letting the overflow pipe drain straight into the storm-water system.

You can also connect your overflow to a lower lying tank that can be utilized for secondary water needs.

You can also connect your overflow to flow into a pool, pond or cistern for livestock or firefighting needs.

Be wary of letting water flow directly onto the ground. Not only does it cause erosion and loss of soil, but it can potentially also damage structure foundations.

Installing an overflow is a simple process and you should follow the manufacturer's instructions. A general installation will look like this:

a. Place a rubber gasket behind the fitting. This will stop water leaking out the fitting joint when the water level is high.

b. Drill a hole in the tank with a hole saw that is ¾" (2cm) larger than the fitting size.

c. Clean the hole by breaking away the small fragments of plastic (burrs) on the surface of the hole drilled. They can cause leaks.

d. Insert your overflow fitting from the outside of the tank. The inside curve must be facing upwards.

e. Screw the overflow fitting into the tank while maintaining pressure on the screws. Tighten the screws evenly: Begin tightening the screws gradually and evenly, working your way around the port. Tighten each screw a little bit at a time until the port is snug against the tank. Don't let the gasket bulge, since it will cause leaks.

f. Add an overflow pipe that matches your inlet pipe diameter, and/or

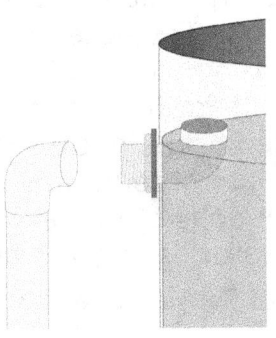

add a screened insert for mosquitoes.

Step 13: Install the drainage outlet

The water of poorest quality is located in the anaerobic zone at the bottom of the tank. The tank must have a drainage outlet and most tanks do come with one pre-installed. Flushing out sludge is part of the maintenance required and should be done biannually. Install the drainage outlet by following the manufacturer's instructions. If you inherited a tank without a drainage outlet then you will have to install it yourself:

1. Choose a location for the drainage outlet: The location should be at the lowest, flat point of the tank, and should be accessible for installation and maintenance.
2. Purchase a drainage outlet kit: This typically includes a bulkhead-type fitting with a ball valve or tap.
3. Drain the tank: Before installing the drainage outlet, you will need to drain the tank of any water that is currently inside.

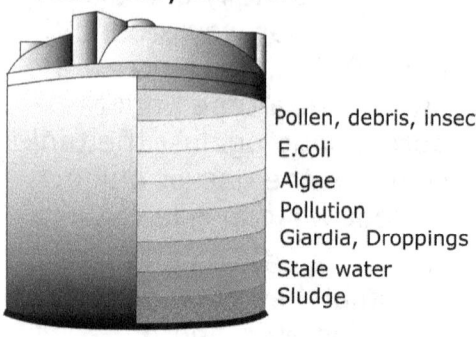

Pollen, debris, insects
E.coli
Algae
Pollution
Giardia, Droppings
Stale water
Sludge

4. Drill the hole: Follow the manufacturer's instructions to install the fitting onto the tank. This will typically involve drilling a pilot hole in the tank and then using a hole saw to make the final hole. The hole should be as low as possible to reach the sludge at the very bottom of the tank.
5. Remove the burrs and make sure the fitting fits.
6. Prepare the fitting and install it in the tank wall according to the manufacturer's instructions. The biggest obstacle will be reaching into the tank to fit the fitting in the hole. This can be a tricky process, but look at the method described on page 146, where we recommend using a Fish Tape to install the water outlet port.
7. Attach a brass ball valve: Use Plumber's tape and attach a ball valve to the fitting.
8. Attach a hose or pipe: Attach a hose or pipe to the ball valve or tap using any necessary fittings. Be sure to use a hose or pipe that is suitable for carrying water.
9. Direct the water to a suitable location: The hose or pipe should be directed to a suitable location, such as a garden or drain. Be sure to take into account any local regulations or guidelines for rainwater harvesting and use.

Step 14: Tank accessories

The main purpose of tank accessories is to give you access to cleaner water. These components should be installed according to the manufacturer's directions. Modern tanks come with a **floating pick-up** (extractor) that floats at the top of the water's surface. This is to guarantee the best water gets used and not the water at the bottom rich in sediment. **Overflow siphons** act as skimmers and when the tank overflows, they skim the pollen and dust from the water's surface. **Calmed inlets** prevent the disturbance of water and sediment when rainwater enters the tank. This means cleaner water.

Step 15: Water outlet

Most tanks come with an outlet pre-installed or included as a separate fitting. Install the drainage outlet by following the manufacturer's instructions. If you inherited a tank without a drainage outlet, then you will have to install it yourself. Installing a water tank outlet involves several steps, including selecting the right location for the outlet, drilling a hole in the tank, fitting the outlet, and sealing it properly. Here's a step-by-step guide on how to install a water tank outlet:

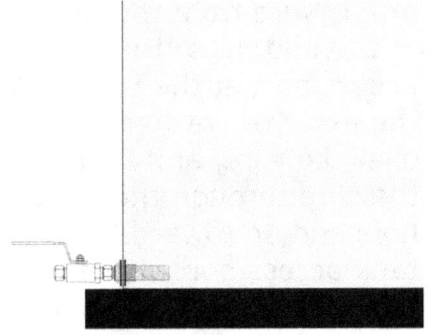

a. Select the location for the outlet: Before installing the outlet, you need to select the right location for it. Make sure that it is at least four inches (10cm) from the bottom, to avoid using the dirty sediment-containing water. This is the anaerobic zone of the tank. Ensure that the location is easily accessible and safe to work on.

b. Drain your tank if it's filled with water.

c. Drill a pilot hole at the outlet location. This should be done on a flat section of the tank wall; four inches (10cm) from the bottom is be enough.

d. Drill a hole in the tank: Use a drill or hole saw to make a hole in the tank at the chosen location. The size of the hole should be slightly smaller than the outlet to ensure a snug fit. Remove any burrs and take care not to drop them in the hole.

e. Check if the outlet fits. Make the necessary adjustments if needed.

f. Remove the tank access lid or strainer basket.

g. Install the outlet: Installing a component in a tank where you cannot reach is tricky. A simple solution is to use a gadget called a Fish Tape which allows you to extend a rigid wire from the hole at the bottom to the tank access port at the top. Slide the external washer and nut over the wire, and then push the wire through the outlet hole and up towards the tank access port at the top. Retrieve the wire from the access port with your hand or a curved stick. Place the internal fitting and washer over the wire and slide it down towards the outlet hole. Use long nose pliers to retrieve the threaded part of the fitting. (Note that if you do not have access to a Fish Tape or similar, the you can use a piece of string that you lower from the top of the tank, tied to a pole, and that you have to "catch" at the bottom through the outlet hole. Once you've done that you can use the same technique to retrieve and connect the fittings.)

h. Remove the Fish Tape.
i. Seal the outlet: Use a waterproof sealant, such as silicone or PVC cement, to seal the joint between

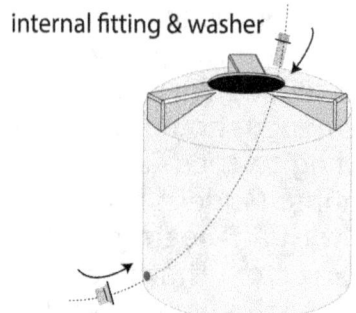

internal fitting & washer

external nut & washer

the outlet and the tank. This will prevent any water from leaking out. Place the washer over the fitting and wrap Plumber's tape over the thread. Tighten the nut.

j. Test the outlet: Once the outlet is installed and sealed, fill the tank with water and test the outlet to make sure it is working properly. Check for leaks or other issues. You can add a valve, spigot or tap.

k. Connect pipes: If required, connect the necessary pipes to the outlet to direct water flow to its intended destination.

Step 16: Air vent

A rainwater tank needs an air vent to prevent the buildup of pressure inside the tank that can cause it to collapse or rupture. When a tank is filled with water, the volume of air inside the tank decreases. If the air inside the tank is not allowed to escape, the resulting pressure can cause damage to the tank or the plumbing system that is connected to it. Additionally, if the tank is not properly vented, it can lead to a vacuum forming inside the tank, which can

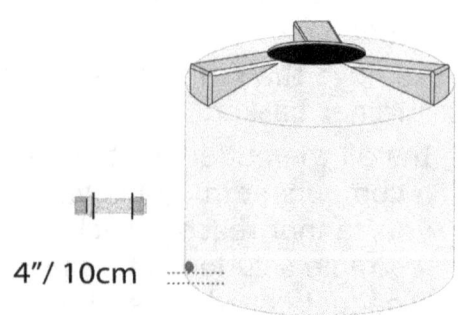

4"/ 10cm

prevent water from flowing out of the tank when it is needed.

An air vent allows air to flow in and out of the tank, equalizing the pressure and preventing any damage or malfunction (see *Vents*, p.132).

Step 17: Test the installation

Turn on your rainwater collection system and check for any leaks around the connections and fittings. Adjust and tighten connections as needed to ensure a proper seal.

Step 18: Regular maintenance

Regularly inspect and clean the tank to ensure it is functioning properly. Clean the inlet to prevent debris from entering the tote, and ensure the water extraction outlet is free from obstructions. Additionally, consider installing a screen or filter at the inlet to prevent debris from entering the tank

Tank Setup

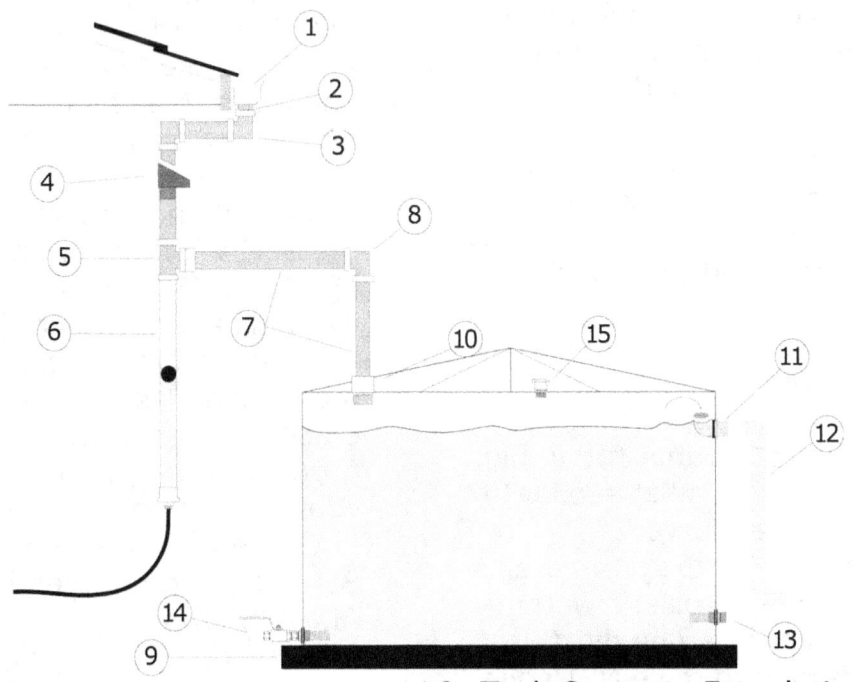

1- Gutter
2- Gutter Outlet
3- 90° PVC Elbow
4- Leaf Trap
5- First Flush Diverter Tee
6- First Flush Diverter
7- PVC Pipe
8- 90° PVC Elbow

9- Tank Concrete Foundation
10- Tank Inlet
11- Overflow (Bulkhead Fitting)
12- Overflow Pipe
13- Tank Outlet
14- Tank Drainage Outlet
15- Air Vent

and potentially clogging the system.

Note: It's important to follow local regulations and guidelines for rainwater harvesting in your area, including any permits or restrictions that may apply.

If you're not confident in your DIY skills, it's recommended to seek the help of a professional plumber or rainwater harvesting system installer to ensure proper installation.

Take Note Of The Following:

- Remember to prepare fittings for secondary applications like irrigation, fire fighting equipment, pond/pool systems, etc.
- If you are using two or more tanks, don't stack them. Normal tanks aren't designed to withstand that much pressure.
- Most tanks curve up from the bottom to the tank walls. Make sure that you place your outlet fitting high enough (on a flat section) to avoid that curve.
- Female PT type ports are common on plastic water tanks and can be directly threaded to male NPT style plumbing and connector fittings.
- A well laid out rain harvesting system can rely on gravity for irrigation, but if your water flow is not strong enough for domestic use, then you should invest in a pump. (See p.194)
- Make sure you understand the importance of the overflow. The overflow should at a minimum be the same diameter as the inlet pipe, since on a big rain day (cloudburst) the overflow will not be able to keep up with the extreme volume of water flowing into the tote. This is where venting becomes crucial to disperse the built up pressure inside the tank.
- When adding PVC cement to PVC fittings, do the female side first, then you can set it down and not get debris on the glued surface while you are doing the male end.
- The brass male by male nipples can cause problems when used with PVC fittings. You should never thread metal MALE fittings, into FEMALE PVC fittings, as metal expands and contracts (due to temperature) significantly more than plastic, and will crack the plastic female fittings. When you make a conversion from metal to PVC, the PVC should ALWAYS BE male, and the metal ALWAYS female.
- Ideally you should spray paint the PVC parts after finishing, as UV light weakens PVC over time, making it brittle.
- Consider PVC unions to simplify maintenance.
- You must vent your tank otherwise it will be damaged during a cloudburst.

Linking Multiple Tanks From The Top

In this section we will build on the knowledge gained in the previous section. We will see that adding another tank (or two) works on the exact same principle, which is water-flow, with the aid of gravity.

Linking multiple water tanks offers greater flexibility in increasing your rainwater harvesting capacity. In fact, an often easier and cheaper way to increase your water storage tank's capacity is to add another tank rather than replace an existing tank with a larger one.

How does it work?

In short, this method fills the tanks in the system one at a time in such a way that water flows from the overflow of one tank into the inlet (top part) of the next. This means that the first tank to receive water will have to fill to the point where the tanks are connected before any water will flow to the next tank in the connection system. Then the second tank will follow this pattern until the container is full. When the second tank is full, it should then also use an overflow to distribute water to a third tank for filling.

Rigid or Flexible Plumbing

The first problem you'll encounter is whether to use flexible or rigid plumbing. **When connecting tanks, you can use either rigid PVC pipes, or reinforced flexible hosing.** However, it's important to note that PVC's rigidness can be a problem; any movement or bulging of the tank may lead to damage to the fittings, pipes, or the tank. **For this poly tank installation, we do not recommend using rigid PVC pipe.** To get a clearer picture of this issue and the problems you'll encounter, see the section on p.125, *Flexible Plumbing vs. Rigid Plumbing*.

Basically you have two options:

- **Rigid plumbing**, such as standard PVC pipes, is suitable when you can ensure that the tank will maintain its rigid structure regardless of being full or empty. This is commonly seen in tanks made of concrete, steel, fiberglass, or wood.
- On the other hand, **flexible plumbing**, like reinforced hosing, flexible PVC, or HDPE pipes is a better choice when you anticipate movement or flexing of the tank's walls. This can be caused by factors such as seismic activity, unstable ground surfaces, or the material of the tank itself. Poly tanks are an example of tanks that may benefit from flexible plumbing solutions.

Connecting multiple tanks via the inlets and overflows at the top is a fairly easy process, but you'll have to keep the following in mind:

- Either a new fitting or an overflow port kit will need to be installed on the tanks. Typically, new poly tanks come with pre-installed fittings for inlet, outlet, overflow, and drainage.
- To connect overflows to inlets, additional piping is required. It is crucial for both parts to match in size to ensure a smooth flow of water and avoid obstructions or leaks. Typically, top inlet and overflow port fittings range from two inches (5cm) to four inches (10 cm) in size, which are commonly used in rainwater harvesting.
- Match the tank overflow pipe size to the inlet pipe size. This makes sure the overflow can keep up with the incoming water if your tank is full of water. Example: If you have a 2" inlet to your tank, you'll want your overflow to be 2".
- Many overflow kits feature bell-shaped PVC ends, while bulkheads are often NPT style threaded fittings. These fittings make PVC plumbing installation easy and convenient.
- The principle of linking tanks together from the top works on gravity, which means that the overflow of the first tank in the chain has to be slightly higher than the inlet of the second tank, and so on, in a stair-step pattern.
- When connecting tanks at the top, it is best to install the tanks as close to each other as possible. System simplicity, efficiency, and reliability are the main reasons why.

In this section we will look at an example where we use the top of the tank, at the inlets and overflows, to connect three poly water tanks with flexible plumbing. Every individual's vision is different, but the information provided here will get you on track to create your own, unique system to meet your needs.

How To

Let's get started:

Step 1: Choose the location

Choose a suitable location for the three tanks, preferably on a flat, stable surface for each tank, and close to the area where you plan to use the water.

Step 2: Position the tanks where they will be installed

Keep a space of at least six to twelve inches (15-30cm) between the tanks so they are not touching. Rotate tanks so that the fittings or overflow ports are directed towards each other. Make sure that the distance from tank to tank has enough length to make a secure connection. If all tanks

You Will Need

- 3 x Tank (clean/ food-grade, preferably with a top fill cap)
- Pre-prepared level surface or suitable platform
- Gutters, downspouts, screens as needed
- Tank inlet, outlet, overflow, drain valve fittings as needed
- Ball valves/spigot (as needed)
- Plumber's (Teflon) tape
- PVC primer and cement
- Pipe wrench or pliers
- Drill or Hole saw

Select a **flexible** connection:
Flexible PVC/HDPE:

- HDPE pipes, or Flex PVC (as needed)
- Compatible bulkhead fittings (same size as pipe)

Flexible reinforced hose:

- 4 x 2" Threaded male nipples (goes into bulkhead fitting)
- 4 x 2" ¼ turn PVC ball valves (connects to nipples)
- 4 x 2 " Male hose camlock couplers (connects to ball valves)
- 4 x 2 " Female hose camlock couplers (or 4 x hose clamps)
- 3-½ft (100cm) length of 2" flexible reinforced hose

are the same height (based on the overflows), they have to be placed on an incline and/or on different height foundations to create the stair step formation that facilitates gravity. If the water tanks involved in the system are different heights (which means that their overflows and inlets naturally form a stair step pattern), then they can be placed on the same level of foundation.

Step 3: Make sure that your tanks have inlets and overflow ports

You will need an inlet and an overflow on all three tanks. Most new tanks have these fittings already pre-installed, but if you need to do it yourself then consult the previous section, *How To Install Fittings On A Water Tank*, on page 121.

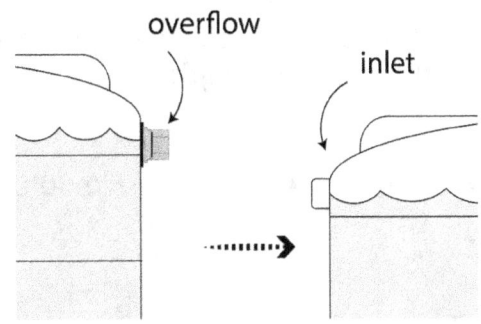

Step 4: Decide which tank will be your primary (first) tank

The downspout from the gutter will flow into the inlet of this tank. This tank should be at the highest level. The principle of linking tanks together from the top works on gravity, which means that the overflow of the first tank in the chain has to be slightly higher than the inlet of the second tank.

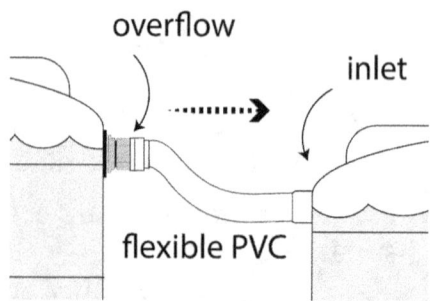

Step 5: Connect the tanks

Use flexible hose or flexible PVC pipe to connect the overflow outlet on the first tank to the inlet on the second tank.

For a **flexible PVC** or HDPE pipe connection follow these steps:

a. Ensure that the first tank is higher than the second tank. Make sure that the inlet port on the second tank is at the correct height to accommodate your pipe arrangement.

b. Prepare your bulkhead fittings and make sure that both ports are clean and debris free.

c. Measure and cut your pipe to the correct length.

d. Arrange your pipe to run down from the overflow fitting on the first tank, to the inlet fitting on the second tank. You might have to bend your pipe beforehand to get the ideal arrangement.

e. Double-check all connections to make sure they are tight and leak-free. Use Plumber's (Teflon) tape and PVC cement where necessary.

f. Test the system by running a hose through the pipe while checking for any leaks. Adjust the connections if needed.

For a **flexible hose** connection follow these steps:

a. Ensure that the first tank is higher than the second

Flexible PVC Connect

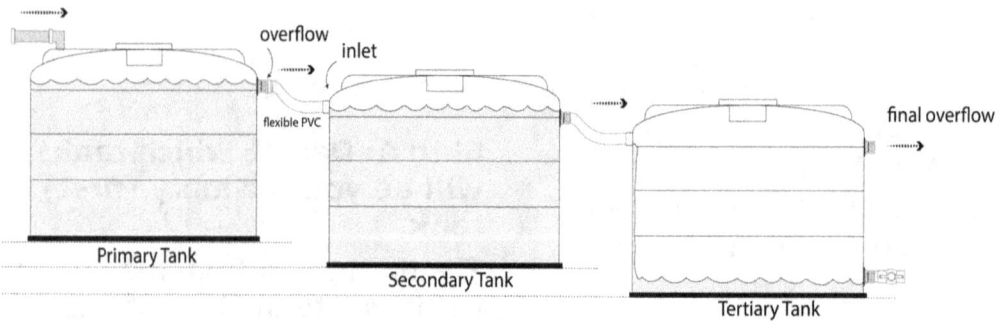

tank. Make sure that the inlet port on the second tank is at the correct height to accommodate your hose fitting.

b. Prepare your fittings and make sure that both ports are clean and debris free.

c. Measure and cut your high-quality, flexible hose to the correct length.

d. Thread one male nipple into the overflow port of the first tank, and also another male nipple into the inlet port of the second tank. Attach a ball valve to each nipple.

e. Attach one end of the flexible hose to the ball valve of the first rainwater tank. This can typically be done using hose clamps. Alternatively you can use a combination of male hose camlock couplers to female hose camlock couplers.

f. Repeat this process when you connect the other end of the flexible hose to the inlet of the second rainwater tank.

g. Double-check all connections

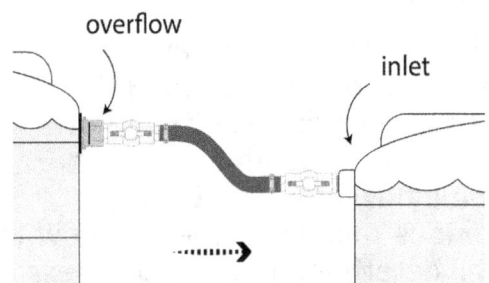

to make sure they are tight and leak-free. Use Plumber's (Teflon) tape where necessary.

h. Test the system by running a hose through the pipe while checking for any leaks. Adjust the connections if needed.

This connection will allow water to flow from the first tank to the second tank when the first tank is full. The ball valves simplify the maintenance process.

Step 6: Continue to connect tanks two and three

Use your preferred hose or pipe to connect the overflow outlet on the second tank to the inlet on the third tank.

Flexible Hose Connect

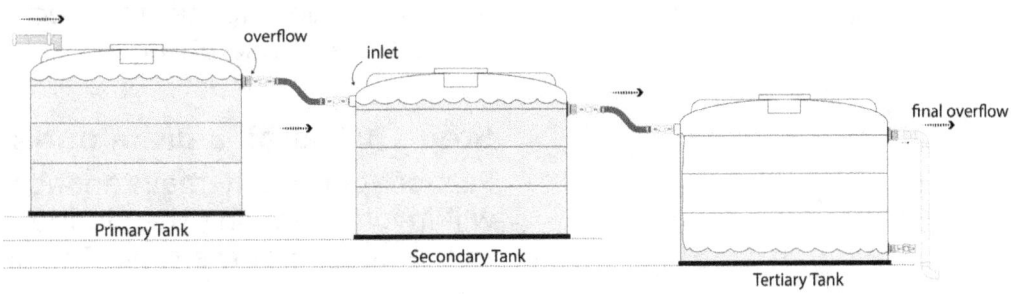

Collection And Storage Of Rainwater 153

Step 7: Install an overflow outlet

The last tank must have an overflow outlet (see p.135). Remember that your overflow outlet must be able to match the inflow of water. During a serious rain event the overflow will determine how much stress the fittings and joints will experience. When it comes to overflows you have the following options:

- You have the option of letting the overflow pipe drain straight into the storm-water system.
- You can also connect your overflow to a lower lying tank that can be utilized for secondary water needs.
- You can also connect your overflow to flow into a pool, pond or cistern for livestock or firefighting needs.

Be wary of letting water flow directly onto the ground. Not only does it cause erosion and loss of soil, but it can potentially also damage structure foundations.

Step 8: Air vents

A rainwater tank needs an air vent to prevent the buildup of pressure inside the tank that can cause it to collapse or rupture. When a tank is filled with water, the volume of air inside the tank decreases. If the air inside the tank is not allowed to escape, the resulting pressure can cause damage to the tank or the plumbing system that is connected to it. Additionally, if the tank is not properly vented, it can lead to a vacuum forming inside the tank, which can prevent water from flowing out of the tank when it is needed. An air vent allows air to flow in and out of the tank, equalizing the pressure and preventing any damage or malfunction (see *Vents*, p.132).

Step 9: Install an outlet

Every tank should still have an outlet port to be used when needed. This will allow you to be selective about which tank to drain, especially when some of the tanks are empty during a dry spell. Add a faucet to direct water for end-use purposes.

Step 10: Install a drain outlet

Part of your yearly maintenance will involve draining the sludge at the bottom of the tank. Install

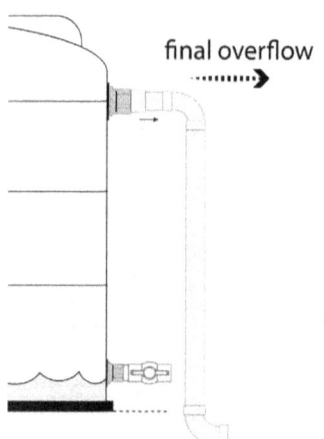

final overflow

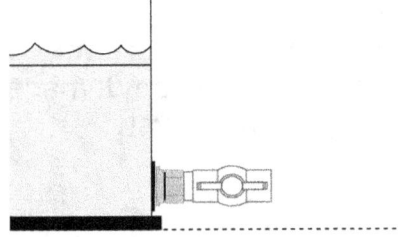

a drain valve to take care of this (see p.144).

Step 11: Install a pump

If you plan to use the water for irrigation or other purposes that require pressure, you may need to install a pump to move the water from the third tank to the point of use. The pump should be installed near the third tank and connected to the outlet using a pipe or hose (see *Pumps* p.194).

Step 12: Test the system

Once the tanks are connected and any necessary pumps or other components are installed, test the system to make sure it is working properly. Fill the first tank with water and observe the flow of water to the second tank and any connected pumps or pipes.

Step 13: Regular maintenance

Regularly inspect and clean the tank to ensure it is functioning properly. Clean the inlet to prevent debris from entering the tote, and ensure the water extraction outlet is free from obstructions. Additionally, consider installing a screen or filter at the inlet to prevent debris from entering the tote and potentially clogging the system.

Note: It's important to follow local regulations and guidelines for rainwater harvesting in your area, including any permits or restrictions that may apply. If you're not confident in your DIY skills, it's recommended to seek the help of a professional plumber or rainwater harvesting system installer to ensure proper installation.

Take Note Of The Following:

- To get enough water pressure you will have to elevate your system, or add a water pump.
- Water can be drawn from any tank in this series but the water level between tanks won't necessarily level out.
- Isolation valves will proof beneficial for maintenance.
- Flexible, hosing should be considered. Although there isn't much flex and bulging at the top section of a water tank, having a flexible hose will guarantee no unforeseen leaks.
- Remember to prepare fittings for secondary applications like irrigation, fire fighting equipment, pond/pool systems, etc.
- Female PT type ports are

common on plastic water tanks and can be directly threaded to male NPT style plumbing and connector fittings.

- Make sure you understand the importance of the overflow. The overflow should at a minimum be the same diameter as the inflow, since on a big rain day the overflow will not be able to keep up with the extreme volume of water flowing into the tank.

  This is where venting becomes crucial to disperse the built up pressure inside the tank.

- When adding PVC cement to PVC fittings, do the female side first, then you can set it down and not get debris on the glued surface while you are doing the male end.

- The brass male by male nipples can cause problems when used with PVC fittings. You should never thread metal MALE fittings, into FEMALE PVC fittings, as metal expands and contracts (due to temperature) significantly more than plastic, and will crack the plastic female fittings.

  When you make a conversion from metal to PVC, the PVC should ALWAYS BE male, and the metal ALWAYS female.

- Ideally you should spray paint all the PVC parts after finishing, as UV light weakens PVC over time, making it brittle.

- You must vent your tank otherwise it will be damaged during a cloudburst.

Benefits Of Joining Tanks At The Top:

- By connecting storage tanks at the top, you can increase the overall capacity and liquid volume available for use.

  This setup allows for individual containers to be positioned freely, eliminating the need for large containers that take up significant land area.

- Additionally, top-connected tanks enable independent draining and drawing from each tank. When water is drawn from one tank, only that specific tank releases and flows water, while the water levels in the other tanks remain unaffected. This flexibility allows tanks to be used for separate tasks or interconnected for a unified system like irrigation. It also facilitates better management and monitoring of water levels and consumption.

- Moreover, the unique drainage style of top-connected tanks enables a designated tank to serve as an emergency reserve supply during periods of low water levels or limited availability, such as during a drought.

Linking Multiple Tanks From The Bottom

Building on the knowledge gained in the previous section we already know that linking multiple water tanks offers greater flexibility in increasing your rainwater harvesting capacity. Connecting water tanks at the **bottom** is usually done at the **tank outlet**. Most water tanks are pre-plumbed and fitted with one (or two) bulkhead fitting type outlet port(s), that can be used to connect multiple tanks together.

How does it work?

The outlet-to-outlet method is different from the overflow-to-inlet method in that water rises and falls evenly. All tanks in the system will fill evenly and they will also drain evenly. This means that the water level between tanks will balance itself out to maintain a steady, equal volume.

If we look closer we see that once the first tank is filled with water up to the point where it connects to the second tank, it stops filling until the water level in the second tank matches that of the first tank. This process continues until all tanks in the system are filled to the same water level. When the tanks have equal water volume, the water level in all tanks rises and falls evenly as water flows in and out, respectively. The rate at which this happens depends on the speed at which water is added to the tanks and the size of the plumbing used.

Keep the following in mind:

- Water storage tank outlets are often only 1 or 2 inches (2.5-5cm) in diameter, making for a smaller plumbing connection and pipe. This can increase the possibility of sediment or debris to build up and cause blockage, which would then need to be cleaned. An optional recommendation would be to replace the smaller diameter bulkhead fitting with a larger one to allow for larger diameter plumbing to be installed.

- It is important to avoid using rigid connections between plastic tanks in a system. As the tanks fill with water, they expand and contract, creating significant bulging at the bottom. This can cause damage to both the tanks and the pipes, leading to ruptures

You Will Need

- 3 x Tank (clean/ food-grade)
- Pre-prepared level surface or suitable platform for placement
- Gutters and downspouts as needed
- Gutter screens, guards and fasteners as needed
- Tank inlet fittings
- Tank overflow fittings
- Tank outlet fittings
- Tank drain valve fittings
- Ball valve/spigot
- Plumber's (Teflon) tape
- PVC primer and cement
- Pipe wrench or pliers
- Drill or hole saw

Flexible tank connection kit:

- 4 x 2" Threaded male nipples (goes into each bulkhead fitting)
- 4 x 2" ¼ turn PVC ball valves (connects to nipples)
- 4 x 2" Male hose camlock couplers (connects to ball valves)
- 4 x 2" Female hose camlock couplers (or 4 x hose clamps)
- 3-½ft (100cm) length of 2" flexible reinforced hose (connects couplers)

and leaks. To prevent this, it is recommended to use reinforced, flexible connections, such as hoses, which can move freely as the tanks fill and empty. This allows for the necessary expansion and contraction without causing damage to the tanks or the connections. There are many flexible tank connector kits on the market and they come with a variety of fittings. Make sure that you opt for a kit with a ball valve type fitting that allows you to isolate your tanks for maintenance. A type of quick connect coupling (camlock) is also very handy for maintenance, but if money is an issue then a simple hose clamp will do the job.

- If your tank outlets came equipped with a valve or a faucet, then they may have to be removed or modified to be compatible with flexible plumbing which will be used to join the tanks together.
- Since we will be using the outlets, or specially installed fittings, the overflow ports will not be used (except on the last tank) and should be capped to keep the water tank closed.

In this section we will look at an example where **we use the bottom of the tank, at the outlets, to connect three poly water tanks**. Every individual's vision is different, but the information provided here will get you on track to create your own, unique system to meet your needs.

How To

Let's get started:

Step 1: Choose the location

Choose a suitable location for the three tanks, preferably on a flat, stable surface for each tank, and close to the area where you plan to use the water.

Step 2: Position your tanks

For this kind of connection it's best to have your tanks closer to one another. Tank one is your primary tank, tank two is your secondary tank, and tank three is your tertiary tank (last). Fifteen to twenty inches (38-50cm) in between the tanks should be enough.

Measure the length between the outlet ports and make sure that you have fittings for your style of outlet port.

Step 3: Make sure that all your tanks have outlet ports

You will need an outlet on all three tanks.

Some rainwater tanks are pre-plumbed with two outlet ports. Make sure that your secondary tank has two outlet ports at the bottom, one to receive water from the primary tank, and the second one to deliver water to the tertiary tank, which is last in line.

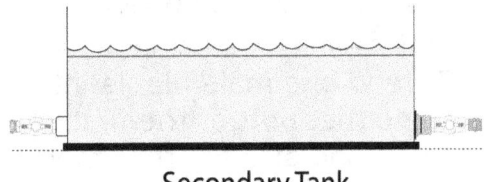

Secondary Tank

Step 4: Decide which tank will be your primary (first) tank

The downspout from the gutter will flow into the inlet of this tank. This tank's overflow should be capped.

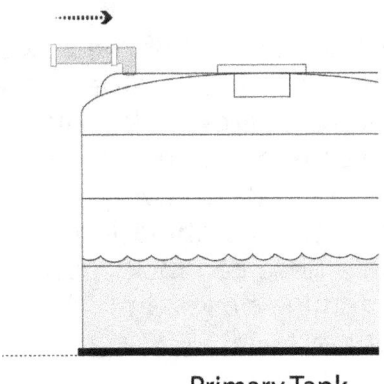

Primary Tank

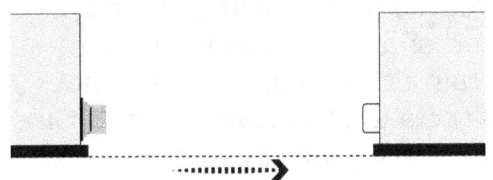

Step 5: Fit the flexible hose

Fit the hose between the two outlet ports of the primary and secondary tanks.

Follow these steps:

a. First, you will need to measure and cut your high-quality, flexible hose to the correct length.

b. Ensure the tanks are level and positioned correctly, allowing for easy connection.
c. Clean the tank outlets to remove any debris or obstructions that could interfere with the hose connection.
d. Thread one male nipple into the outlet port of the first tank, and also another male nipple into the outlet port of the second tank. Attach a ball valve to each nipple.

Valve Fitting

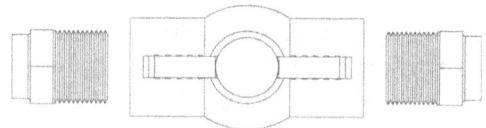

e. Attach one end of the flexible hose to the ball valve of the first rainwater tank. This can typically be done using hose clamps. Alternatively you can use a combination of male hose camlock couplers to female hose camlock couplers.

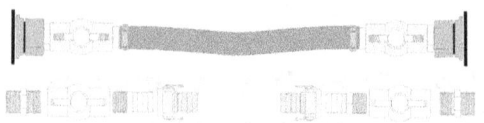

f. Repeat this process when you connect the other end of the flexible hose to the outlet of the second rainwater tank, again using appropriate fittings and ensuring a secure connection.
g. Double-check all connections to make sure they are tight and leak-free. Use Plumber's (Teflon) tape where necessary.
h. If required, support the flexible hose at intervals along its length using clamps or brackets to prevent sagging or undue strain on the connections.
i. Test the system by turning on the water supply and checking for any leaks or issues. Adjust the connections if needed. Make sure that the connections are water-tight.

Step 6: Cap the ports

Cap the inlet and overflow ports of the secondary tank.

Step 7: Fit the flexible hose

Fit the hose between the two outlet ports of the secondary and tertiary tanks by repeating Step 5.

Step 8: Cap the ports

Cap the inlet port of the tertiary tank, but leave the overflow port open. This overflow should be treated like a normal overflow and should be connected to a drain or an overflow pipe that directs excess water away from the tanks and the surrounding area.

Step 9: Install an overflow outlet

The last tank must have an overflow outlet (see p.135). Remember that your overflow outlet must be able to match the inflow of water. During a serious rain event the overflow will determine how much stress the fittings and joints will experience. When it comes to overflows you have the following options:

You have the option of letting the overflow pipe drain straight into the storm-water system.

You can also connect your overflow to a lower lying tank that can be utilized for secondary water needs.

You can also connect your overflow to flow into a pool, pond or cistern for livestock or firefighting needs.

Be wary of letting water flow directly onto the ground. Not only does it cause erosion and loss of soil, but it can potentially also damage structure foundations.

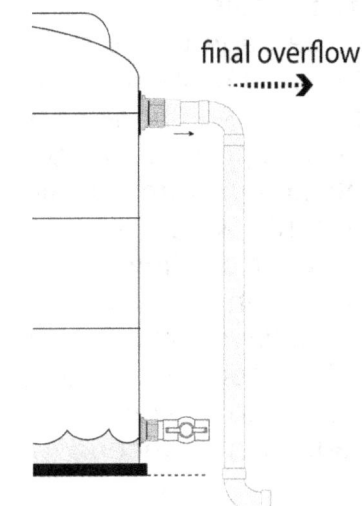

final overflow

Step 10: Add a faucet or ball valve to your outlet

On the tertiary (third) tank you can add a ball valve or faucet on the outlet port to direct water for end-use purposes.

Step 11: Air vents

A rainwater tank needs an air vent to prevent the buildup of pressure inside the tank that can cause it to collapse or rupture. When a tank is filled with water, the volume of air inside the tank decreases. If the air inside the tank is not allowed to

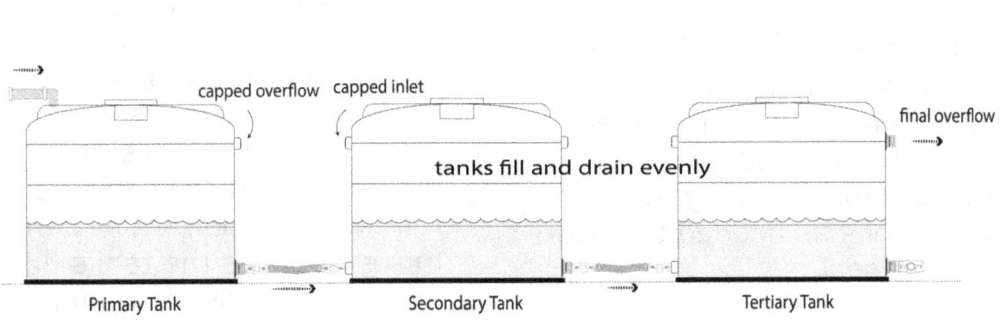

Bottom Flex Connect

Primary Tank — capped overflow, capped inlet — Secondary Tank — tanks fill and drain evenly — Tertiary Tank — final overflow

Collection And Storage Of Rainwater

escape, the resulting pressure can cause damage to the tank or the plumbing system that is connected to it. Additionally, if the tank is not properly vented, it can lead to a vacuum forming inside the tank, which can prevent water from flowing out of the tank when it is needed. An air vent allows air to flow in and out of the tank, equalizing the pressure and preventing any damage or malfunction (see *Vents*, p.132).

Step 12: Install a drain outlet

Part of your yearly maintenance will involve draining the sludge at the bottom of the tank. Install a drain valve to take care of this (see p.142).

Step 13: Install a pump

If you plan to use the water for irrigation or other purposes that require pressure, you may need to install a pump to move the water from the third tank to the point of use. The pump should be installed near the third tank and connected to the outlet using a pipe or hose (see *Pumps* p.194).

Step 14: Test the installation

Turn on your rainwater collection system and check for any leaks around the connections and fittings. Adjust and tighten connections as needed to ensure a proper seal.

Step 15: Regular maintenance

Regularly inspect and clean the tank to ensure it is functioning properly. Clean the inlet to prevent debris from entering the tote, and ensure the water extraction outlet is free from obstructions. Additionally, consider installing a screen or filter at the inlet to prevent debris from entering the tote and potentially clogging the system.

Note: It's important to follow local regulations and guidelines for rainwater harvesting in your area, including any permits or restrictions that may apply. If you're not confident in your DIY skills, it's recommended to seek the help of a professional plumber or rainwater harvesting system installer to ensure proper installation.

Take Note Of The Following:

- It is important to avoid using rigid connections between plastic tanks in a system linked at the bottom. As the tanks fill with water, they expand and contract, creating significant bulging at the bottom. This can cause damage to both the tanks and the pipes, leading to ruptures and leaks. To prevent this, it is recommended to use flexible, reinforced hoses, which can move freely as the tanks fill and empty. This allows for the necessary expansion and

contraction without causing damage to the tanks or the connections.
- Most tanks curve up from the bottom to the tank walls. Make sure that you place your outlet fitting high enough (on a flat section) to avoid that curve.
- Remember to prepare fittings for secondary applications like irrigation, fire fighting equipment, pond/pool systems, etc.
- Female NPT type ports are common on plastic water tanks and can be directly threaded to with male NPT style plumbing and connector fittings.
- It is worth it to consider installing isolation valves for this method, so that if one tank springs a leak, it will not drain all your tanks of their stored rainwater.
- To get enough water pressure you will have to elevate your system, or add a water pump.
- The water in the three tanks moves as one volume of liquid when filling or draining and this is why the levels are always uniform. Air will escape through the final overflow, but you should still add air vents.
- You can place your overflow on your first tank instead of the last tank.
- Make sure you understand the importance of the

It's important to avoid using rigid connections between plastic tanks in a system. As the tanks fill with water, they expand and contract, creating significant bulging.

overflow. The overflow should at a minimum be the same diameter as the inflow, since on a big rain day the overflow will not be able to keep up with the extreme volume of water flowing into the tote. This is where venting becomes crucial to disperse the built up pressure inside the tank.
- If only one downpipe is supplying the system, make sure to close any extra overflows to ensure that water flows from tank to tank without leaking out. In this setup, only the overflow of the last tank in the chain should remain open to handle excess rain or storm-water as intended.
- When adding PVC cement to PVC fittings, do the female side first, then you can set it down and not get debris on the glued surface while you are doing the male end.
- The brass male by male nipples can cause problems when used with PVC fittings. You should never thread metal MALE fittings, into FEMALE PVC fittings, as metal

expands and contracts (due to temperature) significantly more than plastic, and will crack the plastic female fittings. When you make a conversion from metal to PVC, the PVC should ALWAYS BE male, and the metal ALWAYS female.

- Ideally you should spray paint all the PVC parts after finishing, as UV light weakens PVC over time, making it brittle.

The PVC Option

As an alternative, consider joining your tanks with standard PVC, side by side, with the fittings facing forward. This arrangement provides better resilience against tank expansion and bulging when filled with water. PVC pipe and fittings can be used for this design, allowing the tanks to expand outward while accommodating the movement of the fittings.

Alternatively, flexible hose can also be considered as an option.

In this example we used two inch rigid PVC pipe, but you can adjust sizes to accommodate your needs.

How To

Step 1: Choose the location

Choose a suitable location for the three tanks, preferably on a flat, stable surface like a concrete slab.

You Will Need

- 3 x Tanks (Polyethylene)
- 4 x 2" PVC ball valve
- 2" PVC pipe, (as needed)
- 2 x 2" PVC Tee
- 1 x 2" PVC 90° elbow
- Plumber's (Teflon) tape
- PVC primer and cement
- Pipe wrench or pliers

Step 2: Position your tanks

For this kind of connection it's best to have your tanks closer to one another. Ten to twenty inches (25-50cm) in between the tanks should be enough. Orient the tanks so that the fittings face to the front, towards the user.

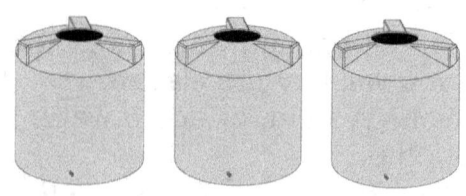

Step 3: Make sure that all your tanks have outlet ports

You will need an outlet on all three tanks. A benefit of this method is that you only need one outlet per tank. The overflow ports of tanks one and two should be capped.

Step 4: Decide which tank

Bottom - PVC Connect

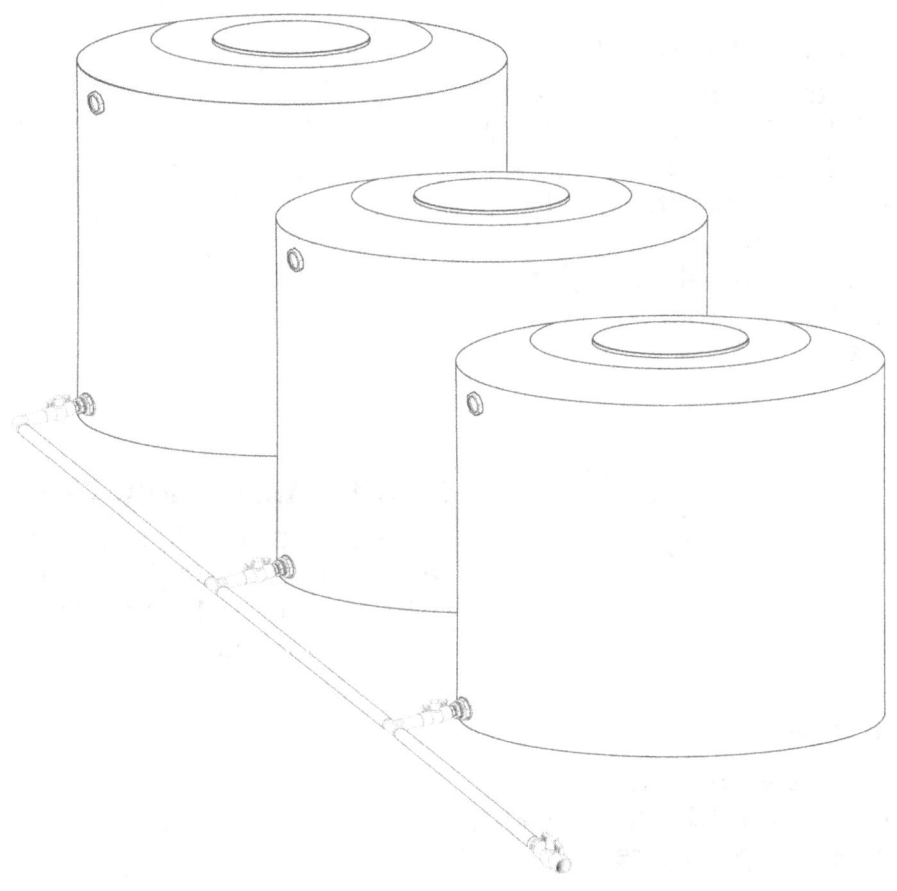

will be your primary (first) tank

The downspout from the gutter will flow into the inlet of this tank.

Step 5: Install the ball valve

Start with the primary tank, and insert the male nipple into the tank outlet fitting. If you are going down to a smaller diameter PVC pipe, then you will have to use a reducer to accommodate the smaller diameter pipe. Connect the threaded PVC ball valve to the male nipple. Use Plumber's tape for a watertight seal.

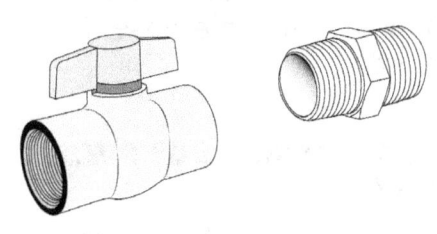

Step 6: Repeat

Collection And Storage Of Rainwater

Repeat this step on tanks two and three.

Step 7: Male PVC adapter

Return to the primary tank. Add a male PVC slip adapter to the threaded PVC ball valve. This adapter has a male threaded end which screws into the female threaded ball valve, and a slip socket end that fits your PVC pipe.

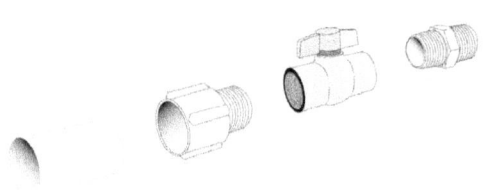

Step 8: Repeat

Repeat this step on tanks two and three. Use PVC cement and/or Plumber's tape where needed.

Step 9: PVC pipe

Add a 4" (10cm) length of PVC pipe to the male PVC slip adapter (or slip PVC ball valve) on all three tanks. This section can also be flexible hose if needed.

Step 10: Install 90° PVC Elbow

Return to the primary tank and attach a 90° PVC Elbow to the 4" PVC pipe. Make sure that the open socket end is facing towards the second tank, to accommodate the PVC pipe

that we will use to connect them both. Use PVC primer and cement.

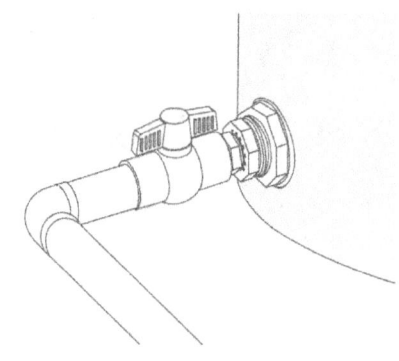

Step 11: Install PVC Tees

Move to the second and third tank and attach PVC Tees to both lengths of 4" PVC pipe. Use PVC primer and cement.

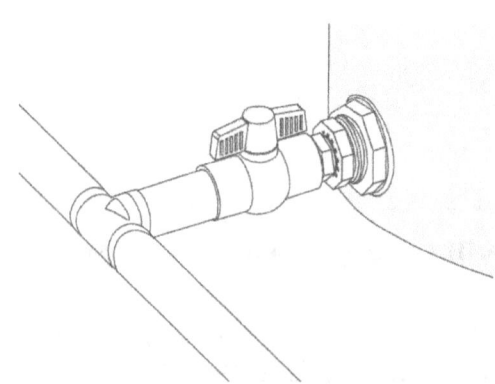

Step 12: Add PVC pipe to tank one

Return to the primary tank, and use a predetermined length of PVC pipe to join the PVC Elbow of tank one and the PVC Tee of tank two.

Step 13: Add PVC pipe to tank two

Use PVC pipe to join the PVC Tee of tank two and the PVC Tee of tank three.

Step 14: Add PVC pipe to tank three

Add a length of PVC pipe of desired length. This pipe will end in a ball valve for end-use.

Step 15: Add a ball valve

There are generally two options to attach a 2" PVC ball valve to a 2" PVC pipe:

Slip Socketed Ball Valve: Use a ball valve with slip sockets, where you can insert the PVC pipe directly into the slip socket of the ball valve. To secure the connection, you would apply PVC cement on the inside of the slip socket and then insert the PVC pipe.

Female Threaded Socket: Alternatively, you can use a ball valve with a female threaded socket.

In this case, you would need to add a male PVC adapter that can screw into the threaded socket of the ball valve. The PVC pipe would then fit inside the male PVC adapter. To ensure a secure connection, you would apply PVC cement on the inside of the male PVC adapter and insert the PVC pipe.

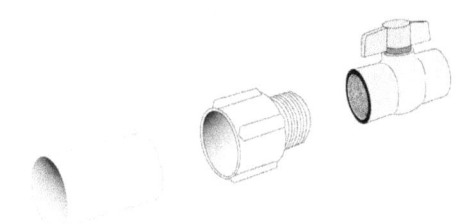

Both options can be effective, and the choice between them depends on the specific requirements of your project and the available components.

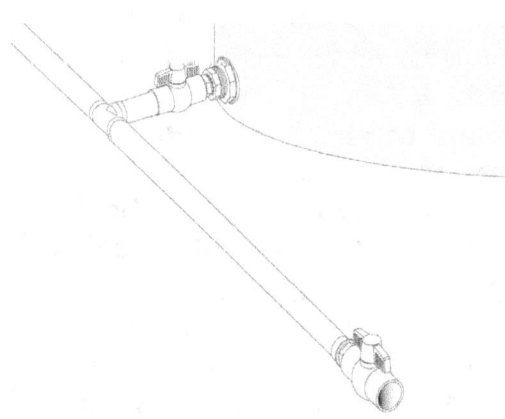

Step 16: Install an overflow outlet

Install an overflow on the last tank. Remember that your overflow outlet must be able to match the inflow of water.

Step 17: Support your PVC pipe assembly

Your horizontal pipe setup (at the tote outlets) will experience stress due to the added weight

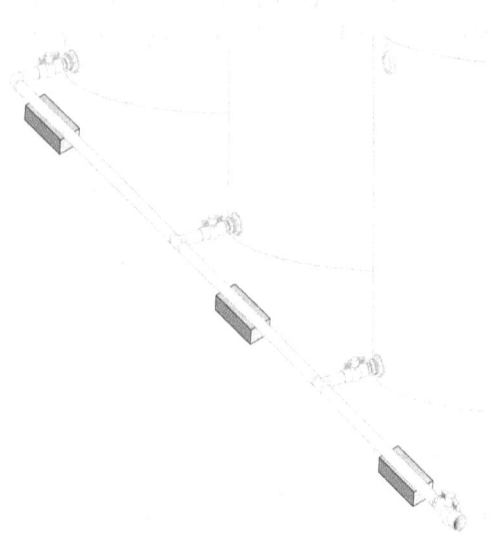

of the water, therefore you should rest your pipes on a platform. Simply place a block or brick under each PVC Tee or elbow.

Remember to use PVC primer and cement on all PVC fittings, and Plumber's (Teflon) tape on all threads.

If you live in an area with seismic activity and pipe fluctuation is a concern, you can substitute the short 4" PVC pipe with similar lengths of reinforced flexible hose.

The flexible hose will guarantee that no harm will come to the pipe connections in case of unforeseen movement or expansion. (Use a male threaded PVC hose adapter instead of a male PVC slip adapter.)

Use hose clamps to secure it to the fitting.

Benefits Of Joining Tanks At The Bottom

- No stagnant volume units of water. When tanks are connected at the bottom, they fill up and empty out uniformly. As water is added to the first tank and reaches the connection point to the second tank, the first tank stops filling up until the water level in the second tank matches that of the first tank. This continues until all tanks in the system are filled to the same level.

  This uniform filling and draining will be affected by the rate of water flow and the size of the plumbing used.

- Bottom-connected tanks offer a significant advantage in that they require only a single tap and overflow port.

  Typically, these features are installed on the tank located at the end of the system. When water is drawn from one tank, the water levels in all tanks are equalized until the entire system reaches the same volume again. This natural balance occurs because the tanks are all connected at the bottom, which creates pressure due to the weight of the water and the influence of gravity.

  This feature makes it easier to manage overflow and access water from the tanks.

Multiple IBC Totes Installation

Installing an Intermediate Bulk Container (IBC) tote for rainwater harvesting is a popular option for collecting and storing rainwater for various uses, such as gardening, irrigation, and other non-potable applications.

Let's take a look at the dimensions of a standard IBC tote used for rainwater collection:

Capacity: 275 gallons (1040 liters)
Length: 48 inches (121.92 cm)
Width: 40 inches (101.6 cm)
Height: 46 inches (116.84 cm)
Weight (empty): approximately 100 lbs (45 kg)

Most IBC totes come with a standard 2-inch National Pipe Thread (NPT) thread for the cap and valve, which is a common thread type used in plumbing and piping systems. However, it's always a good idea to check the specific model and manufacturer of the IBC tote to ensure that it has the correct thread size and type for your intended use.

If your tote came with a buttress thread, then you have to buy a converter which is easy to find. Buttress thread cannot be used with PVC pipe.

Be aware though that clear tanks, like IBC totes, can cause

You Will Need

- 2 x IBC tote (clean/ food-grade, with a top fill cap)
- 1 x 6-½" IBC cap with 2" threaded plug
- 2" PVC pipe (as needed)
- 1 x 3" to 2" PVC reducer bushing
- 2 x 2" PVC 90° elbow
- 1 x 2" PVC MPT male adapter
- 2 x 2" PVC female adapter socket x FPT
- 1 x 2" PVC Tee
- 1 x 2" PVC male slip adapter
- 1 x Threaded PVC ball valve
- Level surface or suitable platform for placement
- First flush diverter kit for 3 inch pipe
- Plumber's (Teflon) tape
- PVC primer and cement
- Pipe wrench or pliers
- Drill/hole saw (if modifications to the IBC tote are needed)
- Hose/faucet for water extraction (optional)
- 2 x 2" PVC 90° elbow (optional for overflow)
- Garden hose adapter barb (optional)

(FPT=Female Pipe Thread)
(MPT=Male Pipe Thread)

Multiple Tote Setup

1- Downspout
2- First Flush Diverter Tee
3- First Flush Diverter
4- 3" to 2" Reducer Bushing
5- 2" 90° PVC Elbow
6- 2" PVC Pipe
7- 2" PVC MPT Male Adapter
8- 6-½" IBC Cap with 2" Plug
9- 2" Bulkhead Fitting
10- Tote Outlet
11- 2" PVC Female Adapter FPT
12- 2" 90° PVC Elbow
13- 2" PVC Tee
14- 2" PVC Pipe
15- 2" PVC Male Slip Adapter
16- PVC Ball Valve

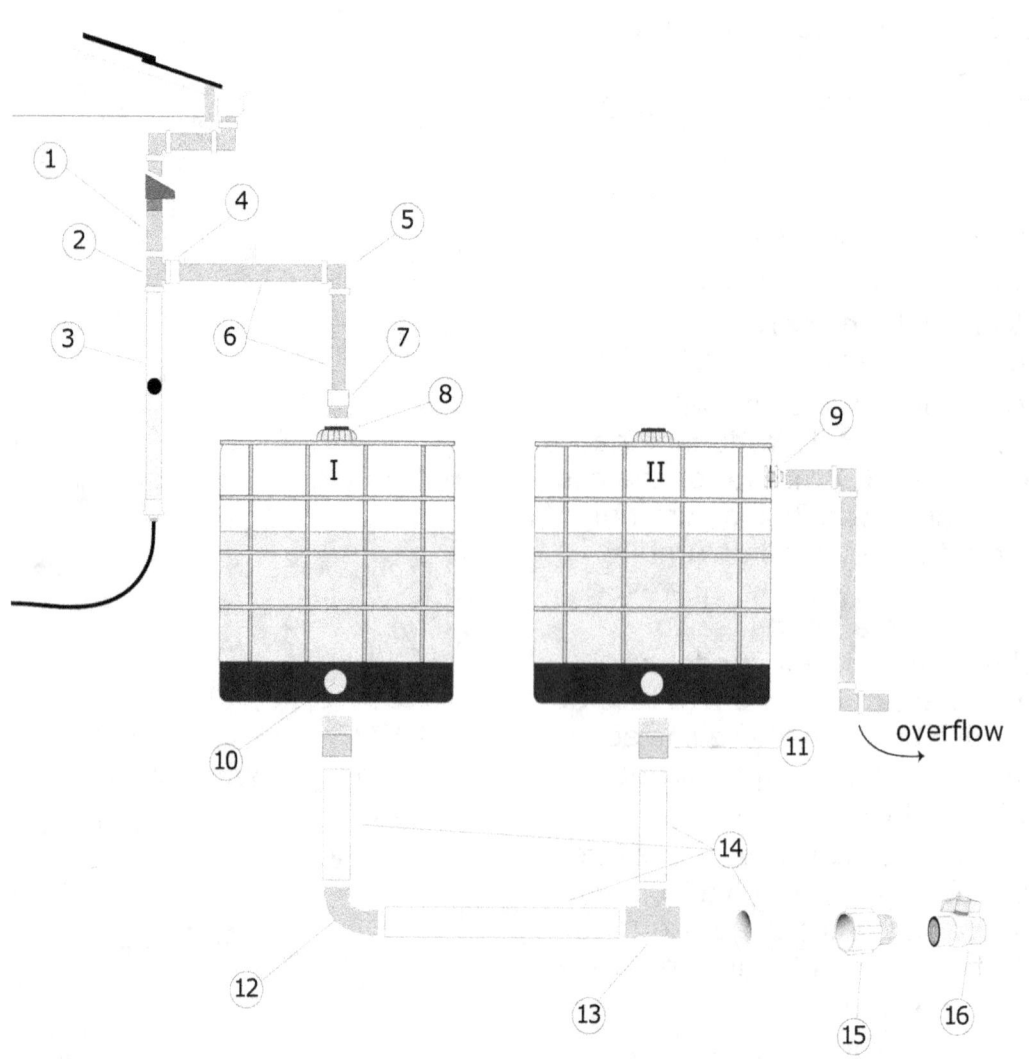

some problems:

- **Light exposure:** If the water tank is exposed to sunlight, it can promote the growth of algae, which can make the water unsafe for consumption and may also clog filters or equipment.
- **Temperature:** A clear water tank exposed to sunlight can also cause the water to heat up, which can cause bacteria to grow.
- **Maintenance:** A clear water tank will require more frequent cleaning and maintenance to keep it looking clear and free of debris.

Some simple solutions to protect you tank against sunlight:

- **Tank Covers:** Tank covers can be made from materials like PVC or fiberglass and are designed to fit snugly over the top of the tank, preventing sunlight exposure, limiting evaporation and providing insulation.
- **Tank Wraps:** Tank wraps are a cost-effective alternative to tank covers. They are made from materials like vinyl and are designed to wrap around the outside of the tank to prevent sunlight exposure and limit evaporation.
- **Tank Paint:** Painting the outside of the tank with a dark-colored paint (marine grade as used on boats) can reduce sunlight exposure and prevent algae growth.

Here's a step-by-step guide on how to link two IBC totes with two inch pipes for rainwater harvesting. You can also use this method to install a single tote or an array of multiple totes. Every individual's vision is different, but the information provided here will get you on track to create your own, unique IBC tote system to meet your needs.

How To

Step 1: Choose the location

Select a suitable location for the IBC tote, preferably on a level surface or a stable platform that can support the weight of a full tote which can weigh around 2,200 pounds/1,000 kg or more. Cinder blocks on a solid surface that allows for drainage (pea gravel) is a good option.

Make sure the location is easily accessible for installation, maintenance, and water extraction.

Step 2: Prepare the IBC tote

Make sure that your tote has a 6-½" IBC cap with 2" threaded plug.

Make sure the cap is securely tightened and sealed to prevent water from entering or leaving the tote except through the designated inlet and outlet.

Step 3: Install the First Flush diverter

Since we are dealing with a

two inch PVC pipe conduit system, it's best to get a First Flush diverter kit for three inch PVC pipe (with a 3 to 2 inch reducer bushing). Connect the downspout, coming from the gutter outlet, to the First Flush diverter. This may involve cutting and modifying the downspout or gutter to fit the diverter tee spout, or using PVC or other suitable piping to connect the two. Follow the manufacturer's instructions for proper installation, and ensure that the connection is secure and watertight.

Remember to align your First Flush diverter Tee so that it is facing in the direction of the IBC tote cap on the first tank.

Step 4: Connect the first flush diverter with your tote

a. Insert the 3" to 2" PVC reducer bushing into the open socket of the First Flush diverter Tee that is facing your tote.
b. Move over to **Tote 1** and locate the **IBC cap**. Your IBC cap has a 2" threaded plug. Thread the 2" PVC MPT Male Adapter into this plug.
c. Cut 2" PVC pipe of the correct length and insert one end into the PVC MPT Male Adapter.
d. Next, take the other end of the 2" PVC pipe and connect the 2" 90° PVC elbow. Align the open socket to face towards the First Flush diverter Tee.
e. Cut 2" PVC pipe of the correct length and connect one end to the 90° PVC elbow, and the other end to the 3" to 2" PVC reducer bushing that you connected earlier to the First Flush diverter Tee.

Remember to use PVC primer and cement for all slip fittings and Plumber's (Teflon) tape for threaded fittings.

Step 5: Connect your overflow

Move over to **Tote 2**, and use your hole saw to make a hole two inches from the top along the side wall. Install a bulkhead fitting (see p.121).

Connect your overflow pipe (2" PVC pipe) and add elbows and pipe as needed. This outlet should be connected to a drain or an overflow pipe that directs excess water away from the tanks and the surrounding area.

Step 6: Connect the water extraction outlet

Locate the outlet valves on the bottom of both totes and install the 2" PVC Female Adapter socket x FPT on each. It should simply screw on to the outlet. Use Plumber's (Teflon) tape to create a watertight seal. This step must be done first, because these adapters screw on to the tote outlet. Once joined permanently with PVC cement to the rest of the pipes, you won't

Totes - PVC Connect

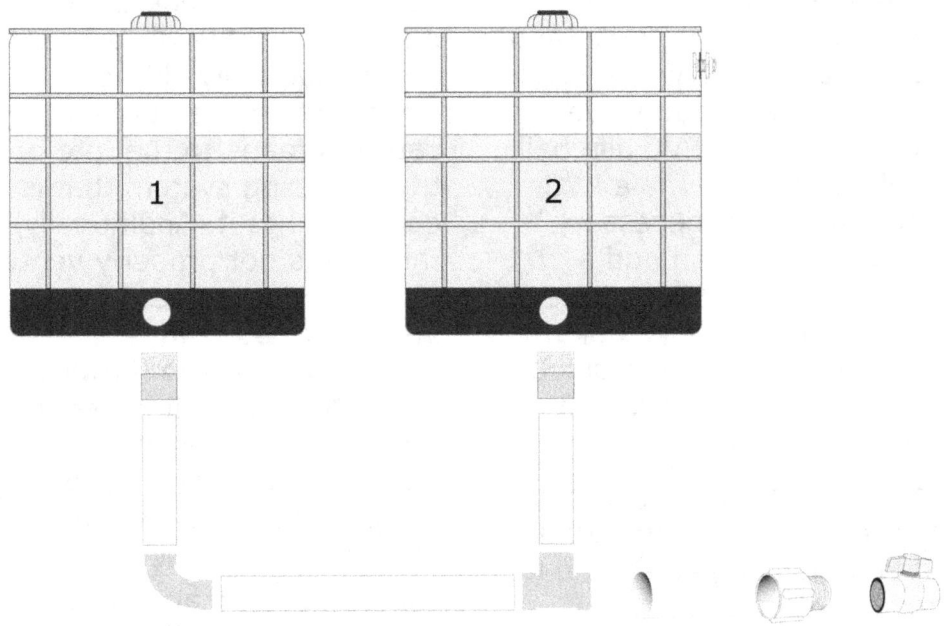

be able to perform the action of screwing (rotating) them into place.

Step 7: Assemble the PVC pipes

The outlets of both Totes 1 and 2 have been fitted with 2" PVC Female Adapter sockets (Step 6).

a. First, focus on Tote 1, and fit a 3-4 inch (10cm) length of PVC pipe to the Female Adapter socket. Then fit a 2 inch PVC 90° Elbow to the other end of this pipe.

b. Next, we move to Tote 2. Here we also fit a 3-4 inch (10cm) length of PVC pipe to the Female Adapter socket. Then we fit a 2 inch PVC Tee to the other end of this pipe.

c. Now check that you have two PVC pipes in front of you; to the left is a two inch pipe with a PVC Elbow and to the right is a two inch pipe with a PVC Tee.

d. Connect the PVC Tee to the PVC elbow with one length of 2" PVC pipe. The length will depend on the distance you allowed in-between the two tanks. This pipe should be around 50 inches (125cm) long.

e. Next we move to the PVC Tee on the right side. Insert a 20 inch (50cm) length of 2" PVC pipe into the right socket of the Tee.

Collection And Storage Of Rainwater   173

f. Next, add a male PVC slip adapter to the PVC pipe. This adapter has a socket end that fits your PVC pipe, and a threaded end which screws into your PVC ball valve. (If you used a PVC slip ball valve, then the male PVC adapter is not necessary.) Later on you can add a garden hose adapter to the PVC ball valve if you plan on having a tap or a spigot. Secure the ball valve with Plumber's tape.

Remember to use PVC primer and cement for all slip fittings and Plumber's (Teflon) tape for threaded fittings.

Step 8: Support your PVC pipe assembly

If you notice that your horizontal pipe setup (at the tote outlets) is experiencing stress due to the added weight of the water, then you can "rest" your pipes on a platform (block or bricks). Simply place a block under each PVC Tee or Elbow.

Step 9: Add a spigot or garden hose

Depending on your end-use goal you can add a spigot or an irrigation line (Garden Hose Adapter Barb) as needed.

Step 10: Add air vents

An IBC tote needs an air vent to prevent the buildup of pressure inside the tote that can cause it to collapse or rupture. When a tote is filled with water, the volume of air inside the tote decreases. If the air inside the tote is not allowed to escape, the resulting pressure can cause damage to the tote or the plumbing system that is connected to it. Additionally, if the tote is not properly vented, it can lead to a vacuum forming inside the tote, which can prevent water from flowing out of the tote when it is needed. An air vent allows air to flow in and out of the tote, equalizing the pressure and preventing any damage or malfunction.

You can buy an air vent and make one yourself (see *Vents*, p.132).

Step 10: Test the installation

Turn on your rainwater collection system and check for any leaks around the connections and fittings.

Adjust and tighten connections as needed to ensure a proper seal.

Loosen the tote caps at the top to release internal pressure.

Step 11: Regular maintenance

Regularly inspect and clean the IBC tote to ensure it is functioning properly.

Clean the inlet to prevent debris from entering the tote, and ensure the water extraction outlet is free from obstructions.

Additionally, consider installing a screen or filter at the inlet to prevent debris from entering the tote and potentially clogging the system.

Note: It's important to follow local regulations and guidelines for rainwater harvesting in your area, including any permits or restrictions that may apply.

If you're not confident in your DIY skills, it's recommended to seek the help of a professional plumber or rainwater harvesting system installer to ensure proper installation.

Take Note Of The Following:

- It's recommended to use clean and food-grade IBC totes for rainwater harvesting to ensure the quality and safety of the harvested rainwater.

- Pay attention to which totes you buy, not all have a 2 inch NPT thread in them, and not all totes have the same connector at the valve (there are at least 4-5 variations).

- Make sure you understand the importance of the overflow. The overflow should at a minimum be the same diameter as the inflow, since on a big rain day the overflow will not be able to keep up with the extreme volume of water flowing into the tote.

   This is where venting becomes crucial to disperse the built up pressure inside the tank.

- To get enough water pressure you will have to elevate your system or add a water pump.

- Remember to prepare fittings for secondary applications like irrigation, fire fighting equipment, pond/pool systems, etc.

- When adding PVC cement to PVC fittings, do the female side first, then you can set it down and not get debris on the glued surface while you are doing the male end.

- Ideally you should spray paint all the PVC parts after finishing, as UV light weakens PVC over time, making it brittle.

- You can make your PVC assembly more versatile by using a quick disconnect fitting straight from the tote outlet. This makes it easier to remove the PVC pipes when doing maintenance.

- Use PVC unions to simplify the maintenance process.

- You must vent your totes otherwise they will be damaged during a cloudburst.

Rain Barrels Installation

Installing multiple rain barrels for rainwater harvesting is a popular option for collecting and storing rainwater for various uses, such as gardening, irrigation, and other non-potable applications.

Let's take a look at the dimensions of a standard blue rainwater barrel used for rainwater collection:

- **Capacity:** 55 gallons (208 liters)
- **Diameter:** 23 inches (58 cm)
- Height: 35 inches (89 cm)
- **Weight (empty):** approximately 20 - 25 pounds (9-11 kg)
- Most 55 gal barrels have two bungs (caps) with a Fine thread and a Course Thread:
  a. 2" White Poly Bung Cap NPT Fine Thread with or without ¾" center tap
  b. 2" White Poly Bung Cap Buttress Course Thread with center ¾" NPT Fine Thread tap

It's important to note that the thread types and sizes on a 55-gallon water barrel can vary depending on the manufacturer and the specific product. Before purchasing a water barrel or any associated fittings or accessories, it's important to check the manufacturer's specifications and ensure that all components are compatible with

You Will Need

- A platform (wood, cinder blocks, steel) to place your barrels on
- 3 x rain barrels (55 gallon) with two bungs (caps) per barrel
- Flexible PVC elbow
- Felt-tip marker
- Spirit level and measuring tape
- PVC primer and cement
- Plumber's (Teflon) tape
- Silicone caulk
- Drill or hole saw (2 inch)
- Jigsaw
- 4 x 2" NPT male plumbing adapter x socket (3 used at outlets, 1 at overflow)
- 2" Conduit lockout nut (for overflow)
- 2 x 2" PVC Tee
- 4 x 2" 90° PVC elbow (2 used at outlets, 2 at overflow)
- 2" PVC pipe (as needed)
- 1 x 2" x 1-½" PVC flush bushing - spigot x hub
- 1 x 1-½" x ¾" PVC Schedule 40 bushing spigot x FPT
- 1 x ¾" MPT brass/PVC hex nipple (in cold areas the brass will contract and crack the pipe, so PVC is a better option)
- 1 x ¾" PVC threaded ball valve

(FPT=Female Pipe Thread)
(MPT=Male Pipe Thread)

each other.

These barrels are fun to work with (light) and there are many iterations that will work:

- You can stack them vertically (placed upright or sideways) **on top of one another**
- You can also stack them horizontally (upright or sideways) **next to one another**

You also have the option of joining multiple barrels :

- With the *Daisy Chain method* where you connect your barrels by linking the bottom outlets. This means that your connected rain barrels will **fill up evenly**.
- With the overflow to overflow method where you connect two barrels by linking their overflow outlets. This means that the **secondary barrel only starts filling up once the primary barrel is full**.

Similar to larger storage tanks, when it comes to attaching the downspout to the rain barrel, you have two options:

- **Connect the downspout** with a flexible elbow to the top of the rain barrel

Multiple Barrel Setup

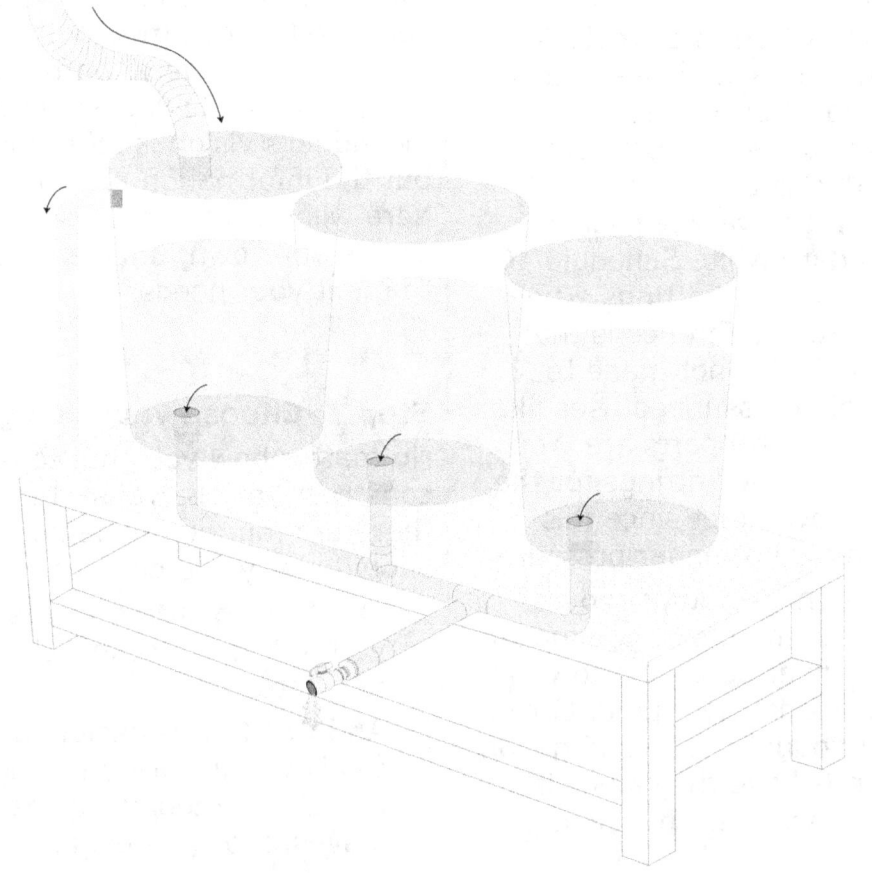

Collection And Storage Of Rainwater

- **The downspout does not enter the barrel**, but rather stops a few inches above the barrel. The top inlet is covered with a screen guard that keeps insects out.

Let's look at a tried and tested example of how to install three rain barrels in a two inch PVC pipe **Daisy Chain configuration**, placed vertically and side by side, under a downspout.

- Keep in mind that gravity is a rain barrel's best friend, and elevating your barrels will aid greatly in providing pressure at end-use and also allow your spigot enough clearance space above ground to fill buckets and watering cans.
- When it comes to material you will have to see what's available at your local hardware store. Schedule 40 PVC is best, but DWV parts look very similar to standard white Schedule 40 PVC pipe and fittings. The only real difference is that DWV PVC is not made to handle pressurized uses like Schedule 40 parts are. You can use DWV fittings for the overflow outlet since this fitting is not under pressure.
- You can also adjust your measurements to go smaller or thinner according to your needs and the type of barrel you bought. Not all 55 gallon barrels have the same size bung holes. In this example we worked with two inch PVC to accommodate the bung hole.
- When sizing your overflow system, keep in mind that you want your overflow rate to match the flow rate of the pipes that enter the tank. This means the overflow has to be the same size or greater than the inlet, so you're not in a situation where there is more water coming into your tank than can get out.

The process of installing multiple rain barrels is a simple one that can be divided into several steps. We will link three 55 gallon water barrels for rainwater harvesting, but you can also use this method to install a single water barrel or an array of multiple barrels. Every individual's vision is different, but the information provided here will get you on track to create your own, unique system to meet your needs.

How to

Step 1: Choose your design

No matter how you choose to construct your barrel system, there are four components that should always be present:

1. The **food-grade barrels** you will use to collect the rainwater
2. An **inlet** that is designed to allow rainwater into the barrel but keep out twigs, leaves, mosquitoes, and

potentially other creatures
3. An **outlet** you can turn on and off to drain your water
4. An emergency **overflow** that directs water away from your home and to an infiltration area when the barrel is at capacity

In this design we are using the standard blue 55 gallon barrels with two bung holes. One of the bung holes must be capped (never to be opened again), and the other will be used to accommodate a fitting. It's important to note that the two bung holes are not the same; one bung has a coarse (buttress) thread and the other one has a fine (NPS) thread.

Step 2: Prepare the barrels

It's advised to thoroughly rinse the barrels inside and out before using them. This helps to remove any debris or contaminants that may have collected in the barrels. Turn the barrels upside down to drain and dry.

Step 3: The bung holes

Take the bung hole cap with the coarse thread (that you will not use), add Teflon tape to its thread to prevent leaks, and then secure it in its hole. Do this will all three barrels. You will leave it in place and it is not to be used again. If your bung came with a rubber O-ring, leave it in place.

Step 4: Drainage outlets

The other bung holes (of all three barrels) will be used as drainage outlets and this is where you can insert the 2 inch NPT Male Plumbing adapters inside.

Remove the o-ring that comes with the bung and place it on your adapter. Remember to use Plumber's (Teflon) tape.

Step 5: Flip your barrels upside down

Select one barrel (referred to as Barrel 1), flip it upside down

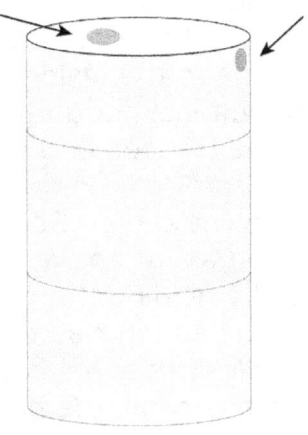

Collection And Storage Of Rainwater

and then use your marker to outline the shape of the opening for the downspout flex-elbow connector. Carefully cut out the marked opening using a jigsaw. This will serve as the entry point for rainwater flowing into Barrel 1 from the roof through the flexible PVC elbow. Make sure that it will be a snug fit.

Step 6: The overflow

Using a hole saw, cut a hole for the overflow near the very top of Barrel 1. This will allow excess rainwater to drain out if the barrel becomes full. Once the hole is cut, ready the last 2 inch Male Plumbing adapter, the 2" conduit lockout nut, and silicone sealing caulk.

Add the caulk to the adapter at the top of the threads. Place the threaded end of the adapter into the overflow hole from the outside to the inside. Take the Conduit locknut and screw it onto the treaded end by reaching through the downspout inlet hole cut earlier. Screw the locknut on as far as possible. Tighten by holding the nut and turning the adapter as tight as possible. Use a wet rag, to wipe off any caulk the has been squeezed out on the outside.

Step 7: The overflow pipe

Add a short section of PVC pipe to the Male Plumbing adapter. Then add your 90° PVC Elbow to the pipe with PVC cement. Put the cement inside the elbow end, fit it over the pipe and twist it sideways as you're pushing it in.

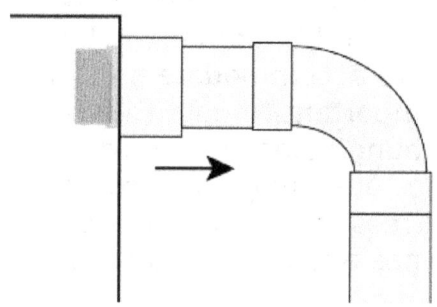

Measure and cut the PVC pipe to the desired length for the overflow (size of barrel plus height of stand). As an extra optional you can attach a 90° PVC Elbow to the bottom of the overflow pipe and add an additional length of pipe to direct water away from the foundation

of the house or platform.

Step 8: PVC pipes

Next we move to the PVC piping section. These pipes and fittings are under pressure and should all be Schedule 40 rated. The PVC pipes will connect the three barrels with one another. Cut the PVC pieces to the lengths your system requires and remember to use PVC primer and cement.

a. Looking at the diagram on p. 182, we can see that the three 2" threaded Male Plumbing adapters are placed inside the three bung holes (see Step 4).

b. Connect a 4"(10cm) length of PVC pipe to each male plumbing adapter.

c. Next we fit a 90° PVC Elbow to pipes of barrels one and three. For barrel two, fit a PVC Tee to the PVC pipe.

d. Connect the 90° PVC Elbow of barrel one to the PVC Tee of barrel two with a 20"(50cm) length of PVC pipe.

e. To connect barrel two with barrel three is a bit more technical (see p.182):

Connect one end of an 8"(20cm) length of PVC pipe to the PVC Tee of barrel two. Fit another PVC Tee to the other end of this pipe. This PVC Tee that sits between barrels two and three should have its open socket end facing the user, and away from the barrels.

Complete this part of the assembly by adding another length of 8"(20cm) PVC pipe to connect this PVC Tee and the 90° PVC Elbow of **barrel three.**

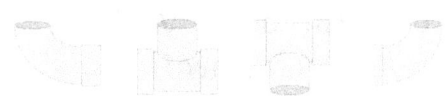

f. Add a 10"(25cm) length of PVC pipe to the PVC Tee that is facing the user and away from the barrels. This is where you will fit your ball valve.

Step 9: End-use fittings

Next comes the ball valve assembly.

Option A: Connect a 2" PVC ball valve with a slip socket, to the 2" PVC pipe.

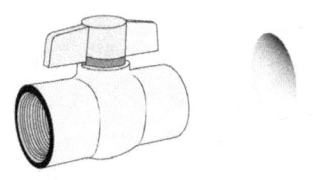

Option B: If you need a fitting where you can connect either a ball valve, or a spigot then attach the 2" x 1-½" PVC Flush Bushing - Spigot x Hub to the 2" PVC pipe. Next, fit the 1-½" x ¾" PVC Bushing Spigot x FPT to the flush bushing. Insert a ¾" hex nipple and then attach your ¾" Ball valve.

PVC Fittings Setup

1- Rain Barrel
2- 2" NPT Male Adapter X Socket
3- 4"(10cm) of 2" PVC Pipe
4- 20"(50cm) of 2" PVC Pipe
5- 8"(20cm) of 2" PVC Pipe
6- 10"(25cm) of 2" PVC Pipe
7- 2" 90° PVC Elbow
8- 2" PVC Tee
9- 2" x 1-½" PVC Flush Bushing - Spigot x Hub
10- 1-½" x ¾" PVC Bushing Spigot x FPT
11- ¾" MPT Hex Nipple
12- ¾" PVC Ball Valve

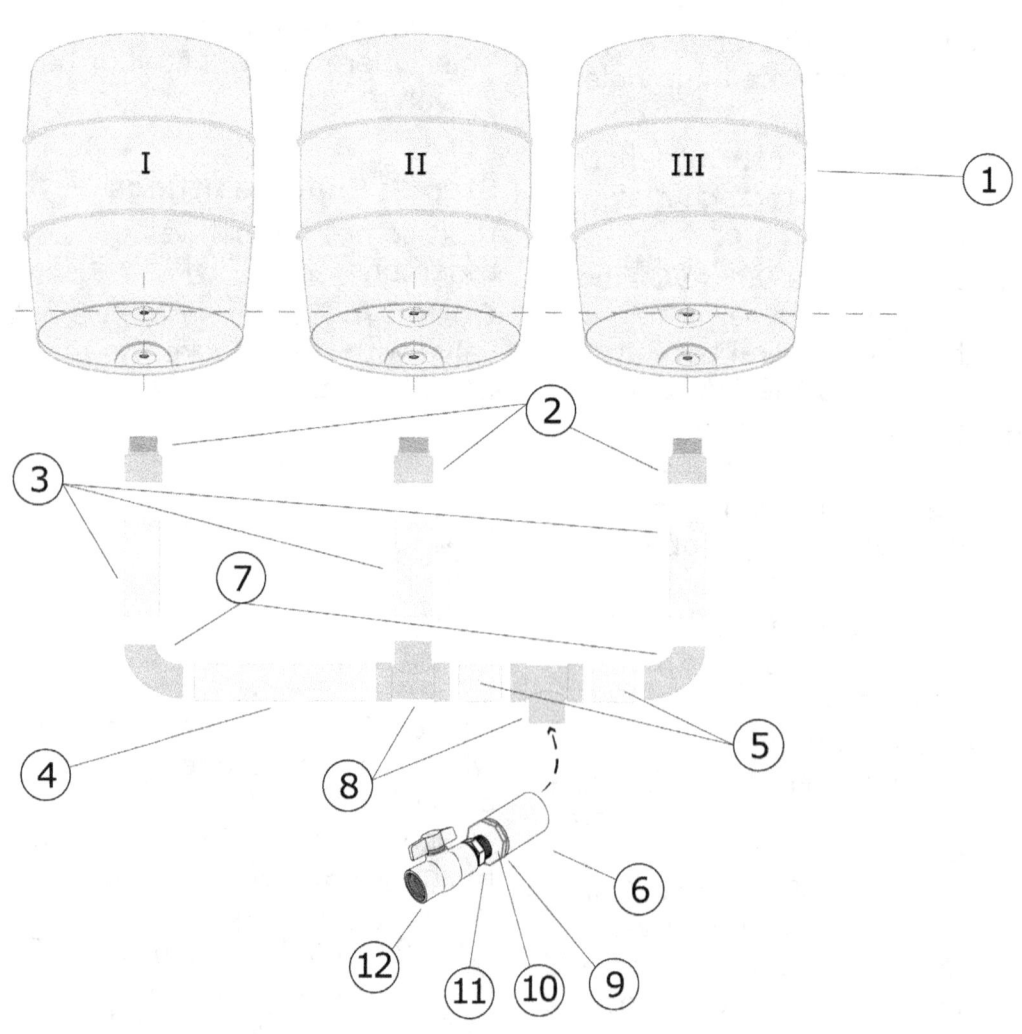

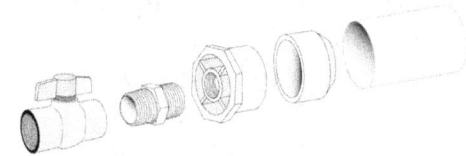

A spigot can fit directly into the ¾" PVC Bushing Spigot x FPT.

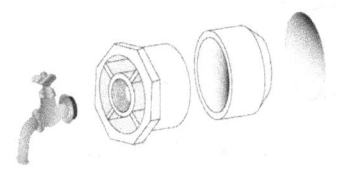

- Do a *dry fit* to make sure that your measurements are correct.
- Remember to position all three barrels so that the fittings are located at the front of your system, facing the user.

Step 10: Secure fittings

Secure Plumber's tape to all threads and screw them into the threaded openings on all three barrels. Glue all other fittings with PVC primer and cement.

Step 11: A platform

You will need a sturdy platform to place your barrels on. Make sure that the platform will not obstruct your pipes and cut or drill holes where needed. The bung holes must be visible and accessible and check for correct alignment of all pipes and outlets.

Step 12: Support your PVC pipe assembly

If you notice that your horizontal pipe setup (at the outlets) is experiencing stress due to the added weight of the water, then you can "rest" your pipes on a platform (block or bricks). Simply place a block under each PVC Tee or Elbow.

Step 13: Downspout inlet

Next it's time to install the downspout elbow. Attach the flex-elbow end inside the hole you prepared. Remember that this connection is not under pressure and sealing the joint with silicone will suffice.

Step 14: Air vents

A rain barrel needs an air vent to prevent the buildup of pressure inside the barrel that can cause it to collapse or rupture. When a barrel is filled with water, the volume of air inside the barrel decreases. If the air inside the barrel is not allowed to escape, the resulting pressure can cause damage to the barrel or the plumbing system that is connected to it. An air vent allows air to flow in and out of the barrel, equalizing the pressure and preventing any damage or malfunction (see *Vents*, p.132).

Step 15: Test the installation

Turn on your rainwater collection system and check for any leaks around the connections and fittings. Adjust and tighten connections as needed to ensure a proper seal.

Step 16: Regular maintenance

Regularly inspect and clean the barrels to ensure it is functioning properly. Clean the inlet to prevent debris from entering the tote, and ensure the water extraction outlet is free from obstructions. Additionally, consider installing a screen or filter at the inlet to prevent debris from entering the tote and potentially clogging the system.

Note: It's important to follow local regulations and guidelines for rainwater harvesting in your area, including any permits or restrictions that may apply. If you're not confident in your DIY skills, it's recommended to seek the help of a professional plumber or rainwater harvesting system installer to ensure proper installation.

When it's time to irrigate, simply connect a hose to the spigot located in front of the rainwater barrels. However, it's crucial to keep in mind that rainwater collected from your roof may contain harmful chemicals and bacteria.

If you intend to utilize the collected rainwater for drinking, it's important to be aware that certain areas may require certification of the filtration system and regular water testing. The most basic and cost-effective type of filtration process involves micro-filtration, which uses gravel, sand, and charcoal. More advanced treatments, such as UV sterilization and ozone treatment, can be more expensive options.

Take Note Of The Following:

- It's recommended to use clean and food-grade barrels for rainwater harvesting to ensure the quality and safety of the harvested rainwater.
- To get enough water pressure you will have to elevate your system or add a water pump.
- Remember to prepare fittings for secondary applications like irrigation, fire fighting equipment, pond/pool systems, etc.
- When adding PVC cement to PVC fittings, do the female side first, then you can set it down and not get debris on the glued surface while you are doing the male end.
- The brass male by male nipples can cause problems when used with PVC fittings. You should never thread metal MALE fittings, into FEMALE PVC fittings, as metal expands and contracts (due to temperature) significantly more than plastic, and will crack the plastic female fittings.

When you make a conversion

from metal to PVC, the PVC should ALWAYS BE male, and the metal ALWAYS female.

- Ideally you should spray paint all the PVC parts after finishing, as UV light weakens PVC over time, making it brittle.
- You can make your PVC assembly more versatile by using a quick disconnect fitting straight from the barrel outlet. This makes it easier to remove the PVC pipes when doing maintenance.
- Use PVC unions to simplify the maintenance process.
- You must vent your barrels otherwise they will be damaged during a cloudburst.
- Make sure you understand the importance of the overflow. The overflow should at a minimum be the same diameter as the inflow, since on a big rain day the overflow will not be able to keep up with the extreme volume of water flowing into the barrel. This is where venting becomes crucial to disperse the built up pressure inside the barrel.

The Flexible Hose Option

The PVC pipe and fittings method is by far the cheapest options and it is a tried and tested method. If money is not an issue, then you can also

You Will Need

- 3 x Rain barrels
- 1 x ¾" Hose Thread Y-splitter per barrel
- 2 x ¾" Female Mender Kits for each barrel
- Flexible hose as needed
- Plumber's (Teflon) tape
- Pipe wrench or pliers

consider connecting rain barrels using Gardener's Poly Tubing, or flexible hoses instead of rigid PVC pipe. PVC pipes provide a rigid and durable option for connecting rain barrels, while Poly tubing and flexible hose offer more versatility and ease of installation.

Both methods can be used effectively depending on your specific needs and preferences. The two systems are both based on the Daisy Chain method, with the main difference being the material and size of the tubing and fittings used.

Let's look at an example of a setup with reinforced flexible hose.

How To

1. Position your three barrels in a line and fit a rain barrel spigot to each barrel.

2. First, begin by attaching the Y valves to the spigot of every barrel. These valves

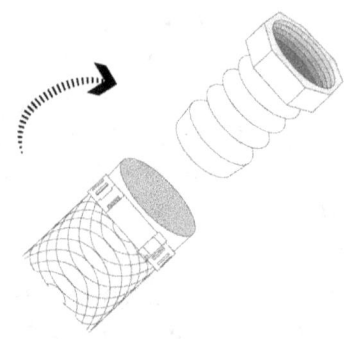

will enable you to split the connection, allowing water to enter from one side and exit from the other side.

4. Now, screw one end of the hose into the Y valve of your primary (first) barrel, and attach the other end to your secondary barrel. Remember to connect the hose to the valves that are closest to each other.

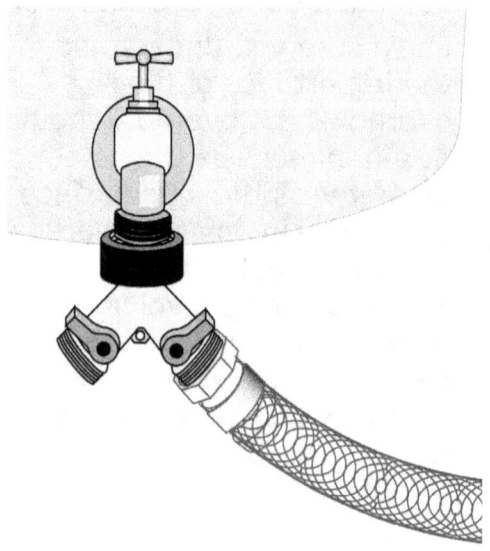

3. Next, cut your garden hose to the desired length. After cutting the hose, connect a female mender to each end of the hose.

Use a screwdriver to securely fasten the clamps, ensuring that the hose remains firmly attached and doesn't slip off.

5. Repeat the aforementioned steps for each barrel in your setup.

Flexible Hose Option

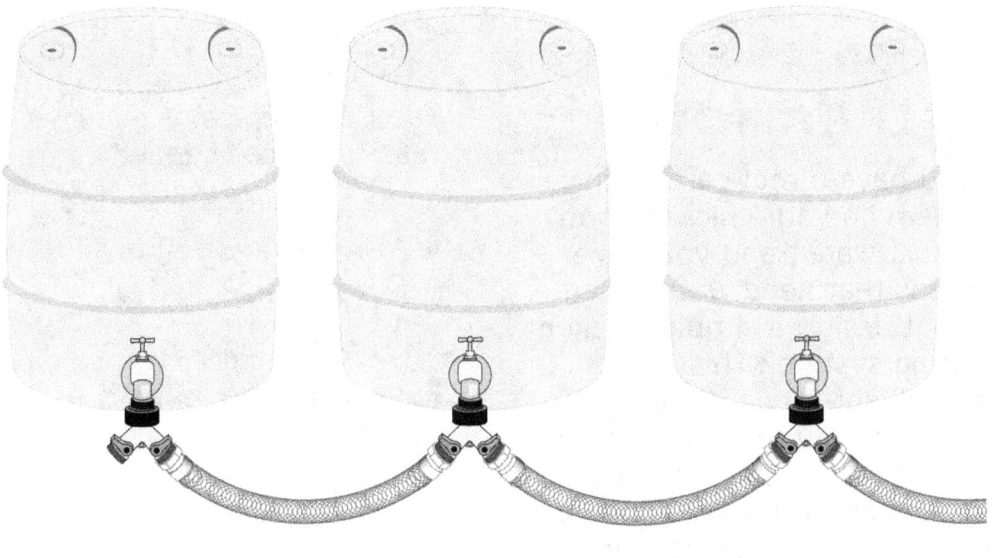

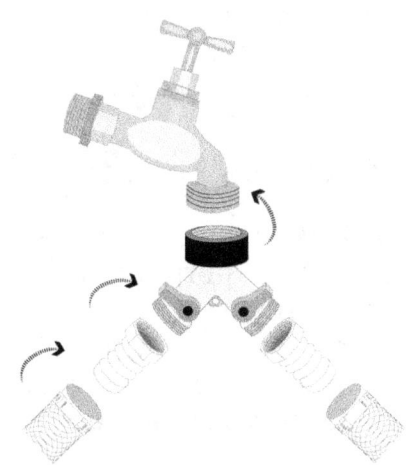

6. For the last barrel, ensure that the overflow valve is clean and clear. Attach another piece of hose to the overflow valve. When all of your barrels are full, the water will flow out through this hose and into a drain or a garden bed.

7. Test your installation and check for leaks.

This simple, yet effective, system will do wonders to supply water to a garden irrigation system.

Remember to still place your barrels on a sturdy platform and to install an overflow and air vents.

All systems require regular inspection and maintenance.

Chicken Waterer System Installation

If you have a roof catchment system on your chicken coop to collect water, and you have chickens that need water, then why not design a simple chicken watering system? This is a very straight-forward concept that allows you to collect water from the roof of the coop and then to store it in a rain barrel. From there the water is guided through a food-safe hose into a PVC pipe. The PVC pipe has a horizontal array of drinking nipples, which the chickens can use as their own little "drinking fountains".

There are several advantages to implementing such a system:

- **Cost savings:** By utilizing rainwater, you can take advantage of a free resource and reduce your water expenses.
- **Maintained water quality:** Due to the elevated position of the pipes, the water remains clean and uncontaminated.
- **Water availability:** By actively monitoring your rain barrel, you can ensure a steady supply of water and avoid running out. If the water level drops, simply refill the barrel as needed.

You Will Need

- Rainwater collection system (e.g., gutter, downspout, rain barrel)
- Watering nipples or poultry water cups (as needed)
- Plumber's tape (Teflon tape)
- PVC primer and cement
- Drill and drill bits
- Saw or PVC pipe cutter
- Screws and mounting brackets
- 2" PVC pipe cap
- 2" PVC pipe (as needed)
- 2" x 1-½" PVC flush bushing
- 1-½" x ¾" PVC reducer bushing spigot x FPT
- ¾" Reinforced flexible hose (as needed)
- Mesh or filter (optional)
- Hose clamps (optional)

(FPT=Female Pipe Thread)

Not all chicken coops are the same, but the information provided here will get you on track to create your own, unique system to meet your needs and to provide for your feathered friends.

How to

Step 1: Install a rainwater collection system.

Set up a gutter system on the roof of the chicken coop to collect rainwater. Direct the

Chicken Waterer

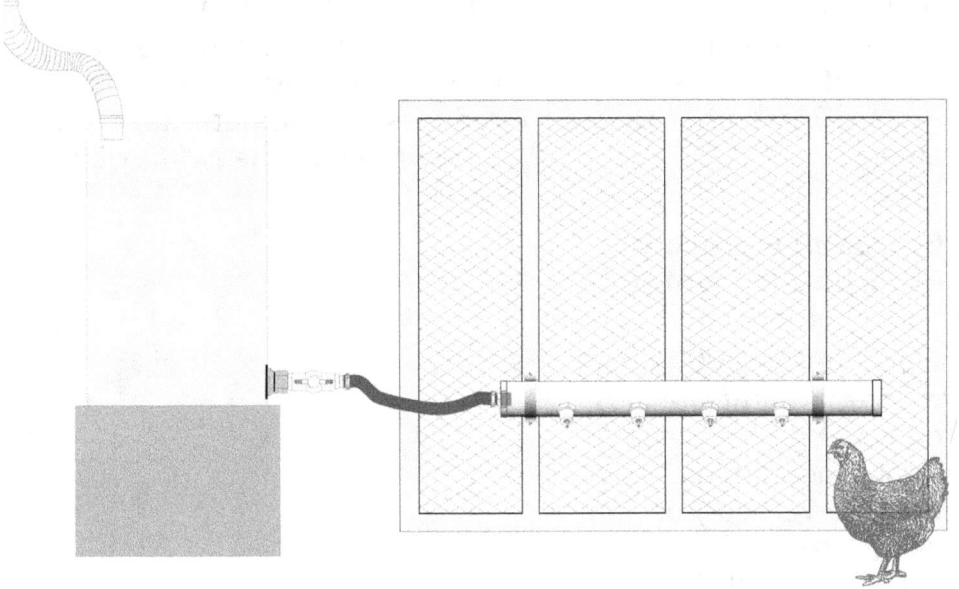

flow of water into a downspout connected to a rain barrel or another suitable container for water storage.

Step 2: Determine the water distribution points.

Identify the locations within the chicken coop where you want to provide water for the chickens. These can include areas near perches, feeding stations, or other convenient spots.

Step 3: Connect your rain barrel.

Add an outlet with ball valve or faucet to your water barrel. Connect one end of the reinforced flexible hose. Remember to use Plumber's tape.

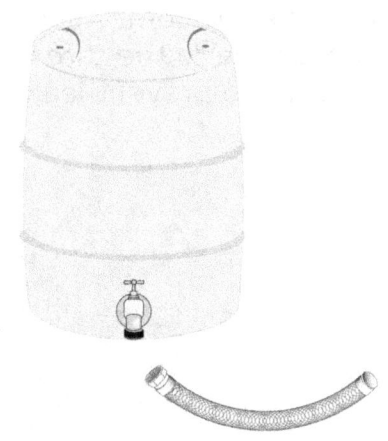

Step 4: Plan the PVC pipe layout.

Design a piping system which, when assembled, will connect the rain barrel to the PVC pipe. Consider the distances, angles, and elevation changes involved in the coop to ensure proper water flow.

Step 5: Cut and assemble the

Collection And Storage Of Rainwater

PVC pipe.

Measure and cut the PVC pipe according to your design, using a saw or PVC pipe cutter. Close off one end of the PVC pipe with the PVC cap. Remember to use PVC cement.

Step 6: Install watering nipples or poultry water cups.

Drill holes into the PVC pipes that you prepared at the selected watering locations. Install watering nipples (horizontal or vertical) or poultry water cups into the holes, ensuring a secure fit. Make sure you space them out to provide ample drinking room. Apply Plumber's tape to the threaded connections to prevent leaks.

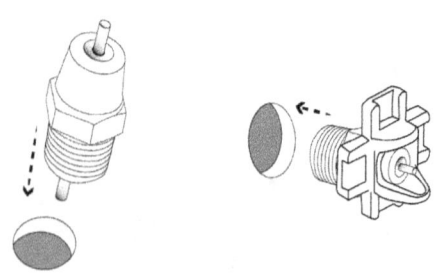

Step 7: Fit the Flush Bushing.

Fit the 2" X 1-½" Flush Bushing in the 2" PVC pipe with PVC cement. This provides a transition from the larger diameter of the pipe to the desired 1½" size.

Step 8. Fit the Reducer Bushing

Fit the 1-½" x ¾" PVC Reducer Bushing into the Flush Bushing with PVC cement. This further reduces the diameter to accommodate the ¾" flexible hose.

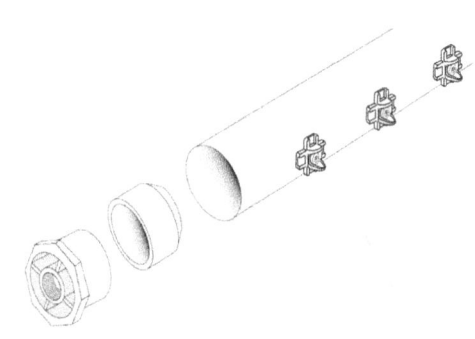

Step 9: Fasten the PVC pipe.

Use mounting brackets to fasten the PVC pipe with fittings to the coop wall. Remember to adjust the height in order to make it convenient for the chickens to drink.

Step 9: Connect the hose to the PVC pipe.

Insert the other end of the ¾" reinforced flexible hose into the ¾" Reducer Bushing. Ensure a tight and secure connection to prevent water leakage. Apply Plumber's tape to the threaded connections to prevent leaks.

Step 10: Test the system.

Fill the rain barrel with water and check for any leaks or drips along the piping system. Adjust and tighten connections as necessary.

Optional

- **Add a mesh or filter:**
  To prevent debris or contaminants from entering the watering system, consider adding a mesh or filter at the entry point of the rain barrel or before the water reaches the PVC pipes.
- **Connect additional hose or tubing:**
  If needed, connect a hose or tubing to extend the water distribution further or to reach specific areas within the chicken coop.
  Secure the connections using appropriate fittings or clamps.

Regularly maintain and clean the system:

- Routinely inspect and clean the rainwater watering system to ensure proper functioning and water quality.
- Clear any blockages, replace worn-out parts, and clean the watering nipples or cups as needed. Clear any potential obstructions underneath the cups.

By following these steps, you can build a rainwater watering system for your chicken coop, providing a convenient and sustainable water source for your feathered friends.

Tanks For Small Spaces

If space is a problem then you should consider a rainwater tank specifically designed for small spaces. These tanks are designed to be compact and fit into tight areas, making them suitable for residential properties with limited space. Here are a few options for rainwater tanks that work well in small spaces:

Slimline Tanks: Slimline rainwater tanks are tall and narrow, allowing them to fit into narrow areas like side yards or along the side of a house. They are available in various capacities and can be installed against walls or fences, maximizing space utilization.

Under Deck Tanks: Under deck rainwater tanks are designed to fit beneath elevated decks or raised platforms. They utilize the space under the deck, which would otherwise be unused, making them ideal for small yards. These tanks are often long and shallow to fit in the available space.

Wall-Mounted Tanks: Wall-mounted rainwater tanks are installed vertically against a wall, making them suitable for areas with limited ground space. They are designed to be slim and can be placed on exterior walls or in narrow spaces, such as between windows or doors.

Modular Tanks: Modular

rainwater tanks are made up of individual modules that can be combined to create a customized storage capacity. They are versatile and can be configured to fit into various spaces, including small ones. The modular design allows for flexibility in adapting to available space constraints.

Underground tanks: Underground tanks come in various shapes and sizes and are simply buried water storage tanks. This hides the tank from obstructing views or using up precious outdoor living space. Since space becomes less of an issue, you can fit larger water storage facilities on your property. Water quality also tends to be better since the tanks are kept dark and cool underground.

When selecting a rainwater tank for a small space, consider the available dimensions, capacity requirements, and all installation options.

Conceal Your Tank

If you want to make your water tanks blend in with the environment, there are several strategies you can employ:

Camouflage: Use paint or artwork to camouflage the tanks and blend them into the surrounding environment. Choose colors that match or complement the natural colors of the landscape. You can also consider painting them with a mural or using stencils to create patterns that mimic the nearby foliage or scenery.

Planting: Use landscaping techniques to create a visual barrier around the water tanks. Plant shrubs, bushes, or tall grasses strategically to conceal the tanks from view. Select plants that are suitable for your climate and that will grow tall enough to screen the tanks.

Screening: Install screens or trellises around the tanks to visually hide them. You can use lattice panels, bamboo screens, or even slatted wooden fences. These structures can be painted or stained to match the surrounding environment.

Natural coverings: Explore the option of using natural coverings to conceal the tanks. For example, you can construct a wooden enclosure with a hinged or removable lid that can be covered with rocks, or moss. This can create the illusion of a natural feature rather than a man-made water tank.

Placement: Position the water tanks strategically in areas that are less visible or where they can be partially hidden by existing structures, such as decks, sheds, fences, or trees. By careful placement, you can take advantage of the existing elements in your environment to minimize the visual impact of the tanks.

Avoid Erosion

When water flows through the outlet at a high velocity, it can exert significant force on the surrounding soil or material. To prevent erosion caused by water tank outlets, several measures can be taken:

Direct water to a safe outlet: Connect a pipe or hose to the overflow pipe to carry the water to a safe outlet. This outlet should be away from vulnerable areas where soil erosion can occur (slopes or bare ground).

Create a splash pad or rock bed: At the outlet point, you can construct a splash pad or a rock bed to help disperse the overflow water evenly. This prevents concentrated flow and reduces the impact on the soil. The splash pad or rock bed acts as a buffer, allowing the water to gently spread and absorb into the ground.

Use erosion control measures: If the outlet area is particularly prone to soil erosion, consider implementing additional erosion control measures. These can include adding erosion control blankets, planting vegetation, or installing gravel channels to help stabilize the soil and to minimize erosion.

Regularly inspect and maintain: Regularly inspect the overflow solution to ensure it remains in good condition and functions effectively. Check for any blockages, damage, or signs of erosion.

Conclusion

The market for storage tanks is constantly evolving with new options being introduced each year. Manufacturers are improving their techniques and utilizing different materials, providing a wider variety of tanks and fittings for consumers to choose from. It is crucial to conduct your own research to select the right material and setup that best suits your specific needs. It's worth noting that storage tanks are not limited to rainwater harvesting alone. They can also be used in conjunction with wells and ponds. In such cases, water is pumped from the well into the holding tank. From there, it can be directed to the house for domestic use or to the fields for agricultural purposes. To ensure the cleanliness of harvested rainwater, employ a combination of first flush diverters, leaf eaters, storage tank maintenance, disinfection, and small micron filters. These components work together to effectively clean the water and make it suitable for various uses within the household plumbing system. Maintenance is a crucial aspect of rainwater harvesting systems. Regular inspection and upkeep of the components are necessary to maintain the system's performance and ensure the quality of the water.

Rainwater Pumps

In a basic rainwater harvesting setup, the water follows a path from the catchment area, down through the gutters and downspouts, into the storage tank.

From the tank's outlet, the water is **drawn into** the pump and subsequently **pushed** towards its intended destination for usage.

When choosing a water pump for your home or garden, several factors should be taken into consideration:

Distance: Evaluate both the vertical and horizontal distance that the water needs to travel between the tank and the intended destination. A greater distance will require a more powerful pump to ensure efficient water flow.

Water pressure requirements: Determine the desired water pressure needed for your specific applications. Different appliances and systems may have varying pressure requirements, so it's important to choose a pump that can meet those demands.

Noise level: Consider the noise generated by the pump during operation, particularly if it will be installed in a residential area or near living spaces. Opting for a quieter pump can help maintain a peaceful environment.

Energy efficiency: Assess the energy usage of the pump to ensure it aligns with your desired level of efficiency. Look for pumps that are designed to minimize energy consumption without compromising performance.

Consult pump suppliers: Seek advice from pump suppliers who can provide expert guidance on selecting the appropriate pump for your specific application. They can also calculate the required flow rate for each indoor appliance, ensuring optimal performance.

Pump Options

When it comes to selecting the right pump for water storage during rainwater harvesting, we generally have two options to work with: **External pumps** and **Submersible pumps**. The main difference between submersible pumps and external pumps lies in their installation and operational characteristics:

Installation: *Submersible pumps* are designed to be submerged directly into the water. They operate while fully submerged.

In contrast, *external pumps* are installed outside of the water.

Design: *Submersible pumps* are self-contained units with the motor and pump combined in a single unit. The motor is hermetically sealed and protected from the fluid being pumped.

External pumps, on the other hand, have a separate motor and pump. The motor is located outside the water and connected to the pump via a shaft.

Efficiency: *Submersible pumps* are generally more efficient since they operate submerged in the fluid, reducing energy losses due to friction in pipes and minimizing priming requirements.

External pumps may have slightly lower efficiency due to energy losses in the suction and discharge pipes.

Noise: Submersible pumps tend to be quieter during operation since they are submerged.

External pumps may produce more noise as they are typically installed above ground.

Maintenance: *Submersible pumps* require less maintenance since they are protected from elements and do not have external components exposed to the environment.

External pumps may require more maintenance due to the exposure of the motor and other components.

Besides considering external and submersible pumps, we should also evaluate the following sub-categories:

Pressure-sensitive pumps: Modern pumps with pressure

Pump Basics

- Use a pre-filter to clear sediment before the water flows into the pump.
- Keep suction piping leading to the pump, as short as possible. This ensures good flow, uniform velocity and good pressure.
- Eliminate elbows mounted close to the pump inlet. This causes "loading" and puts strain on the pump. A straight run of pipe is best.
- Eliminate trapped air from system by keeping water levels high and avoiding unnecessary "high bends" in pipes, which can cause air pockets.
- Don't put any weight on your pump. Your piping arrangement should not put any weight or stress on the pump casing.
- Make sure your pump has a cover to protect it from the elements and to keep all electrical wires contained.
- Make sure your pump is secure and if needed, create a casing that allows you to lock the pump in place.
- Remember: Long pipe, high friction, low flow.
- Short pipe, low friction, high flow.

sensors that activate automatically when a faucet or tap is opened, compensating for water pressure loss. If you want a pump that is on all the time, and you don't have to worry about messing with an electrical outlet by flipping the switch, then it would probably be best to invest in one of these. They are energy-efficient and consume low power. (Also called an on-demand or automatic pump.)

Transfer Pumps:

This type of pump, you plug in when you need to use it and it then pressurizes the water. You cannot let these pumps run constantly. For example, if you are using a hose nozzle with these pumps, you cannot stop the flow of water (turning off the hose) and let the pump keep running. The pump must have water going through it while it is running, or it will overheat. If you want something simple, where you just flip a switch and when you're done using water you can just flip the pump off, then a lower cost economical transfer pump would be a good one to choose.

Inline pump controllers:

Separate devices that monitor line pressure and activate the pump as needed. They can be easily added to existing installations for enhanced convenience.

Solar pumps:

These days solar pumps are very efficient and affordable. A solar pump combined with a pressure tank can provide an effective solution for your water pressure needs.

Booster pumps:

Booster pumps, also known as jet water pumps, are water pumps that operate using electric power and employ a centrifugal-style mechanism. True to their name, booster pumps excel at delivering substantial flow rates and high pressures, often surpassing other available water pump options. This pump variant is also referred to as an aboveground water pump.

Glossary

AC Pump. An AC pump is a pump that operates using alternating current (AC) electrical power and it is designed to efficiently pump fluids in various applications where AC power is readily available

DC Pump. A DC pump is a pump that runs on direct current (DC) electrical power, commonly provided by batteries or solar panels

Ball valve. A ball valve is a type of valve used to control the flow of fluid within a pump system and it consists of a hollow sphere or ball with a hole in the center that can be rotated to control the flow of fluid

Booster Pump. A booster pump is a pump that increases the pressure and flow rate of a fluid in a plumbing or water supply system

Electric diaphragm water pump:

Electric diaphragm water pumps are a type of water pump that utilizes a diaphragm, a flexible membrane, to displace and move water. These pumps are powered by electricity and are commonly used for irrigation and water transfer. The working principle of this pump involves the reciprocating movement of a diaphragm. When the pump is activated, an electric motor moves the diaphragm back and forth. This back-and-forth motion creates suction and pressure strokes, enabling the pump to draw in water and expel it through the outlet. The diaphragm acts as a barrier between the motor and the liquid being pumped, preventing contamination and damage to the pump's internal components. Their self-priming capability means they can automatically remove air from the suction line and begin pumping water without the need for manual priming. They are compact, portable, and durable, making them suitable for both residential and commercial use.

Self-priming pumps: Self-priming pumps have a built-in liquid reservoir and can prime themselves without manual intervention, while regular priming pumps require manual priming before use. As a result, self-priming pumps are ideal for applications requiring frequent priming and are more reliable and efficient in such situations.

Backup pumps:

Having a backup pump in case of emergency is a must. If you lose power, you should have a backup plan. Can your tank setup rely on gravity alone?

It is recommended to consult with a professional to determine the most suitable pump type for your rainwater harvesting system.

Pumps For Rainwater Harvesting

In short, the top recommended rainwater pumps are the **electric diaphragm water**

Glossary

Centrifugal Pump. A device that uses centrifugal force to move fluids by converting rotational energy into kinetic energy

Cavitation. Cavitation in pumps occurs when low pressure causes the formation and collapse of vapor bubbles in the liquid being pumped which can lead to reduced efficiency, damage, and decreased flow

Discharge pipe. The discharge pipe in pumps refers to the pipe or conduit through which the pumped fluid exits the pump and is conveyed to its intended destination or application

Municipal water. Municipal water refers to the water supply system provided by a local government or municipality to meet the drinking water and domestic needs of the

pump, the **submersible water pump**, the **booster pump**, and the **solar powered water pump**.

1. **Electric diaphragm water pump.** Various sizes of diaphragm pumps (３⁄８ and ½ inch) are readily accessible to accommodate standard garden hoses, irrigation pipes, and plumbing systems. Electric diaphragm pumps offer water pressure outputs ranging from 30 PSI to 60 psi, ensuring a suitable range for various applications. The flow rates of these pumps can vary between 1 to 6 gallons per minute (GPM). Most diaphragm pumps are designed for easy operation, with a simple on-off switch that allows users to activate them when needed and deactivate them when no longer required. The cost of diaphragm pumps can range from $100 to $300 or more.

2. **Submersible water pump.** If money is not an issue, then you can consider a submersible rainwater pump. They come in a variety of power ratings, ranging from 1/6 to 2 HP models. When it comes to water pressure output, common submersible water pumps offer a range of 11 PSI to 135 psi, catering to different pressure requirements. The flow rates of these pumps can reach up to 36 gallons per minute (GPM) or even higher, depending on engine power, make, and model. Prices typically range from $100 to $1000 or more.

3. **Booster pumps.** As mentioned earlier, booster pumps employ a centrifugal-style design, utilizing a high-velocity spinning impeller to effectively draw and propel water through plumbing systems. When applications require substantial pressure or flow rates, opting for a booster pump is an excellent decision. Booster pumps are available in a range of plumbing sizes, from ½ to 2 inch piping sizes. Water

---

Glossary

residents

Pressure tank. A pressure tank in pumps, also known as a pressure vessel or storage tank, is a container designed to store pressurized fluid that is used in conjunction with a pump system

Priming. Priming refers to the process of filling a pump or a piping system with fluid in order to remove air or other gases and create a continuous liquid seal

PSI. PSI stands for pounds per square inch and is a unit of pressure commonly used in pumps and other pressure-related applications

Pump switch. An electrical device or mechanism that controls the operation of a pump and it is

pressure outputs for booster pumps typically span from 30 PSI to 145 psi and beyond. Maximum flow rates can vary from 12 GPM to 45 GPM and higher. Prices for booster pumps can range from $200 to $1000 or more.

4. **Solar-powered water pumps.** Solar-powered water pumps typically exhibit plumbing sizes, pressure outputs, and water flow rates that are similar to those of electric diaphragm pumps. These pumps are most suitable for smaller or singular outdoor applications, such as lawn sprinklers, drip irrigation, and garden hoses. Common plumbing sizes for solar-powered water pumps include ⅝ inch and ½ inch, offering compatibility with standard fittings. Maximum water pressure outputs can range from 35 to 55 psi, providing adequate pressure for various outdoor tasks. Flow rates can vary from 1 GPM to 6 GPM, enabling efficient water delivery.

Prices typically range from $100 to $300, but the cost may increase with additional functionalities or higher power output options.

Flexible vs. Rigid Piping

A booster pump can be installed with **both flexible and rigid piping**, depending on the specific requirements and preferences of the installation.

Each type of piping has its advantages and considerations:

Flexible Piping: Flexible piping, such as reinforced, flexible PVC or flexible hoses, offers versatility and ease of installation. It can be bent and maneuvered to accommodate different configurations and tight spaces.

Flexible piping is often used when there is a need for flexibility and easy connection between the booster pump and the water supply or distribution system. Flexible piping can also be used where constant

Glossary

responsible for starting, stopping, or regulating the pump based on specific conditions or parameters

Suction pipe. The suction pipe in pumps refers to the pipe or conduit that brings fluid from the source or reservoir into the pump for the purpose of being pumped

Water filtration. Water filtration for households refers to the process of treating and purifying water to remove impurities, contaminants, and potentially harmful substances, making it safe and suitable for various uses within a home

Water meter. A water meter is a device used to measure and record the amount of water consumed or supplied to a specific location, and typically found in the front yard near the curb of your home

vibration is causing movement of the pump body for some reason. Flexible pipes will prevent pipe or pump damage. However, it's important to select high-quality, properly sized flexible piping that is suitable for the pressure and flow requirements of the booster pump system.

Rigid Piping: Rigid piping, such as PVC, copper, or galvanized steel pipes, provides durability and stability. Rigid piping is commonly used when a more permanent and fixed installation is desired. It offers good resistance to pressure, maintains its shape, and can handle higher flow rates. Rigid piping requires precise measurements, proper fittings, and careful planning for installation, including cutting, threading, and solvent welding (for PVC pipes).

In conclusion, the choice between flexible and rigid piping for a booster pump installation depends on factors such as space limitations, accessibility, pressure requirements, flow rates, and personal preference.

It's essential to ensure that the chosen piping materials and sizes are suitable for the specific demands of the booster pump system.

Pump Accessories

Pressure tank – The tank stores pressurized water to prevent the pump from cycling on and off to meet small demands; it also supplies a constant pressure.

Pressure switch – This device engages the pump when a pressure drop is observed and disengages it when there is no demand. For example, the pump will engage to supply water when a faucet is opened and disengage when no water is

---

Typical Pump Setup

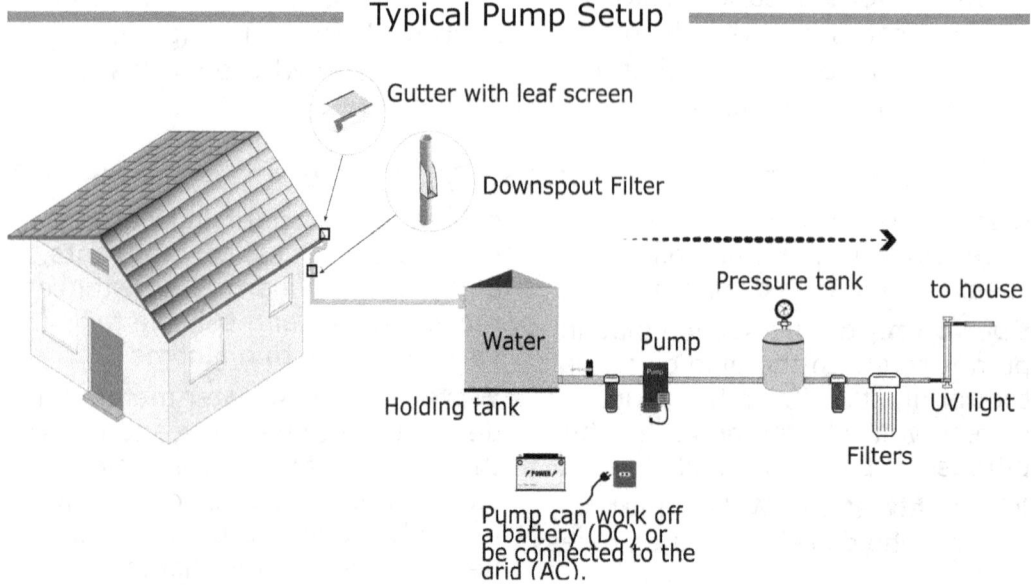

being used.

Check valve – The internal check valve prevents water from flowing back through the pump when it is not running. Some pumps do not have these valves.

Float level switch – This switch, when installed in the cistern, can prevent damage to the pump by disengaging the pump when the cistern runs dry or cistern water falls below a predetermined level.

Throttling valve – This device is intended to control the flow and pressure of water exiting the pump and is typically in the form of a gate valve. Special control systems can be installed to maintain a constant output despite changing water levels within a cistern.

Pressure Tank

Next in line is your **pressure tank** that will provide pressure to the indoor plumbing and minimize pump usage. Pumps work very well in combination with pressure tanks.

Mechanics

Pressure tanks mimic what a municipal water utility does (i.e. pump water to a tank for later use), but on a much smaller scale. A pressure tank has the capacity to store a certain amount of water (about 35 gallons/132.5 liters) under pressure that can be used throughout the day. It has a rubber bladder inside the tank which acts just like a balloon to provide the water pressure. As the pressure in the bladder declines the pump turns on to refill the bladder. This means that the pump is used less often and it also allows you to save on electricity and increases the life of the pump. The pressure tank holds enough pressure to flush a toilet in the house, without turning the pump on.

Tank Pressure

The normal operating pressure ranges for pumps are 20/40, 30/50, 40/60 or even as high as 60/80 psi. This means that pressure is kept constant within a 20 psi range, which sets a boundary for starting and stopping the pump.

A 40/60 ratio is recommended since it's by far the best ratio to keep a constant shower stream. This pressure range indicates when the pump will turn on and off, and this function is controlled by a **pressure switch device**. This is the device that will turn your pump on and off automatically.

Example:

If you have a 40/60 psi setting, the pump will turn on when the pressure in the pressure tank drops to around 40 psi and off when the pressure reaches 60 psi.

Tank Size

The relationship between the

pressure switch, pressure tank and the pump is what allows water to flow through the pipes in your home. **When selecting a pressure tank, you will need to know the performance of the pump in gallons per minute.**

Typically we measure the volume of water required to serve a household, in gallons per minute (gpm). It is calculated by looking at the water consuming appliances in the house and also at the pump's flow rate in gpm.

The gpm of the pump must equal the total number of water consuming appliances.

Example:

5 faucets, 2 toilets, a shower and a washing machine would require 9 gpm.

Therefore, to determine the appropriate size of the tank, match the draw down of the tank to the capacity of the pump. A tank's drawdown volume is the amount of water that is stored between the high and low pressure points, which is usually determined by the pump switch.

"Draw Down" Explained

In plumbing, "tank draw down" refers to the process of water being drawn or consumed from a storage tank or reservoir. When water is used from a tank, the water level gradually decreases or "draws down" as water flows out.

For example, in a rainwater harvest system, there is often a storage tank or pressure tank that holds water. When a faucet or plumbing fixture is opened, water is drawn from the tank, and the water level in the tank decreases. The rate at which the water level drops during usage is known as the "tank draw down."

Understanding the tank draw down is important for sizing and selecting the appropriate size of tanks or reservoirs in plumbing systems to ensure an adequate supply of water is available during peak usage periods without running out.

Another easy way to **size a pressure tank**, is to take the gpm system requirement that you determined for your house, and to multiply it by 3 and then to go to the next largest pressure tank size.

In our previous example:

9 (gpm) x 3 = 27.

Buy a 30 gallon or bigger pressure tank.

Usually, when it comes to pressure tanks, bigger is better. A larger pressure tank will not hurt your pump's performance, but it will give you a larger draw down capacity, which guarantees even more usable pressure in the plumbing system, before the pump needs to come on.

In general, a ½ HP pump is enough for a small house with

one bathroom and a couple of people.

For a larger house, you'll need a ¾ - 1 HP pump.

Take into consideration that a larger pump will affect your electricity bill and that it will also require a larger pressure tank.

If you want a generator as a back-up power source, make sure that the generator can provide enough power to start the pump.

Conclusion

With pressure tanks, it is necessary to buy a pressure sensitive pump that shuts off, when a specific pressure is reached (i.e. 30/50 psi).

Pros:
- Proven technology
- Can be use with lower powered pumps (i.e. Solar systems) or low flow rate pumps
- Easy to service
- Widely available (most large hardware stores)

Cons:
- Requires occasional bladder replacement
- Entire tank needs to be protected from freezing
- Needs a separate enclosure

Having a water meter at this point in the system is a great idea to measure the water usage of the household.

Be conservation minded and keep an eye on consumption per person, and number of water-using appliances.

Check Valve

A check valve, also known as a non-return valve or one-way valve, is a device commonly used with water pumps, which allows water to flow freely in one direction while preventing backflow or reverse flow of water. A check valve is often recommended in applications where the pump needs to maintain its prime, prevent water hammer, or ensure that water does not flow back into the tank.

Mechanics

When installed in the **discharge pipe** of a water pump, the check valve allows water to be pumped out of the system while preventing water from flowing back into the pump when it is not operating. This is important for maintaining the prime or water level in the pump and preventing damage or loss of pressure.

When installed in the **suction pipe (**between the tank and the pump), the check valve prevents the pressurized water from returning to the water tank, and it simultaneously creates a closed system between it and your faucets.

The check valve typically consists of a valve mechanism that allows water to flow in the desired direction and closes to block flow in the opposite direction. It can be a swing check valve, where a hinged disc swings open to allow flow and swings closed to prevent backflow, or a spring-loaded check valve, where a spring keeps the valve closed until the water pressure exceeds a certain threshold, allowing flow in one direction.

Union Couplings Explained

A union coupling in plumbing is a type of pipe fitting used to join two pipes or sections of pipe together and allow for easy disassembly or separation when needed. It consists of three main components: two threaded or slip ends and a central nut or collar.

The threaded ends of the union can be male or female, depending on the specific application and the pipes being connected. The slip ends are designed for pipes without threads and require the use of solvent cement to secure them in place.

To connect the pipes using a union fitting, the central nut or collar is tightened, drawing the two ends together and creating a watertight seal. When disassembly is required, the nut is loosened, allowing the two ends to be separated.

Unions are commonly used in plumbing systems where regular maintenance, repairs, or replacements may be necessary. They provide a convenient way to disconnect and reconnect pipes without the need for cutting or threading. Union fittings are typically made of brass, steel, or PVC.

Check Valve Distance:

The requirement to have a minimum distance of 4 times the diameter of the inlet pipe between the check valve and the suction port of the pump is often recommended for proper pump operation. This distance is commonly referred to as the "straight pipe run" or "straight length" and serves a few important purposes:

- ◇ Flow Stabilization: The check valve creates resistance to the flow of water due to its design and the mechanism that prevents backflow. Placing the check valve a certain distance away from the suction port allows for a length of straight pipe before the valve. This length helps to stabilize and smooth the flow of water entering the pump, reducing turbulence and maintaining a steady flow. Turbulence or uneven flow can cause cavitation, which can lead to damage to the pump and reduced efficiency.
- ◇ Reduced Air Entrapment: Placing the check valve at a distance from the suction port helps to minimize

the likelihood of air being trapped in the pump or the suction line. Air pockets can cause issues with pump performance, such as reduced flow, cavitation, and increased wear on the impeller.

◇ Preventing Vortex Formation: A vortex is a swirling motion of fluid that can occur in the suction line if not properly controlled. Placing the check valve a sufficient distance away from the suction port helps to prevent the formation of a vortex, which can disrupt the flow and potentially lead to air entrainment or cavitation.

By maintaining the minimum distance of 4 times the diameter of the inlet pipe between the check valve and the suction port, these issues can be minimized.

How To Install A Pump

Let's explore a standard installation of a booster pump *(fig 4.26)*, specifically designed to generate sufficient pressure for both indoor and outdoor plumbing systems.

1. Calculations:

   Determine the correct size pump and piping needed for its end-use. One pipe is your **suction pipe** which connects the tank outlet to the pump inlet. The other pipe is your **discharge pipe** which connects the pump outlet to the end-use line.

2. **Gather the necessary materials:**
   - a pump suitable for rainwater systems
   - appropriate piping
   - appropriate fittings
   - an outdoor electrical connection

   Check the manufacturer's instructions for the specific requirements of your chosen pump. Use Plumber's (Teflon) tape on threads and PVC cement on pipe fittings.

3. **Determine the pump's location:**

   Select a suitable location for the pump near the rainwater tank. Ensure it is on a stable, flat surface and protected from the elements.

4. **Install a ball valve:**

   At your tank outlet port you should install a ball valve to isolate the tank water from the pump.

5. **Install the suction pipe:**

   Move across to your **pump inlet** and connect a suction pipe from the pump to the ball valve at the outlet port of the rainwater tank. This pipe should be made of appropriate material, such as Polyethylene, Reinforced flexible hose, or PVC, and sized according to the pump's specifications. A flexible pipe is recommended to accommodate the pump vibrations (movement). Add a union coupling to this pipe to simplify future maintenance.

6. Install a strainer or filter (option):

To prevent debris from entering the pump, you can install a strainer or filter next to the ball valve on your suction pipe line. This will help maintain the pump's efficiency and prevent blockages. Most modern pumps come with a filter add-on that's easy to install.

7. Connect the discharge pipe:

Attach a discharge pipe to the **pump outlet**. The size of the pipe should be determined by the pump's specifications and the intended use of the water. Ensure the pipe is properly supported and secured.

8. Install any necessary valves and fittings:

Install another union coupling for future maintenance work. This **union** should be followed by another **ball valve** to isolate the pump when needed. Depending on the configuration of your system, you may need to install valves, **check valves**, pressure switches, and other fittings as required. These components help control the flow of water and protect the pump from damage.

9. Install a pressure tank:

Depending on the configuration of your system, you may need to install a pressure tank as required. A pressure tank will provide pressure to the indoor plumbing and minimize pump usage.

10. Electrical connection:

Follow the manufacturer's instructions to connect the pump to a suitable power source. Ensure the electrical connection is done safely and according to local electrical codes. If you are not experienced with electrical work, consider hiring a qualified electrician.

Booster Pump Installation

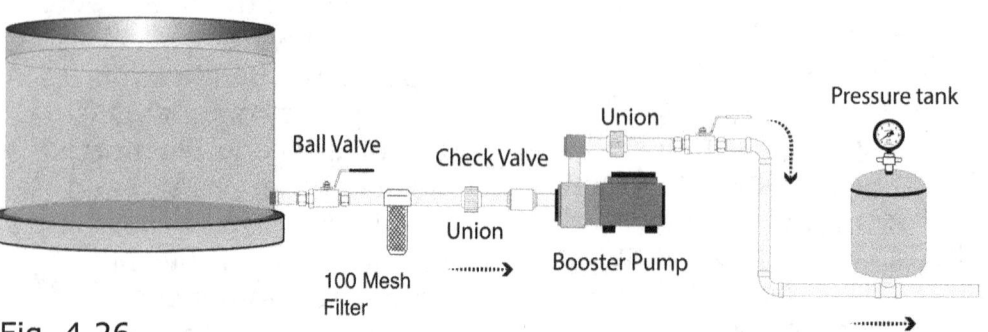

Fig. 4.26

11. **Test the system:**

    Once the pump is installed, fill the rainwater tank and test the system. Check for any leaks, ensure the pump is primed (if required), and verify that it is functioning properly.

12. **Maintain the system:**

    Regularly inspect and maintain your rainwater tank system. Clean or replace filters as needed, check for any signs of damage or wear, and follow the manufacturer's recommendations for pump maintenance.

13. **Protect the pump:**

    Cover your pump with a protective housing which will protect it against the elements.

Irrigation Installation

To utilize tank water solely for a basic irrigation system *(fig 4.27)* instead of indoor use, you can proceed with the pump installation in the following manner:

1. **The right pump:**

    Select the right pump for the job. A booster pump, or electric diaphragm water pump will work just fine. A pump used for irrigation typically relies on a separate device called a pump controller or irrigation controller to determine when to turn on and off. The pump controller is responsible for monitoring and controlling the irrigation system based on specific parameters set by the user.

2. **Flexible hose:**

    Select reinforced flexible hosing for this installation. Cut the hose to the right length. One hose is your **suction hose** which connects the tank outlet to the pump inlet. The other hose is your **discharge hose** which connects the pump outlet to the irrigation line. Use hose clamps or quick disconnect couplers to attach your hoses.

3. **Ball valves:**

    Install a ball valve at the tank outlet, and another ball valve at the pump outlet.

4. **Suction hose:**

    Move across to your **pump inlet** and connect a suction hose from the pump to the ball valve that you installed at the outlet port of the rainwater tank. Follow the manufacturer's instructions and place the check valve and filters where indicated.

5. **Discharge hose:**

    Move to the **pump outlet** where you connected the ball valve, and connect the discharge hose. Use a hose clamp or a quick disconnect coupler.

6. **Power connection:**

    Connect the pump to the

Irrigation Installation

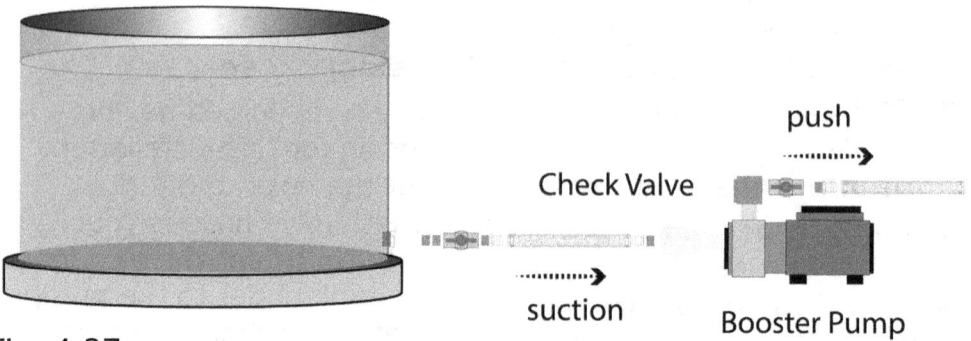

Fig. 4.27

appropriate power source as per the manufacturer's instructions. Ensure that all electrical connections comply with electrical codes and regulations.

7. **Start the pump:**
Once everything is properly connected and in place, start the pump according to the manufacturer's instructions or using the pump controller, if available. The pump will draw water from the source through the suction hose and deliver it to the irrigation area through the discharge hose and sprinklers/emitters. (Some older model pumps may require priming, which involves filling the suction hose and pump with water to create suction. Follow the pump manufacturer's instructions for priming if necessary.)

8. **Maintenance:**
Regular maintenance and monitoring are essential to ensure the pump and hoses are functioning correctly, and the irrigation system is delivering the desired water supply to your plants or crops.

Pump Enclosure

Pump life and longevity should always be a priority and this is where a simple enclosure can be worth your while. An over-the-shelf enclosure is basically a shell made out of Polyethylene and it is UV- and corrosion-resistant. These days you can get simple and affordable enclosures at most hardware stores.

Potential benefits:

◊ **Protection from the elements:** An enclosure provides a barrier against external elements such as rain, snow, dust, and debris, which can damage the pump or its components. Water pumps are typically designed for outdoor use, but prolonged exposure to harsh weather conditions can lead to corrosion, electrical issues, and overall reduced lifespan. An enclosure helps shield the pump from these environmental factors.

- ◇ **Security and vandalism prevention:** Installing an enclosure around the water pump can deter theft and vandalism. It makes it more difficult for unauthorized individuals to tamper with or steal the pump, ensuring the continuous functioning of your water system.

- ◇ **Noise reduction:** Water pumps can generate noise during operation, which can be bothersome, especially if the pump is located near living spaces or in noise-sensitive areas. An enclosure can help reduce the noise produced by the pump, making it more pleasant for nearby residents or workers.

- ◇ **Aesthetic appeal:** If the water pump is located in a visible area, such as a garden or backyard, an enclosure can improve the overall appearance of the space. Enclosures come in various designs and materials, allowing you to choose one that blends well with the surroundings and enhances the visual appeal.

- ◇ **Safety precautions:** Depending on the specific circumstances, an enclosure can add an extra layer of safety by preventing accidental contact with moving parts or electrical components of the pump. This is particularly relevant in environments where children or animals may be present.

Alternatively, you can build a pump enclosure with brick and cement which can provide excellent protection for the pump, particularly against flooding and harsh weather conditions. Using these materials offers durability and strength, ensuring that the enclosure can withstand external forces and provide long-term protection.

Here are some advantages of constructing a pump enclosure with brick and cement:

- ◇ **Structural integrity:** Brick and cement provide a solid and robust structure for the enclosure, ensuring its stability and durability over time. They can withstand heavy rainfall, high winds, and other severe weather conditions, protecting the pump from damage.

- ◇ **Flood protection:** If the pump is located in an area prone to flooding, a brick and cement enclosure can offer a higher level of flood protection compared to other materials.

- ◇ **Security:** Brick and cement enclosures can act as a deterrent against theft and vandalism.

- ⬥ **Insulation:** Brick and cement enclosures can provide better insulation for the pump, helping to regulate temperature and protect it from extreme heat or cold. This insulation can contribute to the pump's longevity and efficient operation.

When constructing a pump enclosure with brick and cement, it's important to ensure proper ventilation to prevent heat buildup and allow air circulation. Adequate access points should be incorporated to facilitate maintenance and repairs when necessary.

Conclusion

Once you have a tank filled with rainwater, you will quickly realize that you have different options concerning the usage of this resource.

To use your water effectively and to maximize usage, you will have to switch between municipal (mains) water and your tank water.

Since the water from your tank is free, you will obviously have to use it first. Once the water level in the tank becomes too low, you can switch to the public water supply. This can be done either manually or with an electronic switch-over device.

The benefit of using this setup is that you will mostly be draining the "free" water from the tank and thus your tank will have space for the next rain event, and it will save you money on your water bill.

- The **manual way** is very reliable and cheap, but it means that you have to be physically available to go outside to open/ close the line.
- Using an **automatic** switch-over device is a bit more expensive and it uses electricity. It is however very convenient and provides a smooth flow of water.

Make sure you check the local government regulations in your area on whether it is required to have a device installed or not.

When it comes to pumps, make sure that you are familiar with the operation of the pump and follow all maintenance and safety guidelines to increase longevity.

A booster pump is a popular choice and can be used to provide pressured water from the storage tank to the household plumbing.

Take note, that before installing a new pump, check the performance rating chart to make sure it can provide you with the capacity and pressure that you require. Typically, 3-4 gpm is a minimum acceptable flow rate per outlet.

Let's take a closer look at the maintenance of a typical rainwater harvest system.

Maintenance Of A Rainwater Harvest System

Regular maintenance is essential for a rainwater harvesting system to ensure its optimal performance, longevity, and to maintain the quality of collected rainwater. Here are some reasons why regular maintenance is necessary:

1. **Water Quality:** Regular maintenance helps maintain the quality of collected rainwater. Over time, debris, leaves, sediment, and contaminants may accumulate in gutters, filters, or storage tanks. Regular cleaning and maintenance prevent the buildup of pollutants and ensure the water remains clean and suitable for its intended uses, such as irrigation or non-potable household uses.

2. **System Longevity:** Proper maintenance practices can prolong the lifespan of the rainwater harvesting system. Regular inspection and timely repairs or replacements of worn-out or damaged components help prevent further deterioration and costly major repairs. By addressing minor issues promptly, you can avoid more significant problems and extend the lifespan of the system.

3. **Safety and Hygiene:** Regular maintenance helps ensure the safety and hygiene of the rainwater harvesting system. By keeping gutters, downspouts, and storage tanks clean, you minimize the risk of bacterial growth, mold, or pests that can compromise water quality and pose health risks. Additionally, proper maintenance reduces the likelihood of accidents or water-related hazards due to system malfunction or blockages.

4. **Optimal Water Supply:** Well-maintained rainwater harvesting systems provide a reliable and consistent supply of water for various purposes. Regular inspections and maintenance allow for early detection of issues such as leaks, clogs, or pump malfunctions, enabling timely repairs and uninterrupted water supply.

5. **Compliance with Regulations:** In some regions, there may be specific regulations or guidelines for rainwater harvesting systems. Regular maintenance ensures compliance with these regulations, preventing any potential legal issues or penalties.

To maintain a rainwater harvesting system effectively,

it is recommended to establish a regular maintenance schedule that includes tasks such as gutter cleaning, filter maintenance, tank inspections, pump checks, and overall system monitoring. Following manufacturer's guidelines, seeking professional advice when needed, and staying attentive to any changes or signs of issues will help ensure the system operates efficiently and provides a sustainable water source. You should know your system inside out and be able to repair, replace or service any of the components at work.

A general inspection of your collection system is a must:

Every three months:

◇ Clean gutters and check screens and guards.

◇ Remove any algal growth if present.

◇ You must have an access hole to see into the tank. Keep the lid of the access hole clean.

Every six months:

◇ Inspect integrity of tank, pipes and the pump system.

◇ Clear overhanging branches and other vegetation.

◇ All tanks must have a drainage outlet to wash sludge and heavy sediment out.

Avoid exposure to sunlight and also to contaminants like external water sources.

Contamination

Regular monitoring of water quality is crucial, especially when using stored water for human consumption. Various factors can contribute to the contamination of stored water, with the quality of the source water being a significant factor, particularly water collected from the roof. To achieve a concentration of approximately 1 to 2 part per million (ppm) of free chlorine, which is commonly recommended for private water systems, follow the sanitizing procedure outlined below. This method allows for the continual use of treated water and **is not intended for shock chlorine disinfection**, which requires higher concentrations of around 200 ppm and may involve draining the treated water.

Procedure

In a clean quart (1 liter) container filled halfway with water, add 1 to 1-½ fluid ounces (2-3 tablespoons) of unscented, non-detergent household chlorine bleach with a concentration of 5.25%, for every 500 gallons of water to be treated. Pour the bleach solution directly into the storage tank and distribute it throughout the tank by stirring or mixing with a clean pole or paddle. Thorough mixing for 2 to 3 minutes is sufficient.

Example solution mix: For 1,700 gallons of water, use 1-¼ fl oz/500 gal, resulting in 4.25 fl oz of chlorine bleach.

Close all tank openings, including the lid, and let the solution stand for a minimum of eight hours before using. If stirring is not possible, let the treatment stand for 24 hours. It's important to note that this disinfection rate is suitable for fresh, relatively clear water. It may not be as effective for turbid or cloudy water containing significant suspended solids. For highly turbid water, consider shock treatment or seek recommendations from your local municipal office.

After treatment, do not consume the treated water until the chlorine odor has dissipated, which typically takes one to two days. Removing the tank lids may speed up chlorine volatilization, but it also increases the risk of subsequent contamination from rodents, birds, or airborne sources. Boiling or cooking chlorinated water usually removes the chlorine, making it tasteless and safe for most people.

Individuals sensitive or allergic to chlorine should avoid direct contact with the treated water until the chlorine odor has completely disappeared. Additionally, some plants may be affected by irrigation with chlorinated water, so it's important to consider this when using the water for such purposes.

Shock Chlorine Disinfection

Shock chlorination is a disinfection treatment recommended when a drinking water system has been contaminated with total coliform or E. coli bacteria. In general, if a tank becomes contaminated, the usual procedure for most people would involve draining, cleaning, and refilling it. However, for those residing in water-scarce regions where access to water is a matter of life and death, the situation becomes more urgent. In such dire circumstances, when having water is absolutely crucial, it becomes necessary to employ a method called shock chlorination. If you do suspect biological contamination like dangerous pathogens (water smells bad and the color is off), you will have to act quickly:

- Get rid of the source of contamination and investigate what's causing the contamination (dead bird or rodent).
- Estimate how long the water has been exposed to contaminants.
- Consider draining the tank, especially if the water has been exposed to contaminants for an extended period of time.
- In case of an emergency, where draining is not an option, you will have to consider boiling and filtering the water or treating the water with chlorine.

Chemicals like chlorine are only to be used in an emergency.

For shock chlorination of contaminated tanks, use 50 to 100 parts per million (ppm) of chlorine and let it sit for 12 to 24 hours.

After adding chlorine, monitor the chlorine levels as they will naturally decrease over time due to factors like water chemistry and temperature.

Test the chlorine residual after 24 hours. If it measures 10 ppm or less, repeat the chlorination process.

If you aim to maintain a safe chlorine residual in stored water, maintain a level of 1-2 ppm.

After shock chlorination, the water becomes temporarily unsuitable for drinking or potable use **until the chlorine levels drop below 2-4 ppm.** The time it takes for chlorine levels to decrease depends on factors such as temperature and water chemistry, typically ranging from a few days to a few weeks.

Use the chart below to find out how much bleach is needed to disinfect a certain volume of water. A general rule of thumb to shock chlorinate and disinfect a storage tank is to mix non-scented NSF-approved household bleach (5.25% chlorine) in the reservoir at the ratio of 1 gallon of bleach for every 1,000 gallons of water (i.e., 1 quart for every 250 gallons of water). This will give a chlorine concentration of 50 ppm, far higher than the 0.5 to 2.0 ppm found in treated city water, and make the water unusable for potable water use until residual drops down to less than 4.0 ppm. To disinfect connected plumbing lines and fixtures, open all taps until a chlorine smell is apparent at each outlet. Close taps and allow the chlorinated water to sit for at least 12 hours to ensure adequate time for disinfection. Do not consume this 50 ppm concentrated solution!

Drain the water storage tank and connected piping to the ground away from plants—not into a

Storage Tank in Gallons	Approx. parts per million of chlorine residual achieved by adding 5.25% chlorine bleach, in the amounts below. Numbers are rounded for easier measuring. 1 Tablespoon = 0.5 ounce.				
	1 ppm	5 ppm	50 ppm	100 ppm	200 ppm
10,000	25.5 oz	1.0 gal	10 gal	20 gal	40 gal
5,000	12.5 oz.	½ gal	5 gal	10 gal	20 gal
2,500	6.5 oz	32 oz	2.5 gal	5 gal	10 gal
1500	3.8 oz	19 oz	1.5 gal	3 gal	6 gal
1000	2.5 oz.	12.8 oz.	1.0 gal	2 gal	4 gal
500	1.3 oz	6.4 oz	0.5 gal	1 gal	2 gal
250	4 teaspoon	3.2 oz	4 cups	0.5 gal	1 gal
100	1.5 teaspoon	1.3 oz	1-½ cups	0.2 gal	0.4 gal

septic system (which can kill the necessary "good bacteria"), stream, or pond. This may kill fish and plant life. It may also be illegal, so be sure to check local and state regulations. Ideally, the chlorinated wastewater should be legally disposed into a sewer network. Refill the cleaned and disinfected tank with potable water.

Next, open the valve to distribution lines and run water from the taps until there is no smell of chlorine. Now you can enjoy safe water from your cleaned and disinfected water storage tanks.

De-Sludge

De-sludging a rainwater tank involves removing accumulated sludge, sediment, and debris from the bottom of the tank. Here's a general guideline on how to de-sludge a rainwater tank:

- **Safety Precautions:** Before starting the de-sludging process, ensure your safety by wearing appropriate protective gear, such as gloves and a mask, to avoid direct contact with the sludge or inhaling harmful particles.
- **Prepare the Tank:** Turn off the water supply to the tank and close any valves or outlets. This will prevent water from entering the tank during the de-sludging process.
- **Drain the Tank:** Open the tank's drain valve or tap and allow the water to drain out completely. You can direct the drained water to a suitable drainage area or use it for non-potable purposes like watering plants or cleaning.
- **Remove Debris:** Once the tank is drained, remove any debris, leaves, or large particles that may have accumulated on the top of the sludge layer. Use a scoop, net, or any appropriate tool to collect and dispose of the debris properly.
- **De-sludging Process:** To remove the sludge at the bottom of the tank, there are a few methods you can consider:
- **Manual Removal:** If the sludge is accessible, you can use a shovel, scoop, or a wet vacuum to manually scoop or vacuum out the sludge. Take caution not to damage the tank lining or structure.
- **Siphoning:** If the sludge is not too thick or the tank has a small access point, you can use a siphoning hose or vacuum pump to suction out the sludge. Insert the hose into the tank, ensuring it reaches the bottom, and use suction or siphoning action to draw out the sludge.
- **Professional Services:** In some cases, it may be necessary to engage professional tank cleaning services who have specialized

equipment and expertise in de-sludging rainwater tanks. They can efficiently remove the sludge and ensure thorough cleaning of the tank.

- **Clean and Rinse:** After removing the sludge, thoroughly clean the tank's interior using a hose or pressure washer. Scrub the walls and bottom of the tank with a mild detergent or cleaning solution to remove any remaining residues or stains. Rinse the tank thoroughly to remove any cleaning agents.
- **Refill the Tank**: Once the tank is clean and rinsed, close the drain valve and refill the tank with fresh rainwater. Ensure the water is clean and free from contaminants before using it for any intended purposes.

Regular maintenance, including periodic de-sludging, helps maintain the quality of rainwater collected in the tank and ensures its optimal functioning. The frequency of de-sludging may vary depending on factors such as the size of the tank, the amount of sediment accumulation, and local environmental conditions. It is advisable to consult tank manufacturers' recommendations or seek professional advice for specific de-sludging intervals for your rainwater tank.

Colder climates

Harvesting rainwater in cold climates with snow and ice is no simple task; although the fundamental basics are the same, the ice factor definitely complicates things. Freezing temperatures can be destructive to a rainwater harvesting system if water in the pump and pipes are allowed to freeze.

For freshly fallen snow, 14 inches of snow will equal 1 inch of water. When the snow is compacted, expect 4 inches of snow to equal 1 inch of water.

Some important maintenance points to consider:

Seasonal variations. Understand the different patterns of precipitation throughout the year, including rainfall and snowfall. Determine the peak snowfall season and assess its impact on the availability and collection of rainwater.

Ice is heavy. All weight-bearing supports will have to be designed to withstand heavier loads. This includes gutter supports, tank stands and roof supports.

Roof design. Using snow guards is essential for gutter protection. Design the collection surface in a way that facilitates the efficient capture of rain and melting snow.

Ice contracts and melts. Prepare for fluctuation of water levels in your tank. A simple, clean tank without complicated internal fittings will make things a lot easier.

Snow removal. Develop a snow removal plan to clear the collection surfaces and access points of the rainwater harvesting system. Snow accumulation can obstruct the flow of water into the collection system, so regularly clearing the snow ensures efficient collection once the snow melts. Use appropriate tools or methods to safely remove snow without causing damage to the system.

Tank type and dimensions. A large, round tank is best simply because of the round curves and the space it provides. More water means that it will take longer to freeze over. Get a tank that is guaranteed (by manufacturer) not to crack in sub-zero temperatures. In areas with heavy snowfall, get a tank that can withstand heavier loads and with a steep-angled roof (for snow to slide off). If you want to put your tank in the ground, follow the manufacturer's instructions.

Use insulation. Check with your local hardware store what works in your area. Insulation materials such as foam, or thermal jackets for tanks can be used to minimize heat loss and maintain the quality and availability of the collected rainwater. Spray-on polyurethane foam provides good insulation for tanks, and valves and pumps can be placed in insulation boxes.

Pipes. Pipes are prone to cracking in freezing conditions. Pipes should be drained when not in use and can also be insulated. Bury your pipes below the freeze line when drained of water. Use pipes that are suitable to the conditions in your area. Standard PVC pipes are not recommended. Valves should be placed in insulation boxes.

Trace heating. There are a few ways that you can prevent pipes from freezing over. Trace heating involves taping a heated element to the side of the pipe. The tape used is insulated and can be used for gutters as well.

Extras. To prevent the water in the tank from freezing over, you can consider an aerator that keeps the water moving in the tank, or even a heat pump that recirculates hot water through the system.

Overflow and drainage. Incorporate proper overflow and drainage mechanisms into the rainwater harvesting system to handle excess water during snow-melt periods. This helps prevent damage to the system and potential flooding. Consider installing additional channels or gutters to direct the excess water away from the collection area.

Mosquito Control Measures

In a complete rainwater harvest system, there are several

components that can be utilized to prevent mosquitoes from breeding and accessing the harvested water. Here are some measures you can consider:

- **First Flush Diverter:** A first flush diverter is a device that diverts the initial dirty flow of rainwater away from the storage tank. It helps remove debris, dust, and pollutants that might have accumulated on the roof. By preventing the initial runoff from entering the tank, it reduces the organic matter and potential mosquito breeding sites.
- **Leaf Eaters/Leaf Screens:** Leaf eaters or leaf screens are devices installed on the downpipes to prevent leaves, twigs, and other debris from entering the storage tank. These screens should have fine mesh sizes to prevent mosquitoes from accessing the water and laying eggs. Regular maintenance and cleaning of these screens are essential to ensure their effectiveness.
- **Mosquito Screens:** Install mosquito screens or mesh on all openings, air gaps, vents, or overflow outlets of the rainwater tank or system. These screens should have small enough openings to prevent mosquitoes from entering while allowing air circulation. Regularly inspect and clean these screens to remove any debris or

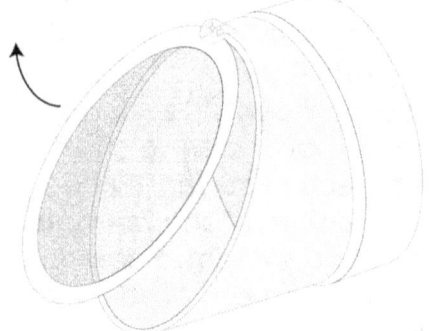

obstructions.
- **Tank Cover:** Ensure that the rainwater tank has a secure and tight-fitting cover. This prevents mosquitoes from accessing the water and breeding. The cover should be designed to allow ventilation while keeping mosquitoes out. Additionally, consider using a cover made of solid materials rather than mesh to minimize potential mosquito entry points.
- **Larvicides and Biological Controls:** Consider using larvicides or biological control agents to prevent mosquito larvae from hatching in the stored rainwater. There are specific larvicides available that target mosquito larvae but are safe for other organisms using the water. However, it's important to carefully follow the instructions and dosage recommendations for any chemical treatments.
- **Regular Maintenance:** Regularly inspect and maintain the rainwater harvest system to prevent

stagnant water and potential mosquito breeding grounds. Check for any leaks, blockages, or areas where water may accumulate. Ensure that gutters, pipes, and tanks are properly sealed to avoid water seepage or standing water.

By implementing these components and measures, you can significantly reduce the risk of mosquitoes breeding in your rainwater harvest system, ensuring the water remains clean and safe for use.

Mosquito Screens

Mosquito screens can be installed in various locations within a rainwater harvest system to prevent mosquitoes from accessing the water. Some specific areas where mosquito screens can be installed:

Inlet Points: Install mosquito screens on the inlet points of the rainwater harvesting system, such as the gutter downpipes or rainwater diverters. These screens will prevent debris, including leaves and twigs, from entering the system and potentially creating breeding sites for mosquitoes.

Overflow Outlets: Mosquito screens should be installed on overflow outlets to ensure that mosquitoes cannot enter the system through these openings. The screens will allow excess water to flow out while preventing insects from getting in.

Air Gaps/Vents: Air gaps or vents in the rainwater tank or storage system should be fitted with mosquito screens. These openings provide ventilation to prevent the buildup of gases, but screens are necessary to prevent mosquitoes from entering and breeding inside the tank.

Inspection Points: Any inspection points or access panels on the rainwater tank or system should have mosquito screens installed. These points are important for maintenance and inspection, so it's crucial to ensure that mosquitoes cannot enter through these openings.

Inlet Filters: If your rainwater harvesting system includes an inlet filter or leaf eater, ensure that it has a fine mesh or mosquito screen to prevent insects from entering the system. This will help keep the stored water free from debris and mosquitoes.

Remember to regularly inspect and clean these mosquito screens to remove any debris or obstructions that may reduce their effectiveness.

By installing screens in these key locations, you can create a barrier against mosquitoes and maintain the quality of the harvested rainwater.

See p.220 on how to make your own mosquito screen.

How To Make A Mosquito Screen

To create a DIY mosquito screen using PVC pipe and a hose clamp to keep mosquitoes out of a water tank, you can follow these steps:

You will need:
- PVC pipe (correctly sized)
- Hose clamp (compatible with the size of the PVC pipe)
- Screw driver
- Marker or pencil
- Fine mesh or mosquito netting (24-mesh non-corrodible screen)
- Scissors or a utility knife

How to:
1. Measure the diameter of the opening you want to cover.
2. Measure (pencil) and cut (scissors) a piece of fine mesh or mosquito netting that is larger than the PVC pipe diameter, allowing for overlap.
3. Place the cut mesh or netting over the PVC pipe, ensuring it covers the entire opening. Leave some excess around the edges to secure it later.
4. Use the hose clamp to secure the mesh or netting to the PVC pipe. Fasten it tightly to ensure that the material remains taut and without any gaps for mosquitoes to enter.
5. Trim any excess mesh or netting around the edges using scissors or a utility knife.

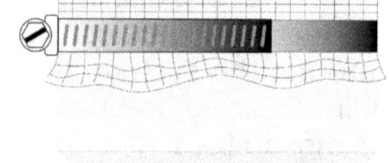

Remember to periodically check the screen for any damage or signs of wear and replace it if necessary.

Operation & Maintenance US Department of Energy

According to the Federal Energy Management Program rainwater harvesting systems require regular maintenance. The components of the system are made to last with regular upkeep, and manufacturers will often provide 15 to 30 years of warranty for storage tanks and pipework, and two to 10 years for pumps. Table 1 lists the recommended operations and maintenance actions for the major components.

Table 1:

#	COMPONENT	DESCRIPTION	MAINTENANCE ACTIONS	SUGGESTED FREQUENCY
1	Collection system	Roof surface and gutters to capture rainwater and send it to the storage system.	Keep clean and clear of excessive debris, especially after prolonged dry periods or after storms. Inspect roof surface and ensure water flows and drains properly as intended.	Weekly
2	Inlet filter	Screen filter to catch large debris.	Clean out filter and replace at regular manufacturer-specified intervals.	Weekly; manufacturer-specified intervals
3	First flush diverter	Diverter that removes debris not captured by the inlet filter from the initial stream of rainwater.	Keep clean and clear of excessive debris, especially after prolonged dry periods or after storms. Ensure the diverter is functioning as intended, diverting only the initial flush of water during rainfall.	Monthly and after prolonged storms
4	Storage tank	Storage tanks composed of FDA-approved, food-grade polyester resin material that is green in color, which helps to reduce bacterial growth.	Inspect tank for cracks or leakage. Infrequent blowdown may be needed to remove sediment from the bottom of the tank. If filters are regularly maintained, sediment accrual should be minimal (2 mm to 2 in. per year).	Annually
5	Overflow	Drainage spout that allows for overflow if the storage tank gets full.	Visually inspect overflow spout to ensure it is clear of debris.	Monthly
6	Controls	Control system that monitors water level and filtration system.	Ensure controls operate as intended, visually confirm response to control commands. Request manufacturer maintenance as needed to repair any controls issues. Check wiring.	Monthly

7	Treatment system	Filtration and disinfection system that treats the water to non-potable or potable standards.	Clean and replace filters at manufacturer-specified intervals. Ensure treatment system dosing intervals are sufficient to meet water quality requirements in the system.	Manufacturer-specified intervals
8	Pump	Pumps move water through the system and to the end use.	Check motor condition. Investigate excessive vibration, noise, or temperature. Perform pump maintenance, such as bearing lubrication, in accordance with manufacturer specifications.	Monthly; Manufacturer-specified intervals
9	Backflow prevention	Backflow preventer to ensure that water cannot flow under instances of negative pressure.	Have an approved professional test annually or at a frequency required by local regulation.	Annually
10	Flow meter	Flow meter (with data logger) to measure water production.	Ensure meter is calibrated per meter manufacturer instructions. Track water use regularly through meter readings automatically (with data logger) or manually with a log book.	Monthly
11	Power supply	Systems may use conventional power sources, or, to improve off-grid capabilities, may use alternative sources such as stand-alone or grid-tied solar systems.	Check power supply and equipment after power outages and ensure no damage to components. Follow manufacturer operation and maintenance guidelines for alternative stand-alone power supplies (e.g., solar photo-voltaic panels).	As needed; Manufacturer-specified intervals
12	Water level indicator	Monitors the water level in the storage tank.	Ensure the indicator is functioning as intended.	Monthly
13	Potable water connection	Make-up water supply (e.g., municipal water) to meet system needs when rainfall is not adequate to meet demand (not pictured).	Inspect potable water supply connection and backflow preventer and ensure that connections are in good condition without leaks. Comply with any regulations for testing required by local ordinances.	Annually

Chapter 5

Extracting Water From The Ground

For thousands of years, man has been digging into the ground, to look for water. Even today, in many developing countries, digging a well is the only way to have access to a potable water source.

For the modern landowner, having a well means independence. It's basically like striking the jackpot of self-sufficiency, for having access to clean, drinkable water takes care of a multitude of land-related problems. By definition, and according to the Oxford Dictionary, a well is "a shaft sunk into the ground to obtain water." When we dig a little bit deeper, we see that wells are fed by natural aquifers of water that are found underground. These aquifers are pockets of water that are located under the layers of ground that we walk on. The ground, which consists of layers of sand, clay, gravel and rock, acts as a natural filtration system. Not only does it filter, but it also adds vital minerals and ions to the water as it seeps through. This is a classic example of the Earth providing and catering to our needs. Once this water has traveled down through all these layers, it accumulates in areas with sand, gravel or rock formations.

In theory, the deeper you have to go to reach the water, the cleaner it should be. Note that this deep water can also be contaminated with natural occurring chemicals that seep into the groundwater. Fortunately, we don't have to go too deep to find water. Professional drill teams often go down to 400-500 feet, but in general, most water is found no deeper than 250 feet down. When we look at domestic drill set-ups, we find that most landowners find their water at less than 50 feet.

Most of it is used for irrigation and livestock, but with a bit of luck, you can get quality water from right underneath your feet. This is great news for the self-sufficient homeowner who is willing to try his hand at well drilling or digging. Having access to a well is something that must be on the minds of all people who want to live close to the land. The water in the cities is transported from reservoirs, rivers, dams and lakes and it is exposed to various pollutants and contaminants. On top of that, it has to be treated for human consumption and that's where (gasp!) even more chemicals are added. Whether unhappy with your current water quality, looking for an independent water source, or if your water bill is starting to hurt your pocket, then get moving, for it's time to get yourself a water well.

Well basics

First, realize that this is a fairly ambitious project and that you need to plan your well very carefully. Let's look at the different methods used to extract water from the ground. Remember that both soil type and depth of the water table will determine which option is best for you.

Self Drilling A Well

Suitable water table depth: Up to 35 feet
Soil: Mostly sand and gravel

This might seem a fairly simple method, but it does involve building and designing a drill from scratch using PVC pipe. The drilling is done by sinking a drill-bit into the ground and creating a hole. As the drill-bit cuts the ground material, the cuttings are flushed out with water pressure. Avoid drilling in rock, but dealing with thin layers of clay in between should be no problem.

Digging Your Own Well

Suitable water table depth: Up to 25 feet
Soil: Sand, clay and gravel

This method involves simple tools and physically digging till

---
Well Drilling Glossary

Aquifer. A geological formation that contains or conducts groundwater, especially one that supplies water for wells and springs.

Auger. A hand-held tool consisting of a twisted rod of metal (drill bit) attached to a handle, used for making holes in the soil.

Groundwater. The water that collects/ flows beneath the earth's surface, like porous spaces in soil, sediment and rock.

Standing Water Level. SWL refers to the level of the water in a well, in a normal rest position when undisturbed and under no-pumping conditions.

you reach water. Dug wells are mostly found in areas where the water table is known to be located at a shallow depth. Personally, I would not recommend digging deeper than 25 feet. This is a time-consuming and very dangerous excavation technique. Shafts often cave-in and there is also the danger of asphyxia. You will need an expert and helpers to assist you with this job. This is not a recommended method!

Using An Auger To Dig Your Well

Suitable water table depth:
Up to 20 feet
Soil: Sand and gravel

An auger is a tool used to cut into the ground and once deep enough, you simply extract it to haul the dirt out. You can add extensions as you move along. Once you reach the water table, you lower a well screen to filter out the sediment. It is a safe and simple method, used in soft soil.

Driving A Well Point

Suitable water table depth:
Up to 25 feet
Soil: Sand and gravel

This method involves the use of a sledge hammer and driving a well point into the ground. As you go deeper, you add extensions to your drill pipe. The well point has a well screen that filters the water as it enters the well.

Hire A Professional

When your water table is located under 50 feet, you can use one, or a combination of the above mentioned methods. If you discover that your water table is deeper, then there is only one option available. Get a local well driller to come over and to give you a quote. Expect it to be fairly expensive. These days people pay around US$4,000-10,000 for a well. You are charged by the foot, so the deeper they go, the higher the costs.

The Soil

You will have to take a good look at the type of soil in

Well Drilling Glossary

Water table. The upper level of an underground surface in which the soil or rocks are permanently saturated with water.

Well screen. A filtering device, placed at the bottom of the well, that allows water to enter from the aquifer, but prevents sediment from entering the well.

Well point. A well screen with a forged-steel point and a threaded pipe shank at the top end. Usually 1.25-2" thick. Driven into the ground with a sledgehammer.

Well casing. A tubular casing made of plastic or metal and placed inside a well bore-hole to maintain the well opening.

your area. The dry area of soil above the water table is called the unsaturated zone. This is an area where the soil particles are filled with air and where water droplets are just passing through on their way down to the water table. Very little water is found in this area. The possibility of digging or drilling through this area depends on the layers of soil present.

Sand. On paper, having fine, white sand sounds like a good type of soil to explore for water, but the fact of the matter is that this kind of sand is just packed too loosely. It will collapse on the well pipe and it will do it often. This makes it an extremely complicated medium to drill through. Coarse, yellow sand with large granules and gravel in between is best. (Finding the odd layer of fine sand is not a problem and you can simply go through it.)

Gravel. Drilling through gravel, or a combination of gravel and sand, is the way to go. These layers are filled with air and are thus easy to "push" around. Gravel is very permeable and good for drilling and finding water.

Clay. Clay particles are tightly packed and do not hold much water. Thick layers of clay is very hard to drill through. Your drill instrument will get stuck in the clay and it can turn into a nightmare very quickly. Drilling or digging through the odd layer

Shallow water wells go down to about 35 feet and are best drilled in soil with layers of sand and gravel.

of clay is not a problem; it all depends on how much clay is found in your area.

Rock. Finding layers of rock can be a problem. If it's solid and thick rock, you will have to start a new well. If it's a thin layer with broken rock, then you can try to break it up and to remove or to push through it with your drill.

The Water Table

The area just below the water table is called the saturated zone and this is where we find the wet soil. This is where the water comes to rest and where the best quality water is found. All the way at the bottom of the saturated zone, we find the bedrock area which is where fractured rock is found with rock aquifers.

When drilling or digging your own well you must realize that these are "shallow" wells that we find at the Standing Water Level (SWL). A shallow well will not reach all the way into the rock aquifer.

Consider the big picture when thinking about wells and water. The groundwater is fed through precipitation. Once you have a well up and running, you should

Water Bearing Layers

Sand. Coarse sand is absolutely perfect for finding water. Normally, this sand will be slightly yellow in color. Water moves easily through this type of sand and when you hit a combination of sand and gravel, it's even better.

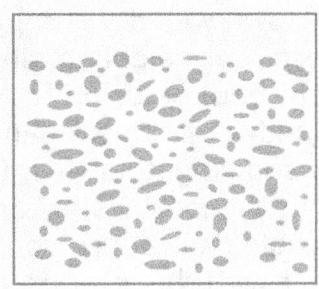

Gravel. Gravel is very much the same as sand in that the air in between the pieces allows the water to flow easily. Gravel is coarse and does not cake together. It is very permeable, just like sand and thus perfect for finding water.

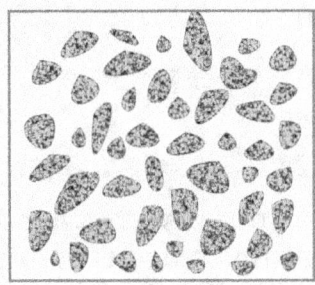

Fractured Rock. Water found in porous rock with fissures can be of very good quality. You will be very lucky to find this type of water. Most of the time, you need to drill very deep to reach these rock aquifers and to do this you will need to consult a professional.

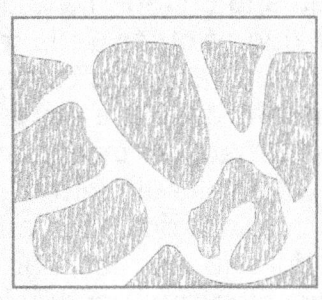

Clay. Clay particles are very small. They are also sticky and "cake" together to form solid layers that are impermeable and that the water cannot penetrate. If the soil in your area is know for its clay content don't waste your time. Consult an expert to get a definitive answer.

take care not to pump it dry. The groundwater relies on precipitation during the wet seasons to maintain its level. Pumping your own well dry can possibly affect your neighbor, since you might both be using the same groundwater. This means that you should gather as much info about the local water table as possible. Make sure you are familiar with the local precipitation patterns in your area. That will require some research on your end.

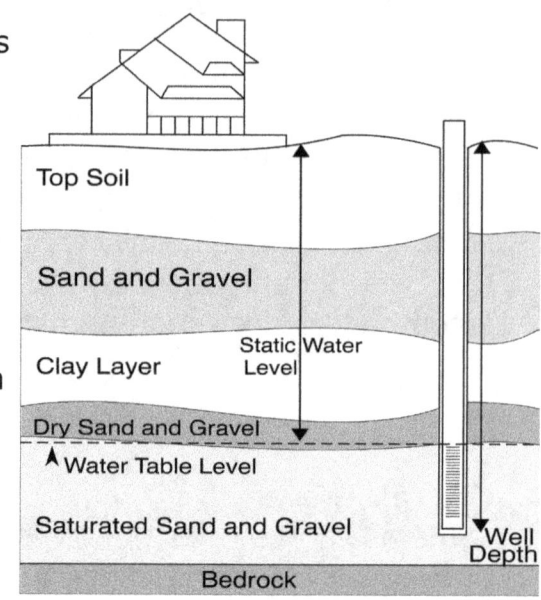

Planning

Step1. Research

• Do your research and read up as much as you can about your local environment. Ask the locals. They know what's been tried and tested in the area. Looking at other wells means that you can get a clear indication how deep the water table lies. Other well owners can also provide valuable information concerning rock and sand formations in your local area. Make sure that you know what your land was used for in the past. Contaminants remain in the soil for a long time.

• Look at geological survey records that indicate the location of the local water table. These records are quite informative and they will highlight other wells in the area. Aquifers are also indicated on these surveys and can be found at the same depth as the water table. If the local water table is deeper than 50 feet, you will have to ask a professional contractor to drill a well for you.

• Look at topographic maps. These maps indicate the topography of the land and as a rule of thumb, the lower the elevation, the easier to find water. Look at the slopes in your area and on your property. Will the runoff from rain contaminate your well? Avoid putting a well near a slope and look for a flat area with minimum risk of floodwater contamination.

• In selecting a suitable location, you will have to look at your immediate environment.

a.) Be sensible and avoid areas with enclosed livestock, septic tanks, fuel tanks, waste disposal or anything that can seep into the ground and contaminate your well. A good rule of thumb is to stay uphill of these areas and at least 150 ft/ 50m away. Your local health department will have guidelines concerning this.

b.) Stay at least 30 feet (10m) from streams and ponds.

c.) In addition, wells should not be located in extremely wet areas with water-logged soil.

• If you have the means, hire a professional consultant. They will do an analysis of the area and provide you with the relevant information.

Step 2. Get Legal

Deal with the local government. In most areas the government requires permits and permissions. Once your well is up and running, you do not want some government representative to come round and rain on your parade.
Make sure that all boxes are ticked and that you are legal and complying with local laws.

Step 3. Plan The Type Of Well You Want

Once you've done your research, established the depth of the local water table and checked with your local government office, you can decide which kind of well will suit you. Shallow water wells can be drilled yourself where the distance to the water

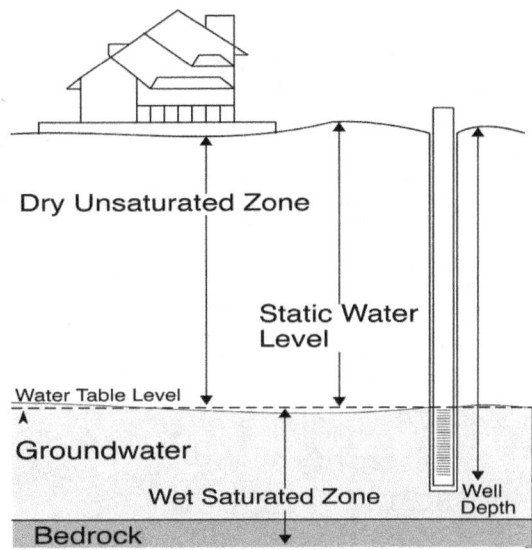

table is no more than 35 ft./ 10.5m. Some home drillers are able to go considerably deeper, but that takes experience and know-how.

A shallow water well can only be used for drinking once it's been tested and deemed safe for human consumption. Water coming from a shallow water table only traveled a few feet through the soil and it has not been filtered properly. This is information that is not to be ignored. People tend to get excited when they see the clear liquid gushing out of the ground, but keep in mind that there are various natural occurring chemicals in the ground that can affect the quality of your water. A deep groundwater well must be drilled by a professional. It can easily be 250 ft/ 76m deep and, in theory, the water can be of very good quality (not a guarantee).

How To Drill And Construct A Well Using PVC Pipe

Once you've determined the size and location of the local water table, you can consider building a drill. This will be used to drill through the dry, unsaturated zone which consists of multiple layers of soil. We will find water once we reach the wet, saturated zone. This is where we can extract water from the water-bearing sand.

Manual Drilling

Shallow water wells are the only wells that can be drilled manually using a DIY method. Generally speaking, **shallow water wells are not considered safe for human consumption**. This should not dissuade you from drilling for water. Manual drilling is cheap compared to a machine-drilled well and the water is ideal for irrigation and other domestic applications. Depending on your location, it is possible to find potable water, but that will depend on the quality and depth of the water table.

We will look at a tried and tested method of simple PVC pipe drilling. This method is also called "jetting" of a well. Expect to spend at least of 20-25 hours on this project. These days you can buy a PVC home drill kit online or at select hardware stores. They include parts, fittings and instructions on how to drill for water. The following setup is based on this method and will give you an idea how to approach basic drilling for water. If you prefer to do things yourself, know that with some basic DIY skills you can build and design your own drill, drill bit and well casing. It is a relatively easy and inexpensive project to take on by yourself. The soil that you are drilling in must be drill-friendly. Soil differs from region to region. As a general rule, expect to drill through layers of top soil, dry sand, clay, rock, water bearing sand and gravel. You will be aiming to strike water in the layers with sand, gravel or even rock (seldom). Ground with coarse sand and gravel is considered best for the DIY driller attempting a shallow water well. **CAUTION.** Take note that you will exert yourself physically. It's hard, manual labor, so wear the right clothes and shoes and take a rest when needed. It's noteworthy to mention that if you have any health issues, you should first check with your doctor.

Let's take a look at the basics of well drilling and the different materials that can be used. Note that this is a project that should be started early in the morning. Once you start "drilling" and have water flowing you cannot remove the drill since the inside of the hole will collapse and if you leave it in the hole, the drill pipe will get stuck. Plan carefully and prepare material and equipment in advance. This is preferably a two man job. When looking at the diagram we can see the basic drilling method in action. Water is pushed through a PVC pipe with a drill bit at the bottom. This pipe initially functions as the drill pipe and later it will be used as the well casing pipe. The water is forced down the drill pipe to soften the ground at the bottom, and as the drill tip cuts away at the soil, the water flushes all the cuttings up to the top along the outside of the drill pipe. The action is flush, cut, flush and it is repeated continuously. This PVC pipe will also prevent the ground, surrounding it, from caving in and therefore we can also call it the well casing. Once the hole is deep enough

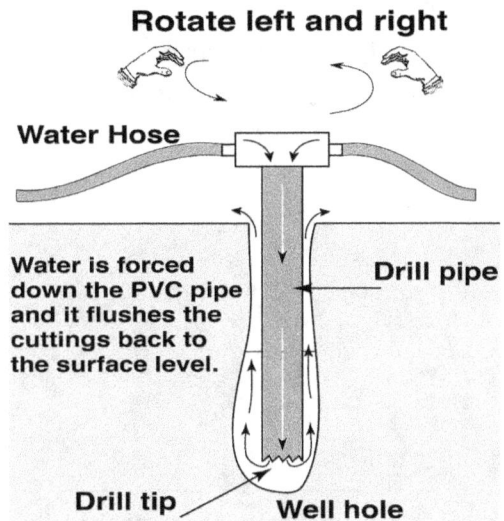

Drilling Method

and you've reached the water table, you can insert a well screen all the way to the bottom of the well. A well screen consists of two parts. The longest part consists of pieces of PVC pipe glued together with inside PVC slip couplings (long). At the end of this PVC pipe you connect the well screen that is 4-6 feet long. A well screen can filter the sand particles and allow the water to enter the well pipe. This complete well screen is just as long as your well casing.

The Use Of Well Water

- Once filtered, can be used indoors for washing, bathing and cleaning
- For drinking purposes, get it tested to see if it is actually safe to drink
- Outdoors it can be used for pools, ponds and recreation
- Can be used for fire fighting
- Perfect for irrigation
- Good for livestock
- It is perfect for small scale mining and manufacturing
- Having a well means that you don't need to store gallons of water for emergencies

Well Drilling Basics

1. Start the hole by rotating the drill left and right. Allow the water to flow and to saturate the soil.

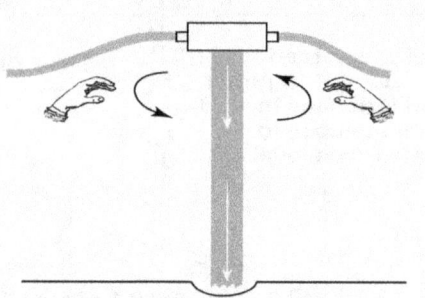

The water, flowing from the two hose pipes, is softening the ground and flushing the cuttings to the surface level.

2. Drill down through all the layers of soil till you reach the water table. Add PVC pipe (with couplings) as you go deeper.

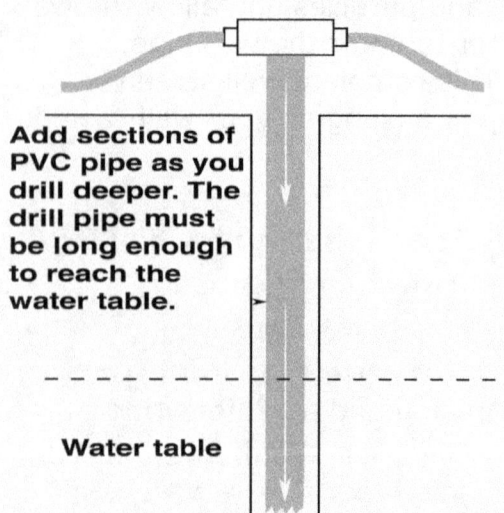

Add sections of PVC pipe as you drill deeper. The drill pipe must be long enough to reach the water table.

Water table

3. Remove the drill top and leave the drill pipe inside the hole. This pipe now acts as the well casing.

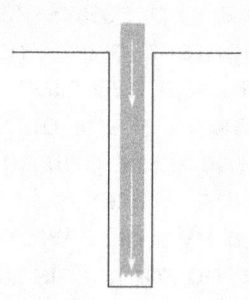

4. Prepare your well screen by adding sections of PVC pipe (or to cut one long piece) till it's just as long as the well casing.

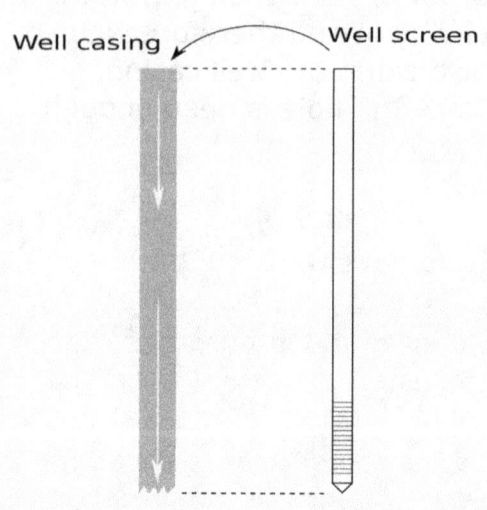

Well casing   Well screen

Well Drilling Basics

5. Slide the well screen into the well casing. The well screen should reach inside the water-bearing sand at the bottom.

Well screen is inside the well casing.

Water table with water-bearing sand.

7. Pour pea gravel down the side of the well casing. It should cover the length of the well screen.

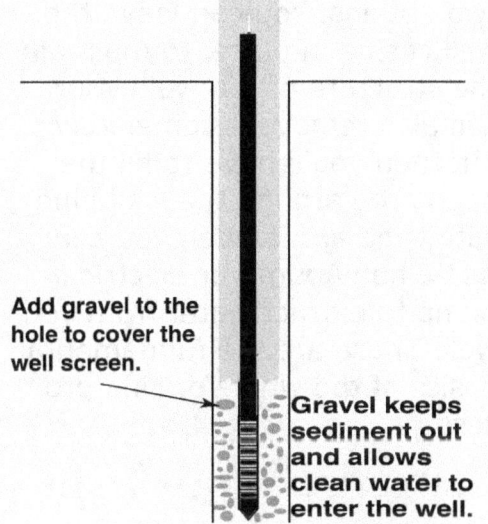

Add gravel to the hole to cover the well screen.

Gravel keeps sediment out and allows clean water to enter the well.

6. Pull the well casing up till it's above the well screen. This is a simple method to expose the slits in the well screen to the water-bearing sand.

Pull up the well casing to expose the well screen to the water-bearing sand.

Water table with water-bearing sand.

8. Leave the well casing in place; this is your well. You can also pull out the well casing and fill the annular space around the well pipe with sand.

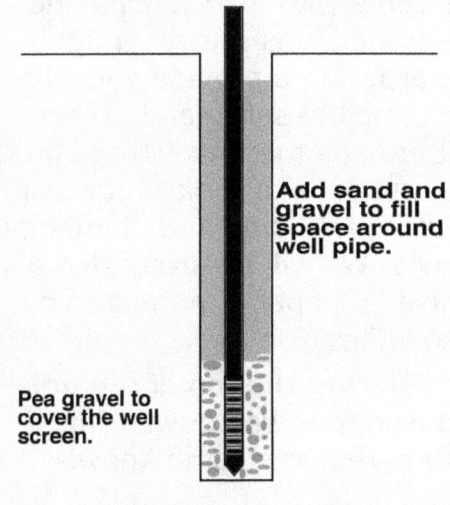

Add sand and gravel to fill space around well pipe.

Pea gravel to cover the well screen.

Once the well screen is in place, you can slowly pull the well casing pipe up to expose the slits of the screen. You want the water-bearing sand to collapse around the well screen, because that is where the water will enter your well. You now have two options. You can leave the well casing in place, to maintain the structure of the well. You can also remove it completely, but then you'll have to fill the open area around the well pipe with sand and gravel. You can add a hand pump or electrical pump to extract water from your well. These are the fundamental basics of the whole drilling process.

Construct Your Drill

Step 1. The Drill Top

Before you can drill into the ground to make a bore-hole, you will need to construct a "drill". You have an option of buying a PVC home drill kit online or from select hardware stores. These kits come with instructions and the assembly is pretty straightforward. Some provide you with the complete setup and others just provide the drill T-top. The best option is to make your own, by using PVC pipe and plumber's fittings. We will be discussing a simple DIY option, perfected by Mike Willis.

Let's start at the top of the drill and work our way down. This section requires some knowl-

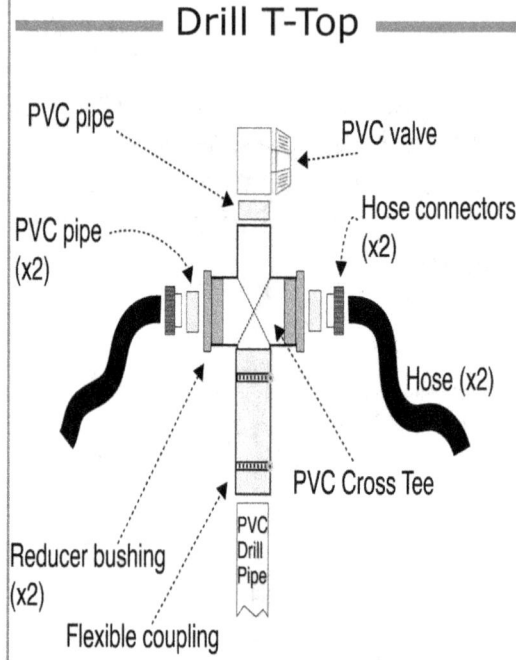

Drill T-Top

edge of joining plumber's fittings. Remember to use PVC cement on all joints. Be creative and adapt to your situation. If you cannot find the exact part or fitting, then you have to improvise. Remember to use Schedule 40 PVC pipe; you can use 2", 3" or even 4" PVC pipe. The larger the diameter, the more flushing power you will require from the hose pipes.

Starting with 2" is a good safe size for your first attempt at drilling. If your drill pipe is 2" you can adjust all fittings to accommodate this size. Make sure that the water from the hoses flow freely and without obstruction. You need as much flushing power possible. Before you start, make sure you look at all the material you will need on page 235.

To Construct A Drill, You Will Need The Following

- **PVC cement and primer.** To join all fittings and PVC pipes together.

- **Hosepipes x 2.** To flush the hole and to push the cuttings to the top.

- **PVC Cross tee x 1.** This is where water, flowing from the two hoses, are redirected into the main drill pipe. (2" recommended)

- **PVC valve x 1.** The PVC valve allows for air and water (under pressure) to be released from the drill. Also adds the option of connecting a mud pump. (2" recommended)

- **PVC pipe pieces x 2.** To connect hose connectors to reducer bushings. (¾" recommended)

- **PVC reducer bushing x 2.** These connect the hose connectors to the PVC Cross tee. (2" reduced to ¾" recommended)

- **Hose connectors x 2.** They connect the reducer bushings to the hose pipes. (¾" recommended)

- **Flexible coupling x 1.** Rubber tubing with radiator clamps will also do. (2" recommended)

- **PVC pipe piece x 1.** Connects to the bottom of the PVC Cross tee. See next page. (2" diameter and 5" long recommended)

- **Drill pipe handle bar.** Made of wood. It allows you to control the drill.

- **PVC drill pipes 8-10 feet in length.** With bell- or plain ends.

Once joined, they will be the drill pipe and well casing. (2" recommended)

- **PVC well pipes.** They connect to the well screen. Your well pipe can be one long pipe, or sections joined together.

This pipe will be connected to your well screen. (1¼ " recommended)

- **Well screen.** This is where the water enters the well.

- **PVC Couplings.** These connect the long drill pipes and the well pipes. Use outer couplings for the drill pipes and inner couplings for the well pipes. (2" recommended)

- **Pump.** Once finished, you'll need a pump to extract water from the well. The pump connects to the top of the well pipe.

The Drill

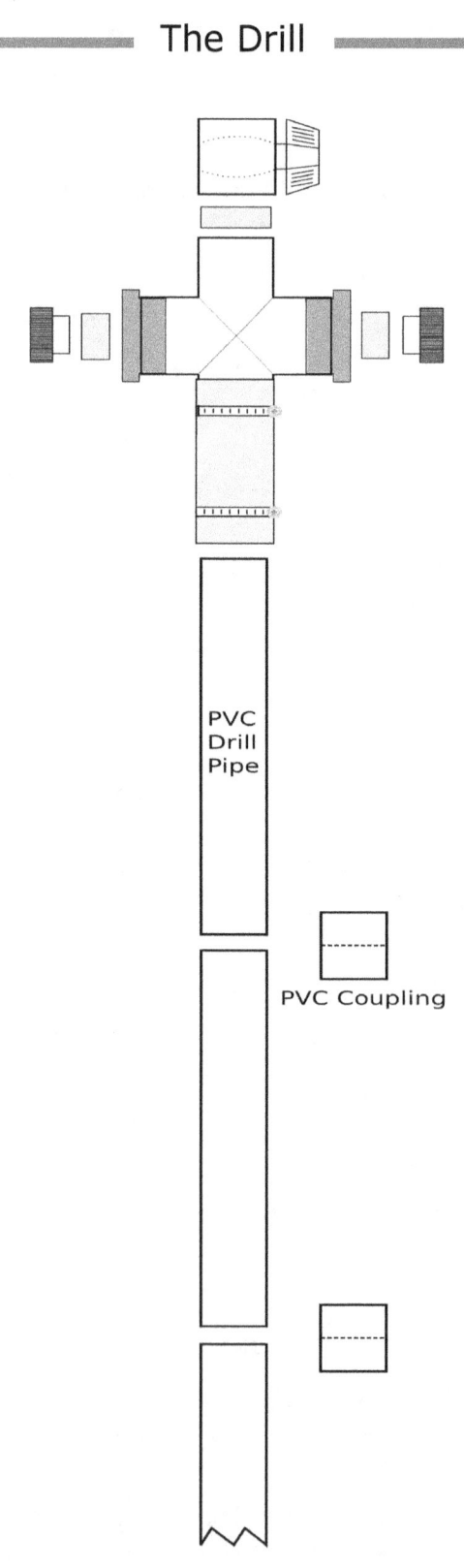

Breakdown of the T-Top upper part of the drill:

- **The PVC Cross Tee** acts as the main facilitator that allows water into the drill pipe. The bottom of the Cross Tee connects with the drill pipe. If your drill pipe is 2", then use a similar sized cross tee of 2". Remember that as you drill deeper, you will have to remove the drill-top to add more pipe to the drill. This means that the design must be practical enough to allow the top to be removed and replaced. The top of the Cross tee connects to the PVC valve. The two sides of the Cross tee connect to the reducer bushings and then to the two hose pipes.

- **The PVC Valve** that goes into the top part of the Cross tee allows you to release air or water from the chamber when needed. You can also connect it to a pump at a later stage. It is a handy option to have. Depending on the fittings you use (threaded or not), you might need a piece of PVC pipe to join the valve and the Cross Tee.

- **Reducer bushings** (x 2) are used to reduce the size of the PVC Cross tee down to the size of the hose connectors.

- **Hose connectors** (x 2) will allow water to flow from the hose pipes through your reducer bushings. Make sure they fit your hosepipe connectors before you buy. To join the hose connectors to the reducer bushings, you will need two short pieces of

¾" PVC pipe.

- **Hose pipes** (x 2) that will comfortably reach your drill hole location and at least 15 feet of extra clearing for when the drill pipe is raised. Check the water pressure from faucets. You might have to consult a neighbor to lend you a hose with water pressure coming from his side. The combined water pressure will be great for flushing the hole. The hose pipes connect with the hose connectors.

- **Flexible coupling** (x 1). If you cannot find a flexible coupling, you can improvise with some rubber tubing and radiator clamps. This coupling connects the drill head to the drill pipe. As you drill deeper, you will reach the end of the first drill pipe's length. This means that you have to remove the drill head from the drill pipe. The next step is to connect another drill pipe to the one in the hole. After you have connected the drill pipe in the hole to the new one, you will reconnect the drill T-top to the top of the new pipe. The flexible coupling will allow you to do all of this.

- **PVC pipe** (x 1) cut 5" long. Your drill pipe and your PVC Cross tee are both the same diameter. To join them together, you will use this section.

- **Use PVC cement with PVC cement** primer to join the fittings. Look for a brand that sets within a fairly short period of time. (Later you will use

The Flexible Coupling

1. The Cross Tee is to be joined to the piece of PVC (grey color). Line the 2 pieces up and add PVC cement and primer. This combined unit forms the top of the drill that will connect to the drill pipe.

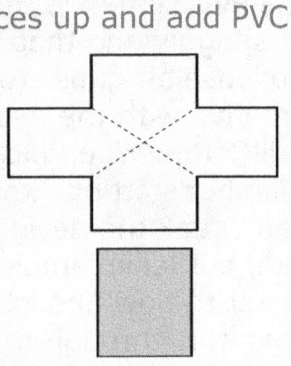

2. Join the two pieces. Later the other end of the PVC pipe will slide into your drill pipe.

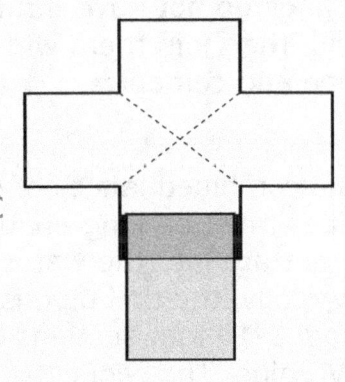

3. The flexible coupling joins the drill top and drill pipe. To change drill pipes, loosen the bottom screw of the flexible coupling and add a new pipe section. No cement needed there, because you have multiple pieces of drill pipe to add later.

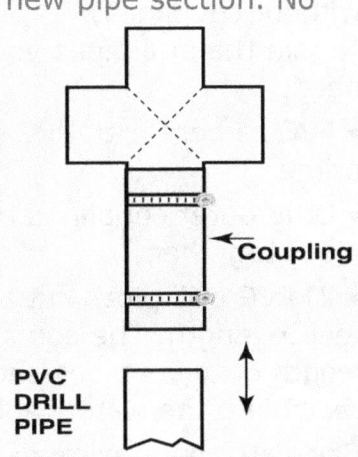

PVC DRILL PIPE

the PVC cement to add drill pipe during the drilling process.)

Make sure that when you assemble the parts (before using the PVC cement), that they all fit snuggly and that there are no loose air gaps. You can get creative with the design of your drill T-top. When picking up your plumber's fittings know that you can tweak the design as you like, but leave ample space for water to flow unobstructed and smoothly through the fittings. If you find that some of your fittings do not have a threaded end, then join them with PVC pipe and cement.

Step 2. The Drill Pipes

The combined length of drill pipes must be long enough to penetrate into the water table. Typically, the drill pipe is made from 8-10 foot sections of 2 inch PVC pipe. The wider the pipe the stronger the flushing power required to flush the cuttings. You can use standard PVC pipes or the ones with the bell endings (which means you can join them without couplings).

To join the drill pipes you will need:

- PVC cement with PVC cement primer.
- Long outer couplings to join pipes together.
- 2" PVC drill pipes cut to 8-10 feet in length. The combined length of these pipes should reach into the water table.

Calculate the maximum depth you are willing to go by looking at the local water table in your area. You will need at least 3 or 4 pieces of pipe.

Draw lines, measured in feet, on every pipe to indicate the drilling depth.

You can add a setting screw on every coupling if you feel all the twisting and turning might loosen the pipes.

Step 3. The Drill Bit

The drill bit is the part that breaks up the dirt at the bottom of the hole. It scrapes the dirt away, breaks it up and then the water flushes it to the top of the hole. To make your drill bit you have three options.

- The first option is to simply cut teeth into the lead PVC drill pipe. Serrated edges are best and you can manage this with an electric grinder or manually with a hacksaw. You can use this option for soft shallow drilling. It is not a recommended option and the teeth will wear down easily.

- The second option is to make a replaceable bit from a PVC coupling. Just connect the coupling with PVC cement and once dry use a saw to cut the triangular teeth for the drill bit. This can be an effective drill bit in soft soil, but as you go deeper the teeth will undoubtedly wear out.

- The third option is the best option. Get a piece of galvanized plumbing pipe. Serrated edges are best and you can manage this with an electric grinder or manually with a hacksaw. This drill bit will allow you to cut more effectively through sediment, roots and debris. You can add a short screw to set this drill bit in place since the constant turning of the drill pipe will loosen the drill bit.

Step 4. The Drill Handle Bar

You can make a handle for your drill from wood. A piece of wood fastened with hose clamps will work just fine. Remember that this handle must be moved up the drill as you add your PVC drill pipe extensions. The hose clamps are easy to unscrew and when adding pipe, it will allow you to place the handle around the new piece of PVC drill pipe.

Preparations

Before you start drilling, lay out the individual pieces and make sure all equipment and tools are ready. Check that hoses are the correct length, fitted with correct connections, and will provide enough water pressure. Double-check your T-top and make sure you have a screwdriver ready for later when needed to loosen the hose clamps. Your drill pipes should be 8-10 feet long with lines drawn to indicate drilling depth and a mark where you will refasten the grip handle. This saves time for when you have to add pipe and reposition the grip handle. If you are connecting your pipes with couplings, then a coupling should be fitted to each extra length of pipe beforehand. When dealing with pipes this long, you need strong couplings with long connections. Get a crate or platform to stand on to make things easier.

A ladder is not a stable platform. **Do not stand on a ladder when drilling.**

How To

When everything is ready, you can start to drill. You should know your water table depth, to give you an indication as to how deep to drill. You do need a fair amount of luck on your side to hit the water table with your first try. Make sure that you have a helper since drilling is very hard to do on your own. Be prepared to dig quite a few holes. Do not prepare the well drilling site by digging any large diameter holes beforehand. Remember that water will be flushed out of the hole for an extended period of time and if not careful you'll soon be standing in a lake of water. Just focus on the hole made by the drill head in front of you and make sure water can drain away from the hole.

If you run out of daylight, pull up the pipe and start again the next day. A pipe left overnight will get stuck.

1. Start the hole

This is physical, hard labor. You must be in good shape and fairly fit. To start the hole, put your drill pipe down and start twisting it side to side.

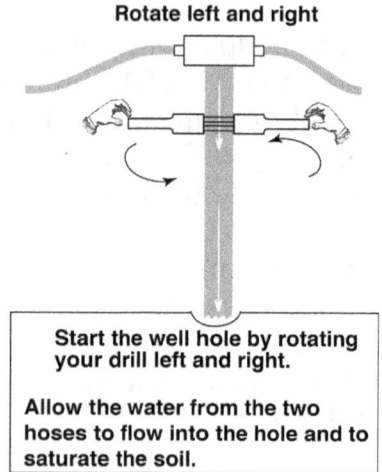

Start the well hole by rotating your drill left and right.

Allow the water from the two hoses to flow into the hole and to saturate the soil.

Turn the water on slowly. Leave the top valve of the drill head slightly open, just enough for air pressure buildup to escape. Rock the pipe side to side and twist it sideways with the aim of making the hole wider. Make sure that there is ample flow of water.

2. Drill deeper

Once you've progressed to about two feet down, you can start to move the drill up and down to flush the hole with water. This is very important. This action, combined with water flow, is essential to keep the pipe from getting stuck.

Constantly keep your eyes on the water coming from the hole. Look for the cuttings and pay attention to the color of the water. Be patient. Do not rush

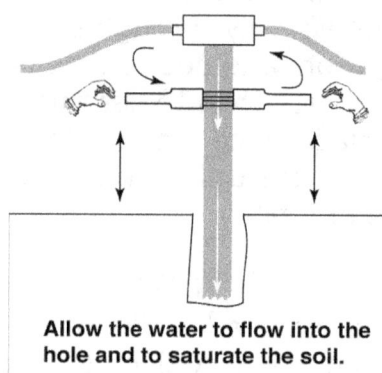

Allow the water to flow into the hole and to saturate the soil.

anything. You will be drilling for a few hours, so settle in and commit to the task at hand.

3. Add extra pipe

Add more lengths of PVC pipe as you move along. You should have the pipes prepared (with their couplings if necessary) beforehand. This is where a second person will come in very handy. Try to change pipes as quickly as possible. You do not want to stop the flow of water down the well. The whole process should look like this:

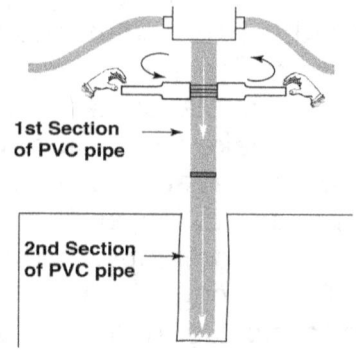

1. Turn off the water.
2. Take off the T-top.
3. Take off the handle grip and fit it to the new pipe.

4. Apply PVC primer and cement.

5. Connect the new pipe.

6. Reattach the T-top and turn on the water.

Make sure that the new connection is secure and that the PVC cement is dry. You can add a short screw to every coupling if you feel the need.

4. Add more pipe

Continue drilling and add more lengths of pipe as you go deeper. When you hit a depth of 10 feet, slow down and make sure

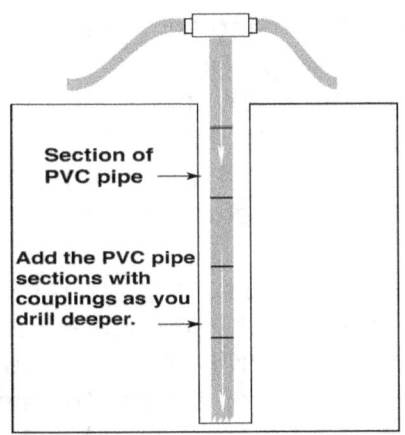

Section of PVC pipe

Add the PVC pipe sections with couplings as you drill deeper.

that the water is flowing steadily. You should pay extra attention to the up and down movement of the drill. This is where the ground becomes harder and you want to make sure you don't get your pipe stuck. Be patient and proceed slowly while making sure the ground stays saturated with water.

5. How will you know when you've hit water?

It's not easy to tell when you've hit water. The cuttings and water are constantly flowing out of the hole and this can create confusion. Look for sand. Water bearing sand is coarse and light in color. You will recognize this sand when you see it. Don't stop. Let it flow and go a few feet deeper till you find the coarser sand. Expect to find it at the water table depth that you have researched for your area. Let go of the handle and watch your drill "settle" in the hole. It should sink deeper onto the hole all on its own, indicating that you have reached the water table.

6. Drill deeper

Continue to drill deeper. You want your well to penetrate nice and deep into the water table and to be surrounded by the coarse, yellow sand. Work that

When Your Pipe Gets Stuck

1. One option is to use a mechanical instrument like a winch or a hydrolic jack
2. Try designing a lever with a long handle that rests on a fulcrum. Move the fulcrum as close possible to the pipe. The lever should push up against the drill pipe handle, which should be reinforced with extra clamps.
3. As a last resort, you can try to sink a smaller diameter pipe (½ or ¾ inch) right alongside your drill pipe to get the soil liquefied again and to loosen it.

drill up and down and side to side and try to make the hole as big as possible. Your drill pipe should feel very loose inside the hole. Go as deep as you can while still in the coarse, yellow sand. If you can still go 8-10 feet deeper, then do so.

You should try to be well under the **Standing Water Level** (SWL) as used by professional drillers. The SWL refers to the level of the water in a well, in a normal rest position when undisturbed and under no-pumping conditions. Once you start pumping the water from the top, the water level will drop. When you turn off the pump, the water will return to the SWL level. Your well screen top should be 8-10 feet under the SWL.

7. The well screen

A well screen is 4-6 feet long, has a pointy tip and hundreds of tiny slits all along the side. These slits keep the sand out

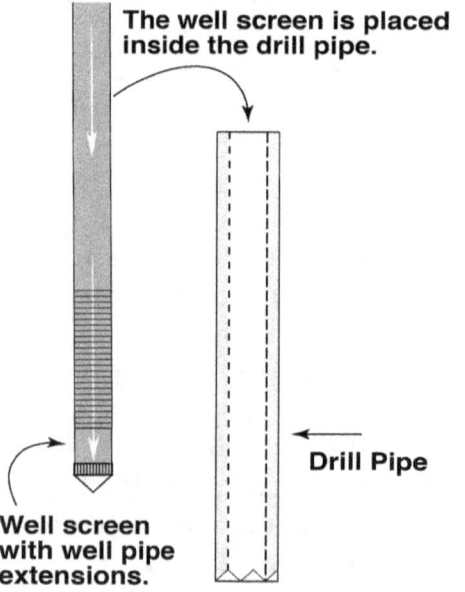

Well screen with well pipe extensions.

The well screen with PVC pipe extensions form your well. The drill pipe forms your well casing.

and allow the water to flow into your well.

This is the point of entry, where water enters your well. You can buy your well screen from certain hardware stores or from an online merchant.

Your actual well pipe (that you will stick inside the well casing pipe) can be prepared in advance.

If you are using sections of 1¼" well pipe (instead of one long well pipe), then join them together with long inside couplings. This will prevent the pipe from getting stuck inside the 2" well casing.

This well pipe connects to the well screen. If you used a 2" well pipe to drill with, then your well consists of a piece (or pieces) of 1¼" PVC pipe with a 1¼" well screen connected at the bottom. The total length of your well pipe should be approximately just as long as the well casing pipe.

Once you've joined the well screen and the well pipe, your well is almost done. At this stage the water is still running. When ready, turn off the water and take off the T-top.

Slide your well pipe with the well screen attached all the way down the well casing pipe. Make sure it has reached the

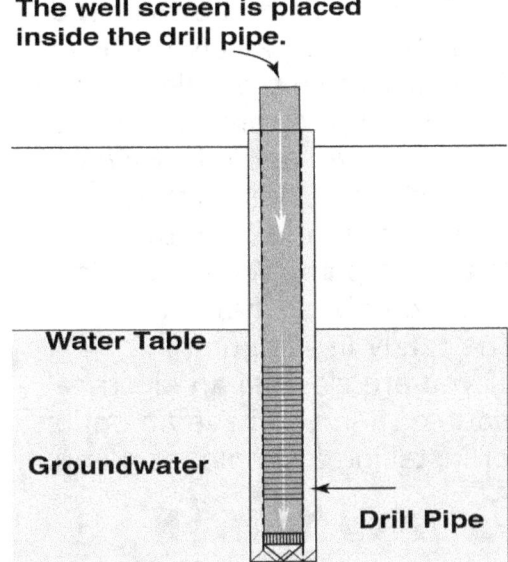

Drilling's hard work. Wear the right clothes and if you have any health issues, check first with your physician.

but it is coarse enough not to clog your well screen.

b.) An easier and better option is to pull it up just far enough for the total length of well screen to be exposed to the water-bearing sand. It will feel strange to pull your well casing back up the hole, but it has to be done. If you leave it in place, you will not get any water from the well.

bottom. If your well pipe is the same length as your well casing pipe, then just a small section of well pipe should stick out the top.

Now comes the pullback of the well casing pipe. You have two options.

a.) You can pull the total length of well casing pipe from the hole and then fill the space around your well pipe with sand and gravel. The sand and gravel will allow the water to pass through,

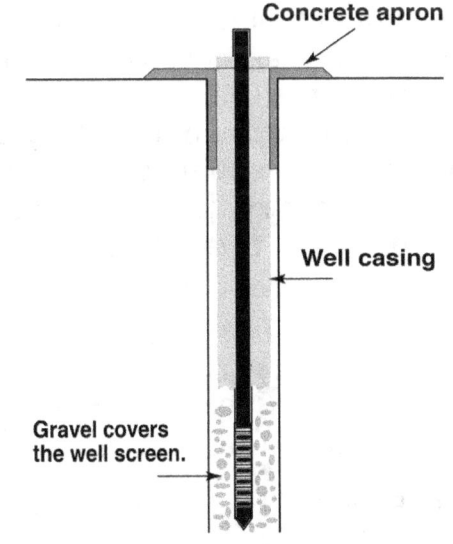

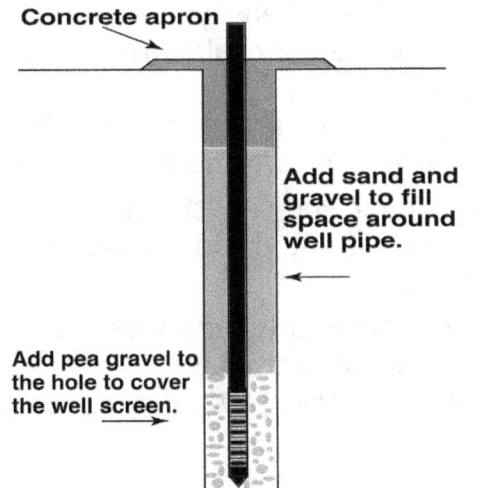

Make sure you are not pulling the well pipe up as well. That's why we suggested inside couplings to be used on the well pipe, so as to not obstruct the well casing pipe when removed. The well screen should be surrounded by coarse sand.

Extracting Water From The Ground

Hopefully the sand has completely covered the well screen. If not sure, you can still pour some sand down the hole. Sand is what you want. It is coarse and prevents plenty of space for water to flow in between and it won't enter your well screen. Pour some cement all around the top of the well to prevent debris and runoff to contaminate your well. Cut the well casing pipe (that's sticking out the top) shorter.

8. The Pump

Install a hand-pump at the top of your well. Use this pump to develop the well and to remove the sediment from the bottom. This could take a while. Remember that off-the-shelf hand-pumps are mostly used for depths up to 20-25 feet. These pumps have to be primed to get the water from 20 feet down all the way up into the pump. If you cannot or don't want to prime your pump, then you will need to install a check valve. It is a one-way valve that allows water to go up towards the pump, but not down and out. Having a check valve means that you don't constantly have to prime the hand pump. The result is that you always have water at the ready in the pump.

If you are close to an electrical source then you have an option of installing a simple jet pump.

Conclusion

When drilling for water, just keep in mind that you are simply extracting water from the water table. Keep things simple as possible and remember to focus on longevity. You want the whole structure to last for as long as possible and therefore you have to use quality material (Schedule 40 PVC) and make sure that all fittings are sealed and tight. Examine the process of drilling and see if your own expertise will allow you to come up with an even better solution. Safety should always be a priority and do not take any short cuts.

Keep animals and people (children) away from the area.

Fill, cover and seal the well opening if you decide to abandon it.

If you have any health issues, you should first check with your doctor before starting this project.

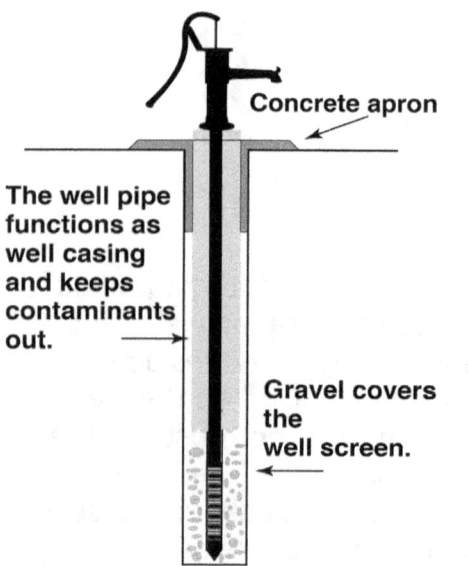

How To Pour Pea Gravel Around The Well Screen?

◇ The **drill pipe** is a 2 inch PVC pipe and once the well is finished, it becomes the **well casing**.
◇ The **well pipe** is the 1¼ inch PVC pipe that we connect the well screen to.
◇ Eventually, the **well pipe** is placed inside the **drill pipe** (well casing).

1. If you are using 2" drill pipe, drill down to the desired depth. *Example: 25 feet.*
2. Move your drill pipe from side to side, up and down, to widen the hole. The bottom of the hole should be considerably wider than your 2" drill pipe. Having a wider bottom space means that we can fill it with plenty of pea gravel.
3. Prepare your 1¼ inch well pipe, with well screen, in advance. Remember that both your well pipe, and drill pipe, should be more or less the same length (in this case, 25'). When you add the 3 foot long well screen, the well pipe will be around 28' long. The 28 foot long well pipe, all glued together, will later be placed inside the drill pipe.
4. Once you feel that the bottom of the hole is wide enough, you can stop the water and remove the T-top. We are working with a 3 foot long well screen, so take some pea gravel and pour 3 feet of gravel down the drill pipe. As you pour the gravel, work the pipe up and down. The gravel will fill the space underneath the pipe and gradually, as you pour, the pipe will move up. When your drill pipe is three feet higher than it was, it means that you have three feet of gravel inside your hole.
5. This is where the pointed tip of the well screen comes in handy. Slide the whole 1¼ inch well pipe down the 2 inch drill pipe. The well screen will hit bottom (the gravel). Both pipes will be at the same depth in the well. Start to turn and twist the well point and try to work it deeper into the gravel layer. The pointed tip will help in getting it through the pieces of gravel. It will take some time, but you will manage.
6. Once the 28 foot long well pipe and the 25 foot long drill pipe is sticking out at the same level, then you can stop. This means that you have three feet of well screen exposed to the pea gravel at the bottom of your well.

(For a 3" drill pipe, first slide the 1¼" well pipe down inside. Pour the pea gravel down the annular space between the 3" and 1¼" inch pipes. Unlike a 2" pipe, the space between a 3" and 1¼" pipe is enough.)

How To Dig And Construct A Well Using An Auger

In suitable soil conditions, using an auger can be a straightforward method used for water extraction. With the auger you "scoop" out the soil till you reach the water table. Place a well screen in the hole and use a pump to extract your water.

Auger Drilling

Another very simple method to extract water is to use an auger post-hole digger for drilling.

Augers can be used in sand and gravel or a combination of both. You'll have less success in clay and absolutely no success in rock. A 6-8 inch auger should be enough. This is basically a fancy drill bit with a handle and it slowly bores into the ground as you turn it. A T-handle will give you more leverage and grip, especially when you get past 10 feet.

Once the auger is filled with dirt, you have to lift it out of the hole and empty it out. You can add extension rods to the auger as you go deeper. This method is obviously useful for digging shallow wells with appropriate soil conditions.

With an auger, expect to drill to a maximum depth of around 20 feet. It will be a slow and energy sapping process. Once you have reached the desired depth, you can fit a PVC well screen inside the hole. The well screen allows clean water to enter your well.

Seal your well and add a handpump or centrifugal pump.

How To

1. Get started. Put the auger in the ground. Start drilling by turning the auger in a clockwise direction. This will ensure that you are drilling straight down.

Do this throughout the process

You Will Need

- An Auger, post-hole digger
- Metal extension rods (20-25 feet)
- PVC pipe
- A PVC well screen
- PVC cement and primer
- A length of rope and PVC bailer
- A pump to bail water
- A hacksaw

of drilling. Take your time and be patient. This is time-consuming and hard work.

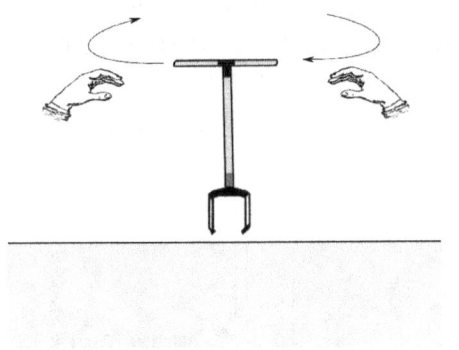

2. Soil. When full, pull it up and empty it. Repeat.

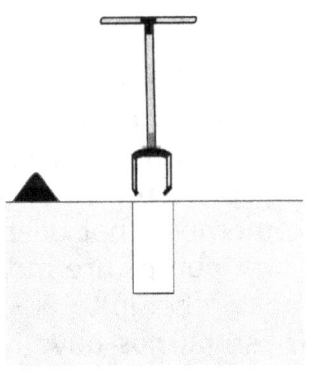

Go as deep as the auger will allow you. Keep the extracted dirt in a heap nearby. You can use it later to seal your well.

3. Extensions. Add extensions when necessary. Make sure that they are tightened securely. Be patient. It will become harder and harder to turn the auger. Get a second person to assist. As you go deeper and deeper, keep an eye on the depth that you have drilled. Once you bring up more water than dirt, you've hit the water table. You should know the depth of the water table in your area beforehand.

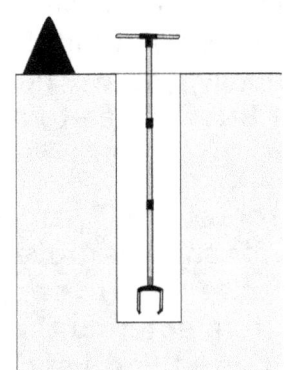

4. Water table. Once you reach the water table, you still have to go deeper.

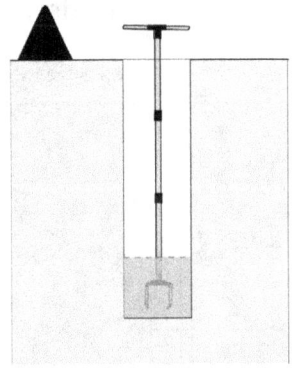

Don't get over-excited. You need to be nice and deep inside the water table to have a working well with steady water flow. Measure the depth and aim to go at least another 8 feet deeper.

5. Water. Bail the well when water seeps in. You can try to do

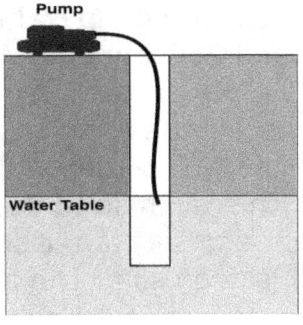

Extracting Water From The Ground

it manually, but it's always more effective to use a pump. If you do it manually, you will need a **PVC Bail Bucket** tied to a rope (p.249).

6. Well screen. Well screens are used for sand or gravel aquifers where we want to prevent particles from entering the well. Once connected, it functions as part of the well pipe and it works like a sieve that extends into the water table and that lets the water into the well pipe while filtering out sand and debris.

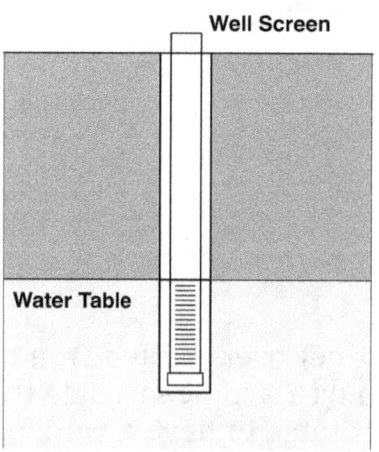

You can buy a well screen online or make your own from PVC pipe. (See page 250.) For a standard well dug with an 8" auger you can use a 4-6" well screen casing. Cap the bottom of you well casing pipe with PVC primer and PVC cement. Make sure it's a perfect seal, nice and tight. Fill the annular space around the well screen with pea gravel. This will help to keep sediment and silt from clogging the well screen slits. The area above the well screen can be filled with the dirt that you excavated earlier. Pack it in nice and tight. Fill the top three feet with concrete and make a nice apron all around your well.

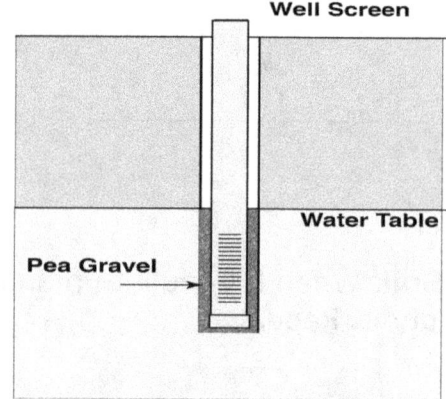

7. Pump. Install a hand-pump at the top of your well, about 3 feet above ground level (to protect your water from flooding). Remember that off-the-shelf hand-pumps are mostly used for depths up to 20-25 feet. These pumps have to be primed to get the water from 20 feet down all the way up into the pump. Not priming the pump will cause damage to the leather cup on the inside. If your water is deeper than 8' down, consider

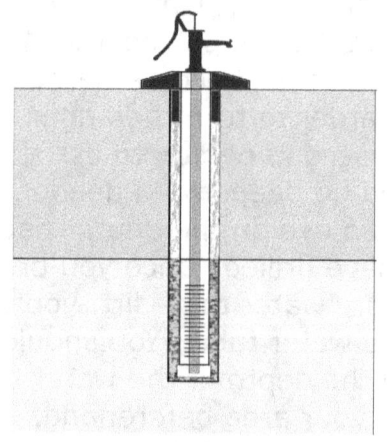

installing a foot valve. This is an inlet valve, connected to the well pipe, that is positioned about 6" from the bottom of the well. It is a one-way valve that allows water to go up to the pump, but not down and out. Having a foot valve means that u don't constantly have to prime the hand pump, since the line is filled and you have water at the ready in the pump.

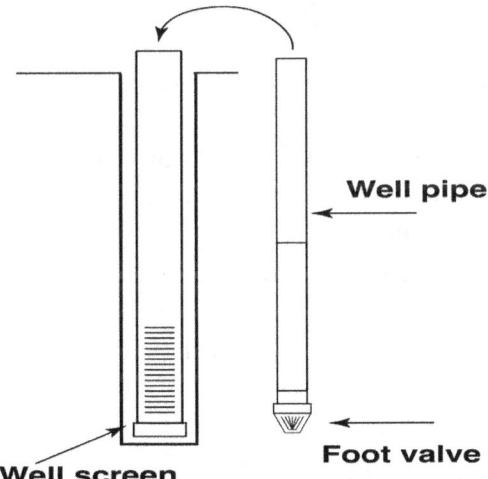

Well screen **Foot valve**

CAUTION! Note that in cold weather, the water in the pipe will freeze and expand. This can split your pipe or foot valve. Consider removing the pump or foot valve in freezing temperatures, or draining the pump and suction line.

For deeper wells, a deep well hand pump will work. If close to an electrical source consider installing a jet pump. See the section on pumps (Chapter 6). If you have any health issues, you should first check with your doctor before starting this project.

Make A PVC Bail Bucket

Mechanics: You will need a long cylindrical bucket, with an opening at the bottom. A simple ball check-valve assembly allows the water to flow into the pipe; when full the ball seats against the reducer and seals it.

Material:
• A 2-3 feet PVC pipe to fit inside the well hole; a 4" diameter pipe should fit most holes (this is your bucket)
• A PVC pipe coupling glued to the bottom of the bucket
• A reducer bushing (4" x 2")
• A solid rubber ball that seats against the reducer
• A stainless steel bolt
• A rope long enough to reach the bottom of the well

Directions:

1. Use PVC cement to attach the coupling to the bottom of the PVC pipe.

2. Glue the reducer bushing to the bottom of the coupling. The ball seats against the reducer opening.

3. Drill two holes about 6" from the bottom and put the stainless steel bolt through. This prevents the ball from going all the way up.

4. Attach the rope to the top for hauling water.

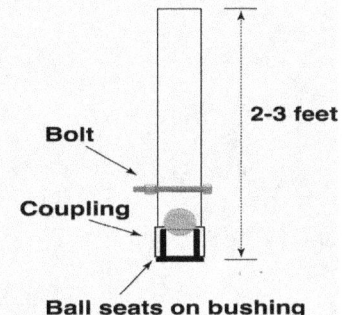

Bolt **2-3 feet** **Coupling** **Ball seats on bushing**

Extracting Water From The Ground

How To Make A Well Screen

The well screen will be used to keep particles and debris out of your well water. It is constructed from PVC pipe and should be an inch or two in diameter smaller than the well hole.
This PVC well screen pipe will also function as the well casing.
For a 6-8 inch hole, use a 4 inch pipe as your well screen and remember that once you are done, to fill the annular space, around the well screen pipe, with pea gravel.
Make the well screen length 4-6 feet long. This means that you should be at least 8-12 feet into your water table. The deeper you can go the better.
You will need a hacksaw to saw slits into the PVC pipe. Use a new, thin blade. You can choose how you want to group the slits together. The main thing is not to compromise the structural integrity of the pipe, by placing the rows of slits too close to one another. For a large diameter pipe, use a grouping of four lines of slits around the length of the pipe. For a thinner pipe, you can use a grouping of three. You can space them out evenly all along the length of the pipe. You can also experiment with your own design.

CAUTION!
Use common sense when making the slits. Their function is to keep debris and small particles out and they should be positioned and grouped to do just that.
Do not cut through the pipe!
Don't group your slits too close together!

Material needed:

- PVC pipe, as required

- PVC cement and primer

- Hacksaw with new, thin blade

- Measuring tape

- Calculator

- Permanent marker

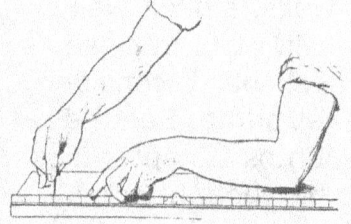

1. Cap the bottom of the pipe. Use PVC cement and primer and make sure it's a secure fit.

5. If you have the patience for it, you can cut a continuous line of slits all along the length of the pipe.

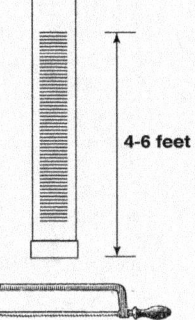

2. Start the well screen around 4-6 inches (10-15cm) from the bottom of the pipe. Use the marker to mark the slits.

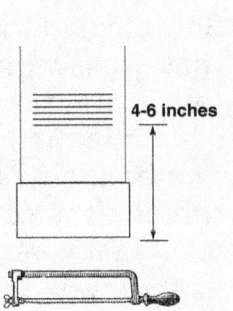

6. For smaller diameter pipes, cut 3 sections of slits all around the pipe.

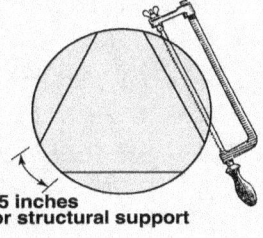

1.5 inches
For structural support

3. Cut the slits into the pipe with a hacksaw. You can group them together in short sections with 3-4 slits.

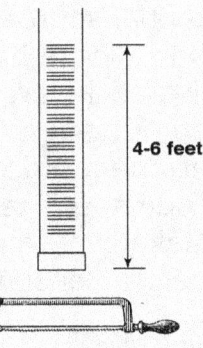

7. For larger diameter pipes, you can try to get 4 sections of slits in.

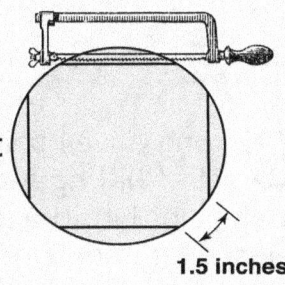

1.5 inches

4. You can also space them out into longer sections with 9 or 10 slits each.

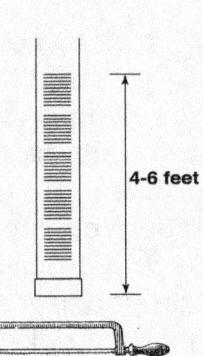

8. For calculating the length of the slits, you need the circumference of the pipe.

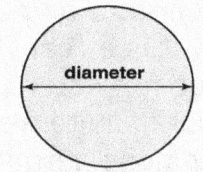

4" diameter = 12.566" circumference
5" diameter = 15.708" circumference
6" diameter = 18.850" circumference
7" diameter = 21.991" circumference
8" diameter = 25.133" circumference

How To Dig And Construct A Well Using A Well Point

Finding water doesn't have to be a complicated process. The invention of the "well point" certainly simplified the process of extracting water for shallow water wells. These days many shallow water tables are susceptible to pollution and the water obtained should really only be considered for irrigation and household chores. Have your water tested before you use it.

Point Wells

Driving a well point is the act of hitting a screened well point into the earth with a sledgehammer. Once you reach the water table, the well point acts as the well screen and it allows clean water to enter the well and to be pumped to the top. It is a really simple technique for water extraction and it works quite well when the soil conditions permit.

• This method is predominantly used in soil that's rich in thick sand or gravel. Don't use it in soil that consists of hard, red clay or rock.

• It is suitable for areas where the water table is 10-25 feet below the surface.

These well points can be driven 30+ feet into the ground if needed, but that's no easy feat. Check with locals and your local water office to compare the success rate for this method in your area.
It will all come down to soil conditions and depth of water table.

Something to keep in mind is that areas with shallow water tables are more susceptible to contamination from land use. Contaminants can easily seep through the soft, sandy soil into your well, even from distances far away. When planning a "point well" it is best to be at least 150 feet away from potential sources of contamination. These include septic tanks, livestock yards, ponds, sewers, etc. Check with your local government office to get the exact code specifics for your area. Make sure you test your water before consuming it.

How To

Start by collecting all the material that you will need. You can purchase complete well point kits from hardware stores or from an online merchant. When looking at the materials used, you will notice that all the components are made of metal. This makes galvanized steel

You Will Need

- **A well point**, also called a "Sand point". The well point is a heavy spear-like object that is around 3-5 feet long and 1¼" to 2" in diameter. It has a metal screen on the outside to prevent larger particles from entering your well and also a fine screen on the inside for smaller particles.

  If you are planning to go down less than 25 feet, then you can use a 1¼" well point with metal piping and a shallow-well pump.

  If you want to go deeper, you will need a 2" well point with metal piping and a deep-well pump installation.

- **Galvanized steel pipe extensions** around 5 feet long. (1¼" or 2" in diameter)

- **Threaded metal couplings.** Try to get long couplings with extra thread.

- **Metal drive cap.** If the drive cap is threaded, the it must be the same diameter as the pipe extensions. Some setups come with a galvanized nipple which is first threaded into the coupling. The drive cap is placed on top of the nipple. Make sure you follow the manufacturer's instructions.

- **Pipe Wrenches** x 2

- **Carpenter's level or Plumb bob**

- **Plumber's tape/ Teflon tape.** Wrap it clockwise, with the thread!

- **Sledgehammer**

- **Auger or post hole digger**

- **Garden hose**

- **A long wooden pole** (or similar) for surging the well hole.

- **Pump.** A Pitcher pump is suitable for a 1¼" pipe and depth of 25 feet.
  A Jet pump is suitable for a 2" pipe and depth of 25-45 feet.

- **An extra pair of hands.** This is a two-man job.

Extracting Water From The Ground

your best option. Quality material will not bend, break or rust. When selecting your well point, remember that a well point of 1¼" will suit most wells down to 25 feet.

1. Start by making a hole in the ground with an auger or post hole digger. Go down a few feet just to get you started.

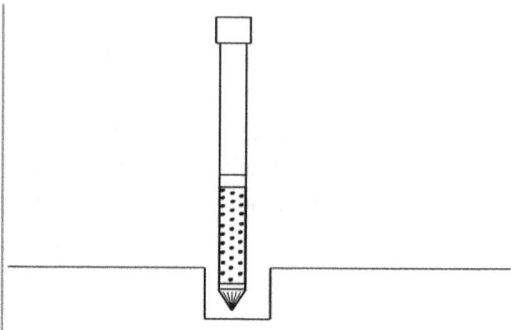

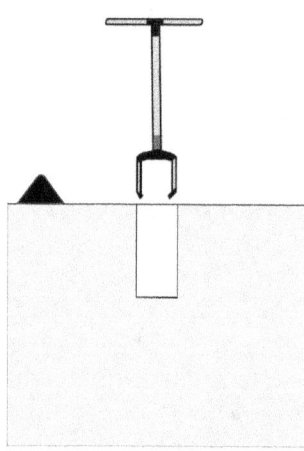

Take the well point and rub a bar of soap over the openings. The soap will clog the holes and prevent debris from entering. Later, the soap will dissolve when you strike water. The soap also makes it easier to drive the point into the ground.

2. Place your well point in the hole.

You will need to add a coupling and pipe extension from the start. Clean the thread first and remember to use Teflon tape (clockwise) and to lock the coupling in place. Use one wrench to hold the pipe in place and the other to tighten the coupling. This is a two-man job.

3. Place the drive cap on the pipe. Use the level or plumb bob to make sure that the pipe is vertically lined up in the hole. Drive it into the ground with your sledgehammer.

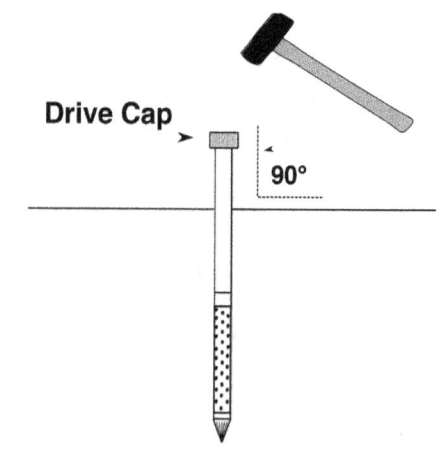

The drive cap protects the pipe threads on the well point and provides a better striking surface. Make sure you strike it evenly. It's best to have a helper to help with keeping the pipe perpendicular.

4. Keep on driving the well point, with pipe extension, into the ground. At first, it will be smooth and easy. When there's about ½ foot (15cm) of pipe left sticking out the ground, you can stop to add another section of extension pipe. Repeat this sequence as you go deeper.

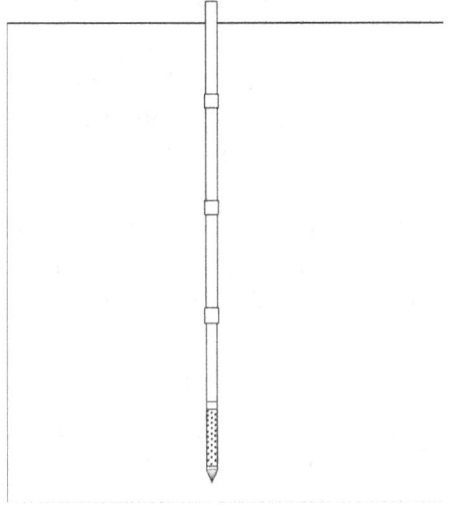

Pour water at regular intervals down the pipe. Wet soil is easier to penetrate.
Remember to use the drive cap when adding the extension pipes.
Check couplings and remember to use Teflon tape.
You can use the plumb bob to try to keep things level and perpendicular to the ground. You are dealing with a long length of pipe. Do your best to send it straight down into the earth. Repeat till you reach the water table.
If you hit rock, don't try to break it with the well point. Depending on the depth, you can try to move the pipe side to side and to wiggle it to create some annular space around it. If un-

Make sure that the new pipe goes straight into the coupling. Threaded pipes and fittings are prone to leaks if not straight and sealed.

successful, pull up the pipe and move to a new location. No use destroying your well point.

5. Reaching the water table means that you are in water bearing sand. There are a couple of ways to test if you've reached the water bearing sand.

• When hitting water bearing sand, you will suddenly become aware that it is easier to drive your well point through this section.

• Also, listen for a variation in sound when you strike the cap. If you suddenly start to hear a hollow sound like made by a dinner gong, then you are probably in fine, water-bearing sand.

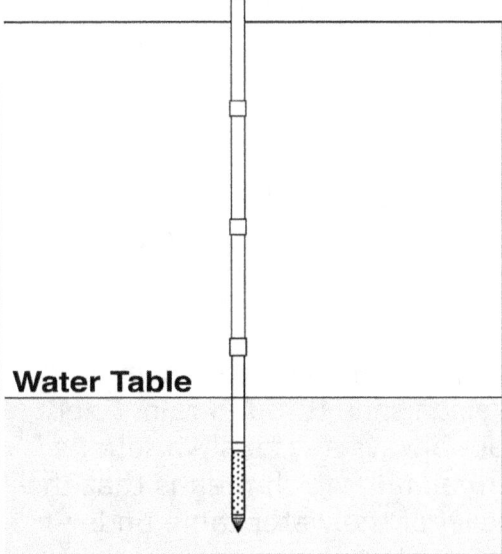

Water Table

Extracting Water From The Ground   255

- You can pour water into the top of the pipe. If it sinks down within a short period of time (2 mins), you are in the water table. If it stays at the top, it means that you are still in the soil layer.

If you sense that you have reached the water table, check your notes and compare with your local water table depth that you researched beforehand. Take a line with a small weight attached and drop it down the pipe. Measure how deep you are into the water table.

6. When you have reached the water table, continue for at least another 8-10 feet if possible. The deeper the better for you want the well screen to be totally submerged in water.

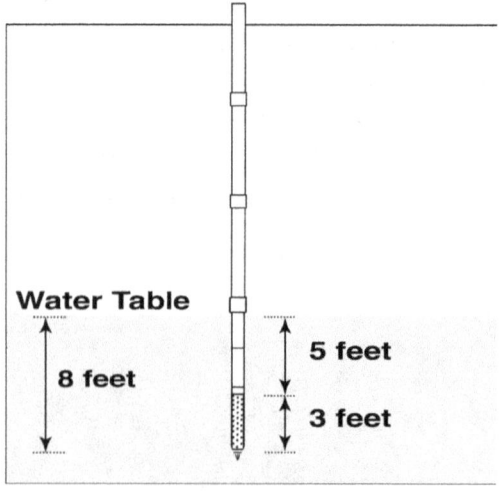

Remember that groundwater is replenished through rain. There are always seasonal variations in rainfall, which means that the level of the water table under-ground, will also rise and fall. The deeper you go into the water table, the better off you'll be in the dry season.

7. Once you've reached the correct depth you can start to develop the well. This will improve the yield and also the longevity. You have to clean around the well screen, to remove the fine particles and to leave only the larger particles behind. The well screen will filter the larger particles. You can clean the well by using either a surging technique or jet washing it with a garden hose.

a.) **To surge**, place a long pole down the pipe till below the water level and push up and down to surge the water through the perforated point.

b.) You can push the garden hose all the way down the pipe **to jet** the particles out.

8. The next step is to attach a pump and a concrete apron at the top of the well. The apron should drain away from the well towards a storm drain.

To avoid well contamination during a flood scenario, you should try to keep your pump at least 3 feet off the ground. This applies to both hand pumps and electric pumps.

If your pump pipe diameter and the well pipe diameter don't match up, you can fit a reducer coupling to connect the two.

Operate the pump to further remove debris and particles from around the well screen. This will take a while, but you have to do it till you see clear, clean water coming out the top of the well.

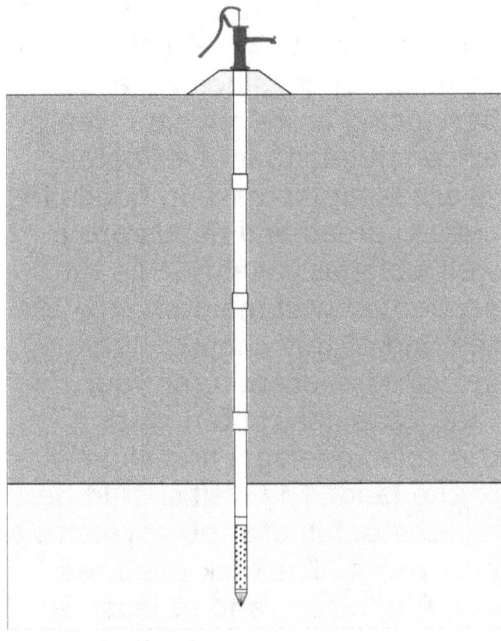

Remember that off-the-shelf hand pumps are mostly used for depths up to 20-25 feet. These pumps have to be primed to get the water from 20 feet down all the way up into the pump.
If you cannot or don't want to prime your pump, then you will need a check valve. This is an inlet valve that is positioned on the horizontal line leading to the pipe. It is a one-way valve that allows water to go up towards the pump, but not down and out. Having a check valve means that u don't constantly have to prime the hand pump. The result is that you always have water at the ready in the pump.

Your pump should be installed a minimum of 3 feet above ground, to prevent contamination during a flood event.

If you are close to an electrical source then you have an option of installing an electric Jet pump.

Conclusion

Using a well point is a very simple and straightforward method for water extraction. Try not to break the well point by forcing it through the rocky patches. Just move to a new location and start a new hole.

People tend to get excited once they hit water. Take your time and make sure that the well screen is completely submerged in the water. The deeper the better.

Be smart and do not drink water from the ground, except if you are 100% sure about the quality. It's always best to test your water before consumption.

All wells need a pump to transport the water to a place of storage. Tanks are perfect vessels for water storage.

Make sure you take a look at Chapter 4 where we take a closer look at holding tanks, and also at Chapter 6 where we discuss well pump basics.

How To Dig And Construct A Well

The most basic method for extracting water involves digging, with basic tools, through the dry, unsaturated zone. The depth of the water table will decide how deep you have to dig. You have to dig into the water table and line it with a porous material for clean water to enter the well. Line and seal the upper part of the well to prevent cave-ins and contamination. This is a very dangerous method that should be done with the aid of a professional.

Dug Wells

To dig a well by hand should be a last resort, for desperate times. This method is dangerous, slow, unpredictable and not feasible as a long-term solution. It is also open to contamination which is problematic, to say the least. A shallow water well should be at least 100 feet from any source of possible contamination. It is this writer's opinion that 150 feet (45m) is a safer distance, since we are experiencing more adverse weather conditions than ever before. There is an increase in floods in certain areas and floods are a well's biggest enemy. See the section on well maintenance at the end of this chapter. When selecting a location for your well, consider contamination, the soils, geology, and slope of the land. The well should be located uphill of septic systems, barnyards, livestock pastures, and fuel tanks, and at least 30 feet (10m) from streams and ponds. In addition, wells should not be located in extremely wet areas.

The following design will introduce the basics of well digging and it will give you an idea of all the dynamics at work. Dug wells work on a simple prin

---

Well Drilling Glossary

Apron. A reinforced concrete floor all around the top of the well. It prevents contaminants from entering the well.

Caisson ring. A watertight retaining structure used on the foundations of a bridge pier, for the construction of a concrete dam, or for the repair of ships.

Concrete. A mixture of cement, sand and gravel.

Windlass. A type of winch used to lower buckets into and hoist them up from wells.

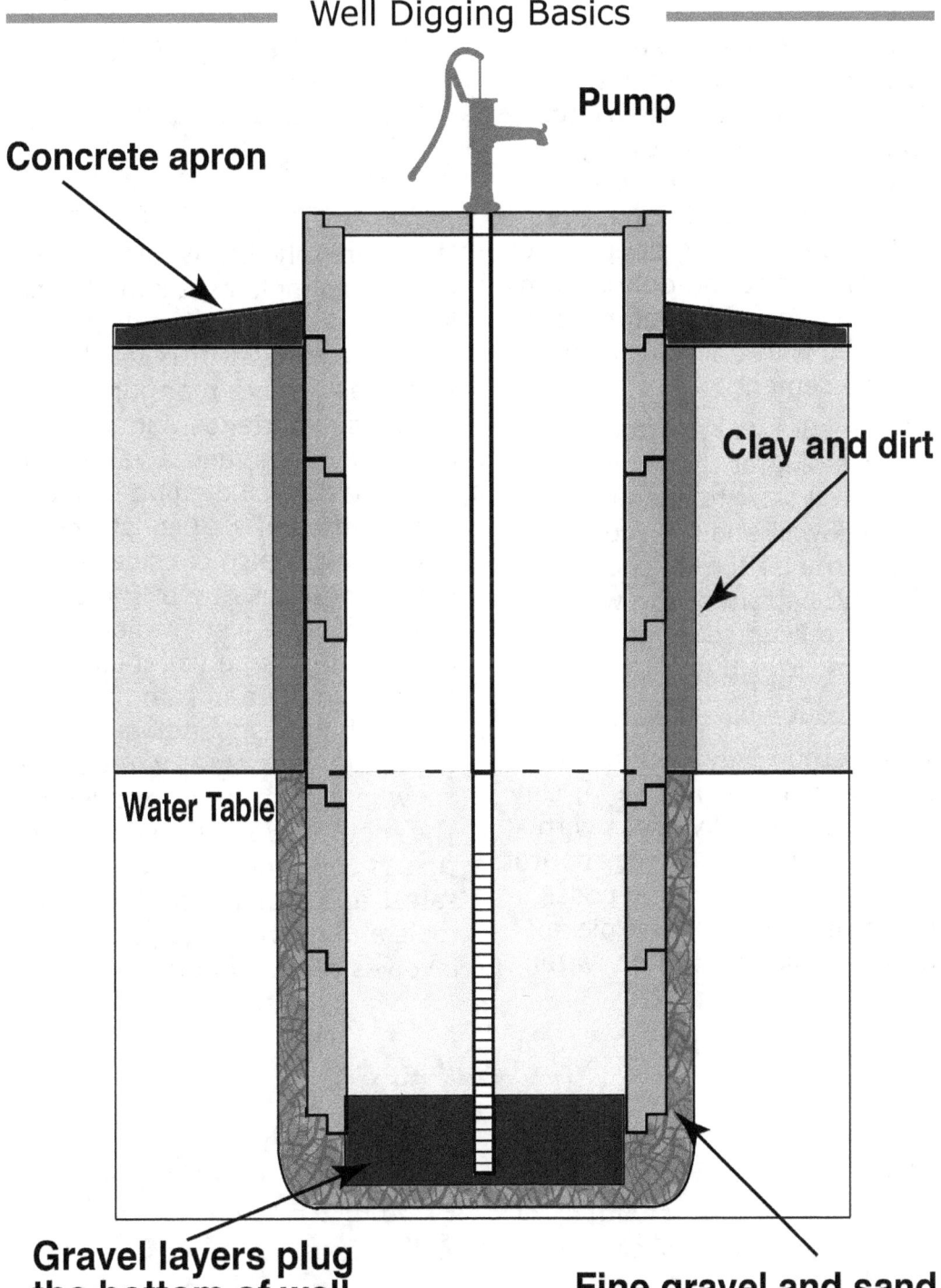

Extracting Water From The Ground

259

ciple:

- Find an aquifer close to the surface and dig a round hole that extends below the water table level. Never ever dig a square hole. It will cave-in!
- Remember to reinforce the well wall as you dig deeper and to remove dirt and debris from the hole. You can reinforce the well wall with bricks or with poured cement.
- Make sure you are below the water table, for this is where water will start to seep through the well wall and to accumulate in the bottom of the hole. Plug the bottom of the well with gravel to keep sediment and large particles out.

That's your well.

This method only works in areas with suitable soil like sand, clay or gravel. In rocky areas with boulders you can forget about it. On paper, it's definitely possible to use a shovel and pickax to dig a hole and to find water.

The width of the hole is determined by the soil conditions. Loose soil, smaller hole, firm soil wider hole.

Unfortunately, the whole process is not as simple as it sounds and it is not recommended doing this on your own. It is, at the very least, a five-man job that requires experience, lots of common sense, and a variety of tools. Basic well digging is a low cost method of well construction, but it is also considered very dangerous due to the possibility of cave-ins. Traditionally these wells for personal use are 4-5 feet in diameter and are dug by hand. Depending on the ground in your area, you can go anywhere from 8-20 feet deep. The method we are going to discuss is considered the easiest, safest and requires fairly simple tools and moving equipment. It involves reinforcing the inside

You Will Need

- A shovel and pickax for digging
- Simple waterproofing plaster
- Lengths of rope or cable used to lower and haul equipment
- A winch with head frame to haul heavy objects from well
- Pulleys for simple lowering and hauling tasks

- Bailing buckets for hauling dug material from the well
- Hammers, nails, wood
- Hard hats
- Level, plumb bob and plumb line
- Ladder (rope or rigid)
- Eye protection
- Medical kit

of the well with a concrete well lining, also called a Caisson ring. If digging in unstable ground, you run the risk of the well wall collapsing.

The best and simplest method is to use precast concrete well rings as your well lining (Caisson rings). These are precast and are about 2" (5cm) thick and 3-5 feet (90-150cm) in diameter. They are similar to the manhole cement rings we see on road construction sites. They should have joints that allow them to "fit" into place. You will need to construct a windlass (winch on tripod or head frame) above your well to assist with the moving of dirt, rocks, concrete rings and bailing of water.

Preparations

There are various ways to dig a well. The use of precast concrete rings has proven to be a very efficient method, but it does involve a fair amount of preparation.

• Prepare your concrete rings. You will have to do some research to see what's available in your area. Discuss in detail what you need and the implementation of the rings. They must be wide enough for a person to stand in. Around 4-5ft (1.2-1.5m) will do. Remember to determine the depth of the water table beforehand. This will give you an indication of how many rings you require. You will also need professional lifting equipment.

Dug wells should only be attempted as a last resort, when all options have been exhausted and only if it's a dire emergency.

Do not attempt to move or lift a concrete ring without proper lifting equipment as prescribed by the manufacturer!

• These rings are heavy and it will be an arduous and a dangerous task to move them around. Each ring must be safely position over the well hole and to do this, you will need a forklift and a truck. There is no way around this. A standard ring has at least two "lifting holes" (through the side) which are used for hoisting and lowering. Make sure you get rings with lifting holes on the inside or with holes that go straight through the sides. Use the correct lifting straps and lifting keys/ hooks. Don't rig any ropes around or on the underside of the ring. Once your well is finished, remember to seal off all holes where potential contaminants can enter.

• You should have a sturdy platform constructed around the opening of the well hole. When you are ready to sink the first concrete ring, place two thick pieces of timber over the hole. You can rest your first concrete ring on top. Once ready, lift the ring, remove the timber, and slowly lower the ring into the hole. This part of the process will require patience and teamwork. Make sure that safety is a first priority.

- You will need lifting equipment. Moving material in and out of the well is no simple task. Take care that your lifting equipment is not too heavy, for that can also cause the top of the well to collapse. Use a heavy-duty electric winch with head frame for heavy objects and a simple pulley system (or windlass) for buckets filled with dirt. It is not advisable to lower the cement rings by hand. The process of hauling and lowering material is a very dangerous one. This is where most accidents occur. The head frame can also act as windlass and should have a solid platform underneath (wood) to prevent objects from being knocked into the well hole.

- You will need waterproofing plaster to seal the top ten feet of the well. This is to prevent contaminants from entering the clean well water. It is not necessary to plaster the bottom section of the well. This part is submerged in the water-bearing sand of the water table and you want the water to enter through the openings between the concrete rings.

- Organize your well. Draw up a plan with the layout of the construction areas. This should include the dumping area, no-driving zone, no-contamination zone, washing area, rest area, tool storage, etc.

- You will have to prepare a dumping area well away from the well hole. This is where all the material taken from the well will be dumped. Calculate how much space you'll need by determining the volume of dirt that must be excavated from the well. Make sure that the daily routine of moving the dirt will not cause the well sides to collapse.

- Prepare storage space for construction tools. Tools should be locked away when not in use.

- Create a path system for workers to follow. This creates a safe environment with designated areas for work and rest.

- The construction team should be well trained and everybody should know his or her task(s) and what equipment they are responsible for. Use a system of hand signals when communicating with diggers inside the well.

How To

1. Start by digging a hole wide enough for you to stand in. Remember that your concrete ring has to fit inside this hole. Around 4-5ft (1.2-1.5m) will do. Keep the diameter of the hole in mind and leave the wall curves nice and smooth.

At this stage, nothing should be suspended or hanging above the digger.

2. Go down to about 3 feet (1m) and stop to inspect your hole. The first cutting ring

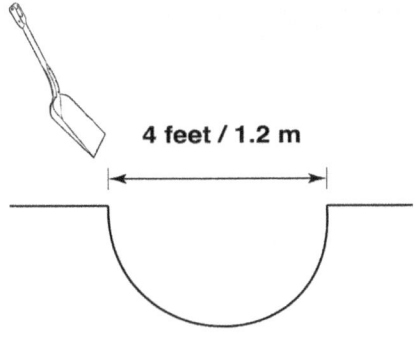

should be able to fit snugly inside. Typically, the first well ring has a "cutting edge" to scrape away the dirt. Dig down to a depth that will allow the first well ring to sit level in the well.

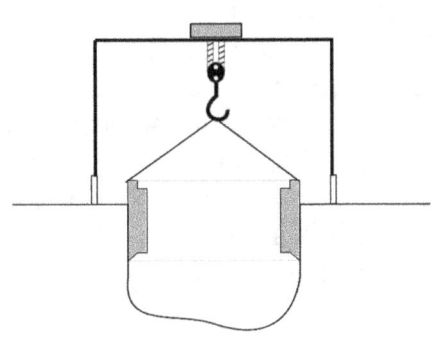

CAUTION! When lowering these rings, no workers should be inside the well!

3. Once the ring has been lowered, a worker can be allowed to go down and to stand inside the hole. While standing inside the ring, start to excavate underneath it; as you remove the dirt, the ring's weight will allow it to sink deeper into the hole. Remove the soil gently and evenly from underneath the ring. This will allow the ring to slide in a level position down the well.

Just allow it to slide down a few inches. You have to leave the upper part of the ring exposed to join the second ring. Gently lower the second ring and make sure that the joints line up.

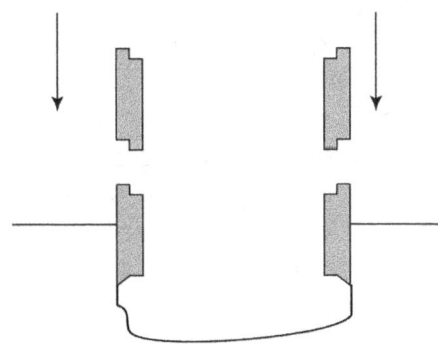

Remember that once the hole is completed, you will have to seal the upper joints of the well with waterproofing plaster!

4. The next step is very important. Connect the second ring to the first ring with metal rods or chains. This can be achieved by placing a metal connection inside the "lifting eye" of the ring. This method will aid in sending the whole column of rings down

Extracting Water From The Ground

evenly and in a rigid, straight line.

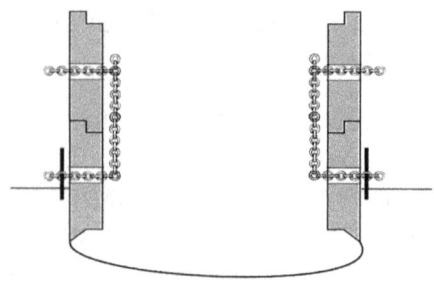

5. Level. Use a level or plumb-bob (a weight on a string) to check your progress and if you are digging in a straight line. Do this routinely as you go deeper. The stack of rings must slide evenly, at a ninety degree angle, down the hole.

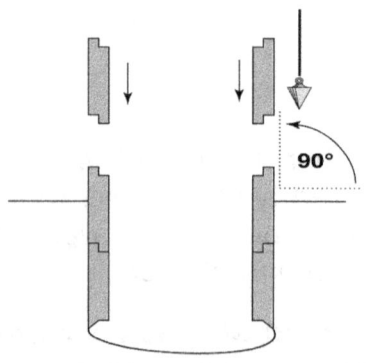

6. Keep on stacking these rings on top of one another till you reach the water table. All the rings should be in a column, tightly fastened with rods or chains. This is no easy feat and you will have to pay close attention to the task at hand. At this stage, the help of an expert will come in very handy. For a novice, this can be considered as on-the-job training and it will require some perseverance.

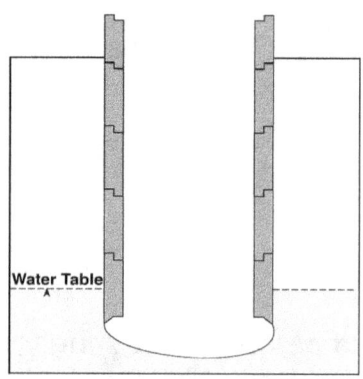

Patience is key and it will undoubtedly take a qualified person to guide you through this process. Upon reaching the water table, you have to stop.

7. Pump or bail. Once you reach the water table, you can expect water to start seeping into the bottom of the well. This is where digging gets hard and where you will get wet. Try to go as deep as you can. At least 5 feet more. You will have to

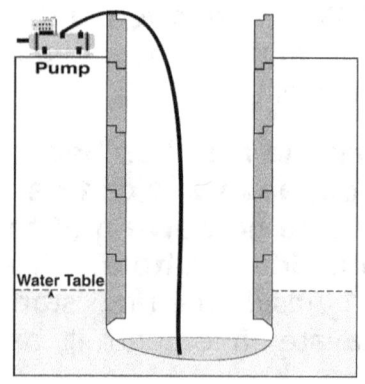

pump or bail the water from the well.

When using a pump and generator, take care that the generator is downwind and that it is not sending fumes down into your hole.

8. Gravel. Once you find that the flow rate of water into the well is faster than the bailout rate, then you have to stop. You are now truly under the water table. (You should aim to be at least 5 feet under the water table.) Fill the bottom of the well with gravel.

9. Protection. Seal the top 10ft (3m) of well rings on the inside with a waterproof concrete sealant to create an effective seal against contaminants that might seep in through the adjacent soil on the outside.

10. Annular space. Fill any openings on the outside of the concrete ring casings (top 10 feet) with clay or concrete. This is to prevent surface water contamination. This is of great importance!

11. Seal the well. At the top of your well, you want the well lining to extend about 3 feet (1m) above ground. This is where you have to consider contamination. Make sure that no contaminants can seep into your well, whether through the well opening at the top or through the adjacent soil.

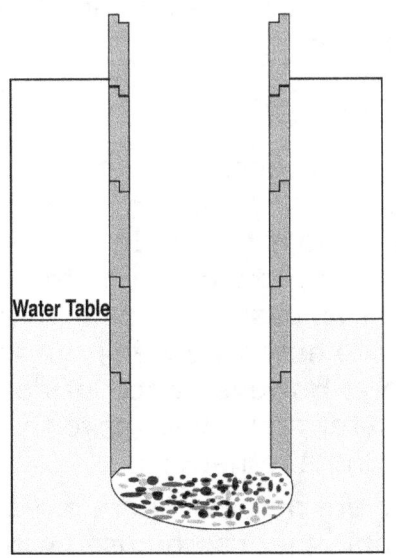

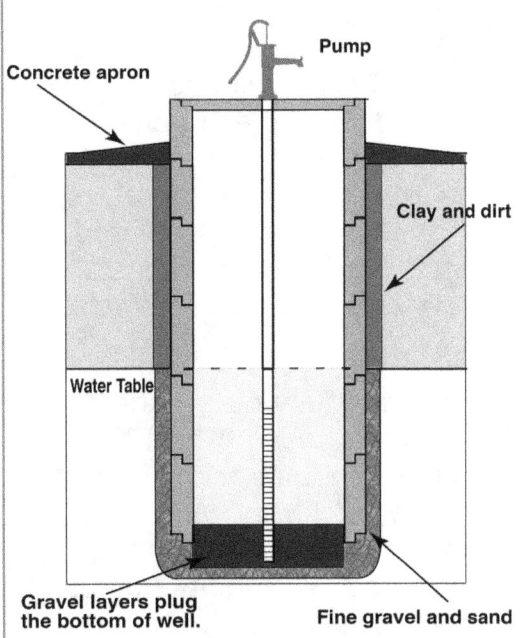

One foot of gravel will do nicely. Use coarse gravel as a top layer and fine gravel underneath. Fill it up properly to prevent clay and other sludgy material from entering the well. Clean water can penetrate into the well from the porous sides and from the bottom filled with gravel.

Extracting Water From The Ground

To seal your well top you can create an apron from reinforced concrete. Make sure that it runs downhill and that it drains water away from the well into a storm drain or another collection area. This apron should be reinforced, strong and wide (6 feet of space all around). This is where people walk and where water containers are filled. Cracks in the apron will cause contaminants to seep into your well.

12. Pumps. Install a hand pump at the top of your well. Remember that off-the-shelf hand pumps are mostly used for depths up to 20-25 feet. These pumps have to be primed to get the water from 20 feet down all the way up into the pump. If you don't want to prime your pump, then you will need a foot valve. This is an inlet valve that is positioned at the bottom, suction end of the well pipe. It is a one-way valve that allows water to go up towards the pump, but not down and out. Having a foot valve means that u don't constantly have to prime the hand pump. The result is that you always have water at the ready in the pump. If you are close to an electrical source then you have an option of installing an electric Jet pump.

How To Lower Concrete Rings In A Level Position?

- Digging method: When scraping the dirt away from underneath the ring, start by making four fist-sized holes in the 3, 6, 9, 12 o'clock positions. Next, make four holes in between the previous four created. Make eight holes in between the existing holes created. Your ring should start to slide down while doing this.

- Join the concrete rings together as you lay them on top of each other. This creates a unit of rings that slide down the wall in a straight line. You can join the rings with bolts, chains or lap joints.

Options

You can consider alternative well casing support. The main issue is safety. In loose soil, the concrete ring method is best. This basic method allows water to enter the well from the bottom. However, once inside the water table, you have the following alternatives:

There are porous concrete rings available that can be used within the water table to let water in. Tiny holes in the concrete wall will keep sediment out and let water in.

Concrete rings with holes (slanted upwards) can also be used within the water table. The holes must be angled upwards from the outside in. This aids in keeping sediment out.

When concrete well casings are not an option, you can consider the following methods:

- **Clay bricks** can be used in stable ground and at shallow depths. You have to dig as deep as possible before you start laying the bricks. This method requires stacking bricks from the bottom of the

Dug Wells And Safety Concerns

- Place a fence around the well to keep ALL animals, no matter how big or small, out of the well. A scorpion or snake can kill you.
- Place a fence or barrier to keep people, especially children, out.
- During the construction period, place a lid on the well opening when the construction site is abandoned.
- Always leave a safety line in the well, in case someone were to fall in. This line should be a sturdy rope, twice as long as your planned well depth.
- Make sure that the digger wears a hardhat or helmet when in the hole.
- Check that all ropes are new and not worn.
- Check that bucket handles are secure and that the bottoms of the buckets have no cracks.
- Get a pair of steel-capped boots to protect your toes. Using shovels in a confined space increases the chance of accidents.
- Double-check all knots.
- Protect the top of the well by placing some wooden planks all along the edge of the well hole. This will prevent workers from wearing away the ground around the well.
- Always keep a safety kit on site.
- Toilets and washing areas should be 75 feet away from your well.
- Plan ahead where vehicles and forklifts will be allowed to drive. Do not disturb soil around your well's edge.
- When using generators, make sure that fumes are downwind of the well hole.
- The attitude of all workers should be professional and alert. Do not joke around a well site and do not allow alcohol or other mind-altering substances to be consumed while working.

Extracting Water From The Ground

well (within the water table) and it means getting wet and bailing water by hand or with a pump. This method is for experts only.

- **Various molds** (Brit. moulds) can be used to pour concrete that can act as the well lining.

  You can dig down to around 10 feet, create a mold from wood or sheet (corrugated) metal, and then pour your concrete. The concrete will set in between the mold and the soil of the well wall. This will also require the help of an expert.

- **In very stable soil**, and with a fairly shallow well, you also have the option of digging the shaft first. Afterwards, you can slide the concrete rings down.

  This is also a dangerous method and not recommended. In stable soil and hard rock, the risk of a cave-in is lower.

Part of being aware of safety and contamination is instructing the excavation foreman to implement strict excavation guidelines.

Conclusion

The construction of a well, using manual drilling/digging techniques is a complicated process. Some of the biggest issues with wells of this nature is **safety** awareness and prevention of **contamination**. A deep, open hole has the potential to cause accidents, not just to workers, but to anybody who ventures onto the

Benefits Of Using Precast Concrete Rings

- They are strong, durable and ideal for use as a well lining.
- Using concrete rings cuts down on manpower needed.
- They can be sealed with waterproofing plaster to prevent contaminated surface water from seeping in.
- They help to cut down on excavation time.

- They will not corrode or rust and will not contaminate the water in the well.
- They come in different sizes are considered safe to use by the industry.
- They are perfectly round, lock into place and provide good structural strength.

construction site.

Wherever humans and machines are involved, we find an immediate rise in pollution levels and also in contamination of the environment.

Extra care must be taken to guarantee a safe, contaminant-free environment. Part of being aware of contamination is instructing the excavation foreman to implement excavation guidelines.

- Before digging/drilling starts a good site has to be selected, where experience suggests that there will be an adequate quantity of good quality groundwater.
- During the digging/drilling process there are a lot of different aspects which require attention to prevent things from going wrong. Besides the practical digging/drilling skills which are executed at ground level, attention also has to be paid to important processes which are happening below ground level during digging. Water used in drilling (working water) could flow away or worse; the borehole could collapse, burying part of the drilling equipment.
- And finally, once the hole has been drilled, the well casing, screen and sanitary seals have to be installed at the right depth, preventing contaminated water from entering, and ensuring a sufficient yield.

With this in mind, it becomes clear that the introduction of manual drilling needs adequate supervision, with proper training and guidance of the drilling teams.

These guidelines will include:
◇ work schedules
◇ safety procedures
◇ emergency procedures
◇ waste disposal
◇ workplace etiquette

Summary

Having a well can be a very rewarding experience, but just how rewarding will depend on a couple of factors.

1. Make sure that your well is **fully developed**.
2. Once your well installation is complete, make sure that all **cuttings and sediment are removed** to ensure that water can freely enter the well.
3. Make sure that your borehole has **no stability issues**. The well should be sealed and the well casing, screen and pump system should be made of quality material. Cracked pipes and screens normally means drilling a new well.
4. Watch for **over-pumping**. This occurs when water is withdrawn at a faster rate than it can be replenished. It is a common problem

Extracting Water From The Ground

with wells and it leads to premature well failure.

5. Watch out for **dissolved gas** (nitrogen, methane), corrosive water and biofouling. Water saturated with oxygen can cause bacterial growth, which causes a biofilm to develop inside the well. This can mean reduced well yield and water quality.

Have a look at the discussions on holding tanks (Chapter 4) and well pump basics (Chapter 6).

My Checklist

√ I had my water tested to see if it is fit for human consumption.

√ I selected the correct well for my needs.

√ I have determined the use of my well water.

√ My well is at the perfect location. It is not affected by surface runoff water and in the vicinity of potential sources of contamination.

√ I checked with local government.

√ I am legally allowed to own a well.

√ I double-checked all installation requirements. Both plumbing and electrics were done (or checked) by a licensed well professional.

√ I know how to maintain my well water system.

√ My well has a well seal to prevent contaminants, and critters from entering.

√ I am not abusing the groundwater in my area.

√ I am only collecting for my basic needs.

√ My well does not pose a safety hazard to people or animals.

√ I have considered all worst-case scenarios. I have a backup plan.

Chapter 6

The Basics of Wells And Water Pumps

Pumps are used to "pull" or "push" water from a water source to a holding tank or to directly feed the household plumbing. Typically, the water source can be a well, spring or a pond. Pump installations should be done by a licensed professional to make sure that you get the most out of your pump. A correctly installed pump will last long, use very little energy and deliver water with optimum flow rate and pressure.

Your well must be fully developed and pumped clean of large particles, debris and fine sand before you install the pump. No use clogging your pump or pipes with sediment or drill cuttings.

To select the appropriate pump for your water volume needs, you will have to consider the following:

a.) The water level of the water in the well.

b.) The diameter of your well pipe.

c.) The number of plumbing fixtures (water consuming appliances) that require water.

d.) The volume of water required in gallons per minute (gpm) during peak water demand.

e.) Your preferred power source. (Connected to the grid or solar.)

These are questions that every well owner should be knowledgeable about.

Different pumps are used for different applications and it's vital to select the correct pump for your needs.

Connecting An Electric Pump

A pump is used to draw water from the well and to move it to a new location. Selecting the right pump for your needs, will depend on the following:

1. The estimated volume of water you need for your property.
2. The power source you will connect to. Pumps can connect to the power grid, batteries, solar panels or generators.
3. The environment. A harsh environment with dry and dusty conditions will require a quality pump with screen and shield protection.

Pump Basics

Depending on the aforementioned conditions, you can consider looking at the following types of pumps:

- **Jet pumps** are, generally speaking, suitable for wells that are 3 inches or less in diameter.

- *Shallow well jet pumps* are often used for shallow water wells which are no deeper than 25 feet. The pump is located at surface level.

- *Deep well jet pumps* work well with wells as deeper as 80-100 feet. The jet is located inside the well, below the water level.

- **Submersible pumps** are long cylindrical pumps (with motor) around 2-3 feet long. They are used in wells 3 inches or more in diameter. Submersible pumps are very efficient pumps with high yield and high delivery pressure.

Check Valves And Foot Valves

One-way valves are a very important component of the pump system. The use of one-way valves prevents the water from flowing back down into the well and it ensures that the pump is primed.

- There are two options for jet pumps:

- Usually, a **foot valve** is placed at the end of the suction line, which is at the bottom-end of the well. They are mostly used in drilled wells.

Having the foot valve at the bottom of the well can be problematic, because it's hard to get to for maintenance purposes.

- A **check valve** is a good option (especially for driven wells) and can be installed in the feed line next to the pump. This assures that the pump line and pump is filled with water when the pump is switched off. If you put your check valve next to the pump, above ground, then you need an additional fitting above ground (at the top of the suction line) to prime the pump.

You can use 1¼ inch male PVC adapters with the valve.

- For submersible pumps we use check valves where needed. Usually submersible pumps come with a check valve pre-installed in the pump that you place at the bottom of the well. If your well is very deep, install a check valve for every 150 - 200 feet (45-60m) of pipe.

You should also have a check valve at surface level to ensure a steady flow of water to your distribution system.

Pressure Tanks

Pumps work very well in combination with pressure tanks.

The function of the pressure tank in a water well system is to create water pressure by using compressed air to push down on the water in the tank.

The pressure tank functions as a reservoir and holds enough pressure to flush a toilet in the house, without turning the pump on.

- The normal operating pressure ranges for pumps are 20/40, 30/50, 40/60 or even as high as 60/80 psi. This means that pressure is kept constant within a 20 psi range, which sets a boundary for starting and stopping the pump.

To measure the depth-to-water in a well, drop a line with a bobber (float) and a small weight down your well hole. Once the line is slack, (the bobber is floating) you have your measurement.

A 40/60 ratio is recommended since it's by far the best ratio to keep a constant shower stream. This pressure range indicates when the pump will turn on and off, and this function is controlled by a pressure switch device. This is the device that will turn your pump on and off automatically.

Example: If you have a 40/60 psi setting, the pump will turn on when the pressure in the pressure tank drops to around 40 psi and off when the pressure reaches 60 psi.

The relationship between the pressure switch, pressure tank and the pump is what allows water to flow through the pipes in your home. When selecting a pressure tank, you will need to know the performance of the pump in gallons per minute.

- We measure the volume of water required to serve a household, in gallons per minute (gpm). It is calculated by looking at the water consuming appliances in the house and also at the pump's flow rate in gpm.

The gpm of the pump must equal the total number of water consuming appliances.

Example:

5 faucets, 2 toilets, a shower and a washing machine would require 9 gpm.

- To determine the appropriate size of the tank, match the draw down of the tank to the capacity of the pump.

Another easy way to size a tank, is to take the gpm system requirement that you determined for your house, and to multiply it by 3 and then to go to the next largest tank size.

In our previous example:

9 (gpm) x 3 = 27.

Buy a 30 gallon or bigger pressure tank.

Usually, when it comes to pressure tanks, bigger is better. A larger pressure tank will not hurt your pump's performance, but it will give you a larger draw down capacity, which guarantees even more usable pressure in the plumbing system, before the pump needs to come on.

- **In general**, a ½ HP pump is enough for a small house with one bathroom and a couple of people.

For a larger house, you'll need a ¾ -1 HP pump.

- Take into consideration that a larger pump will affect your electricity bill and that it will also require a larger pressure tank.

- If you want a generator as a back-up power source, make sure that the generator can provide enough power to start the pump.

Pitless Adapters

In colder climates where water pipes should be protected from freezing, a pitless adapter (fig. 6.1) is used to keep the water,

---

Well Pump Glossary

Cavitation. The formation of bubbles in a liquid, typically by the movement of a propeller through it.

Check valve. A valve that closes to prevent backward flow of liquid.

Foot valve. A one-way valve at the inlet of a pipe or the base of a suction pump.

Pressure switch. A switch that automatically turns water on and off, depending on pressure settings.

Pressure tank. A tank in which a liquid or gas is stored under pressure greater than atmospheric.

This helps maintain water pressure to keep appliances running efficiently.

Depth to water. The distance

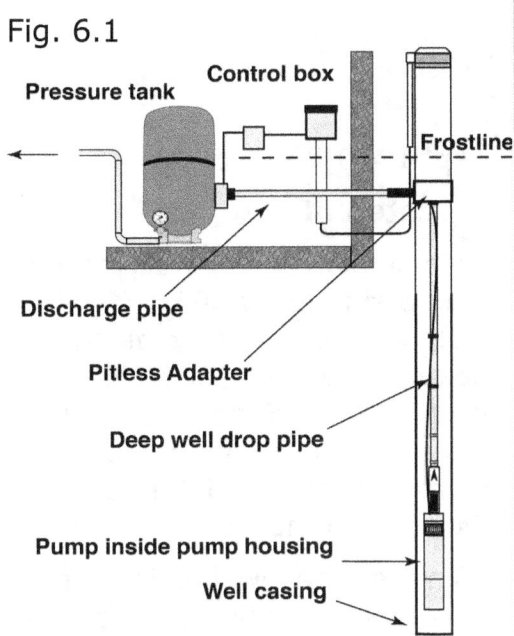

Fig. 6.1

can be used with standard jet pumps, they are commonly used with submersible pumps, where the pressure tank is located in a pump house or basement.

This adapter consists of a special fitting that is installed 6-10 ft (2-3 m) below ground level for protection from frost. A hole is made in the well casing and the water coming up from the well is diverted horizontally at the adapter, and discharged into a water pipe that runs underground to the house or pump house.

The main benefit of these adapters are:

- These adapters are cheap and reduce the risk of contamination in your well.

pumped from your well, under the frost line and in a sanitary environment.

Installing a pitless adapter is a much better option than using the old well pits we still see in certain parts of the world today.

Although a pitless adapter

- A pitless adapter makes it easy to access the well for routine maintenance or emergency service.

Well Pump Glossary

measured from the ground level to the water level in the well when the pump is not in operation.

Draw down. Amount of usable water we can get out of the water pressure tank before the pump has to turn on

Draw down water level. The distance from ground level to the water level inside the well, while water is being pumped from the well.

Well capacity (gpm). The amount of water measured in gallons per minute that the well produces without the water level dropping.

Well total depth. The distance measured from the ground level down to the bottom of the well.

The Basics of Wells And Water Pumps

Typical Shallow Well Jet Pump Setup

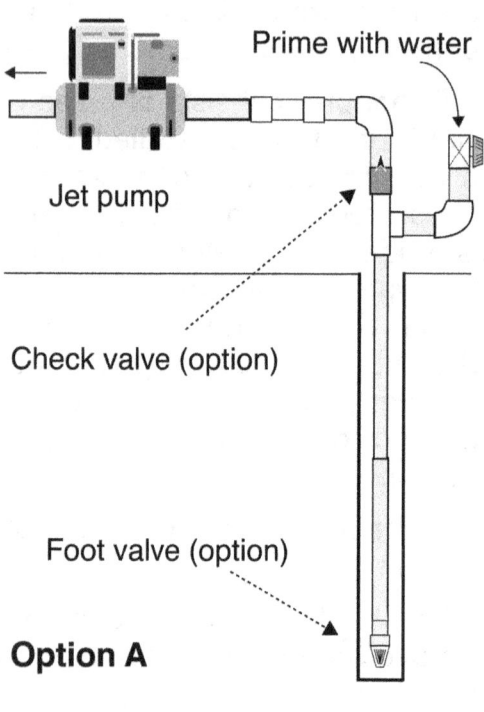

Option A

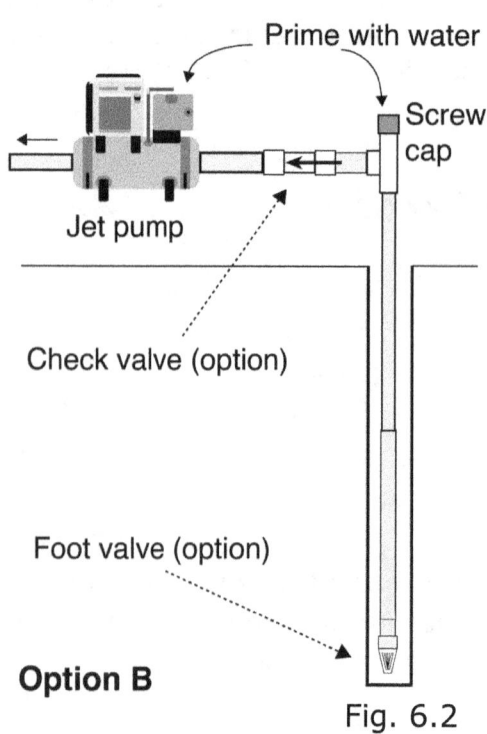

Option B

Fig. 6.2

The Basics Of A Shallow Well Jet Pump

Depth: 10-25 feet

Expected life: 10 years

For a shallow water well you will need a simple jet pump (fig. 6.2). Jet pumps are relatively inexpensive and require little maintenance. Expect your jet pump to last an average of ten years. Some basic information about jet pumps:

• Simple, shallow well jet pumps can draw water up to 25 feet. They do not work with deeper wells. For a deeper well you will need a deep well jet pump.

• Jet pumps can also be used as booster pumps to improve the existing pressure to a higher level.

Mechanics

These pumps are easy to operate and basically "suck" the water out of the well. When we say suck, it actually means that the pressure from our atmosphere is pressing down on the water and pushing it up the well pipe.

When you turn on the pump, it creates suction in the well pipe and with the push from the atmosphere, the water is jetted up the pipe. Inside the pump, the impeller pushes water through a narrow jet which is a cone-shaped device which creates a vacuum with suction. This is the suction that draws water from the well into the pumping system. The sudden decrease in diameter causes the water to shoot through the jet and eventually out the pump at end use.

You can install your pump at ground level, above or next to the well hole.

Installation and priming

Before you install your jet pump, make sure that your well is clean and ready to be connected to the pump.

Your well should have a sanitary seal on top of the suction (well) pipe. This should fit your well diameter.

You can also, if needed, make the well head area frost proof with a concrete containment area.

To prevent air pockets, make sure that the pipe slopes slightly downward from the pump to well.

Basic installation of a jet pump.

1. Make sure there's no debris and large particles in the well.

2. Assemble the pump as per manufacturer's instructions. Make sure that you follow the instructions to the letter.

3. Install a 1¼ inch foot valve at the bottom of the suction pipe if this is an option. This will suit wells that have easy access to the well pipe like a drilled well.

Another option, as seen with driven wells where the well casing is used as the suction pipe, is to install a 1¼ inch check valve in the feed line. You can do it near the pump, in the horizontal or vertical position.

Shallow water pumps generally need to be primed to work. This is the process of flushing and forcing water into the pump in order to create enough pressure for it to work.

1. Make sure that the pump is turned off.

2. Locate the priming plug and unscrew it.

3. Use a funnel and slowly pour water in. The pump head should be completely filled with water. The water primes and lubricates the pump. When priming, remember that you want the pump and suction pipe to be completely filled with water.

4. Replace the plug once the pump head is completely full.

5. Turn on the pump. Check for leaks.

6. Optionally, you can also give yourself the option of being able to prime the well pipe if needed. To connect the well pipe to the pump, use a Tee fitting with a threaded cap on top. This is where you can prime the pipe by filling it with water.

The Basics Of A Deep Well Jet Pump

Depth: 25-110 feet

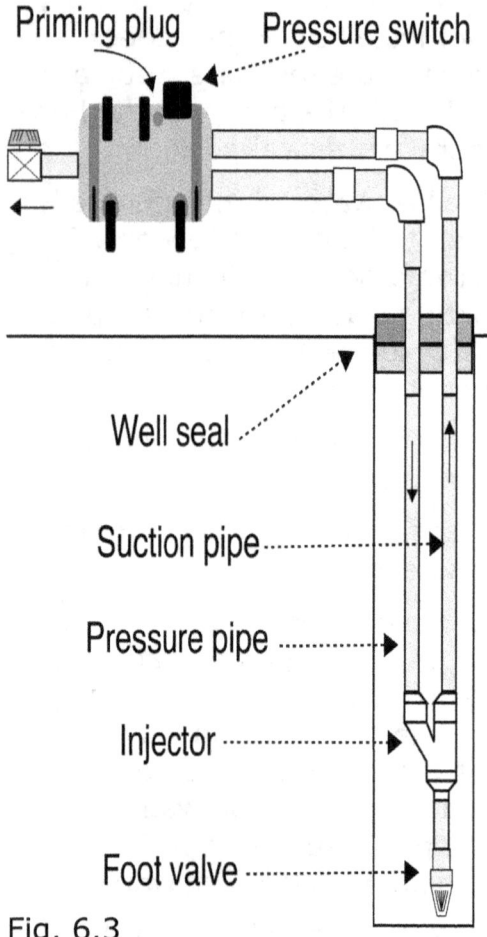

Fig. 6.3

This pump is not as complicated as it looks and uses a very simple method to extract water from a well.

A deep well jet pump works similar to a shallow well jet pump, but the main difference is that the jet is separated from the motor and placed at the bottom of the well. This pump uses a system of two pipes; the one pipe is mounted to the impeller housing and "pushes" water down into the well and into the jet body. The jet is located about 15 feet below the minimum water level.

On the output side of the jet body, the second pipe drives the water from the jet back up to the pump and on to a storage tank or to the household plumbing.

A foot valve is placed at the bottom of the well piping. This foot valve keeps the system primed and prevents draining of water from the pipes.

The jet uses suction and the impeller creates pressure. The combination of these two forces creates a simple application that can produce water lift from 25-110 feet deep.

Expected life: 10 years

If your well's water level is deeper than 25 feet, you can look at a deep well jet pump (fig. 6.3) to "push" your water to the surface.

Deep well jet pumps are slightly different from shallow well jet pumps in that they use two water lines in the well, instead of one.

Expect your jet pump to last an average of ten years.

Mechanics

The Basics Of A Submersible Deep Well Pump

Depth: 25-400 feet

Expected life: 25 years

Submersible pumps (fig. 6.4) are placed inside the well (or water source) and are used to "push" the water to the surface.

Pushing water takes up less energy than pulling water, especially in a deep well.

Mechanics

These are very effective pumps that consist of a pump motor and a series of impellers that are placed inside a cylindrical shape. When turned on, the impellers spin, and this causes the water to be pulled into the pump, which pushes the water to the surface for storage. The pump uses electrical wires that connect to a power source above ground. Note that a pump with three wires plus a GROUND, is a single-phase pump and it requires a control box. The control box is matched to the pump motor. Two wire plus GROUND pump/motor assemblies do not require a control box. Submersible pumps are very reliable and all working parts are sealed within the cylinder. They do not need priming and deliver great volume and pressure. These pumps can deliver water at pretty much any working delivery rate, depending on well depth, pipe diameter and pump horsepower.

Submersible well pumps are typically seen with horsepower ratings of ½ HP, ¾ HP, 1 HP, and 1½ HP.

Pay attention to the following:

• Follow the instructions of the pump manufacturer. Make sure that you are installing the correct pump for your needs.

• When following the manufacturer's instructions, make sure that you get the correct diameter drop pipe to connect to your pipe. The pump's discharge rate in gpm, must match the diameter of the pipe.

Example: A 10 gpm pump can handle a 1" pipe up to 300 ft.

• All electrical wiring, connections and system grounding must comply with local codes and ordinances. If unsure, hire an electrician to assist with the

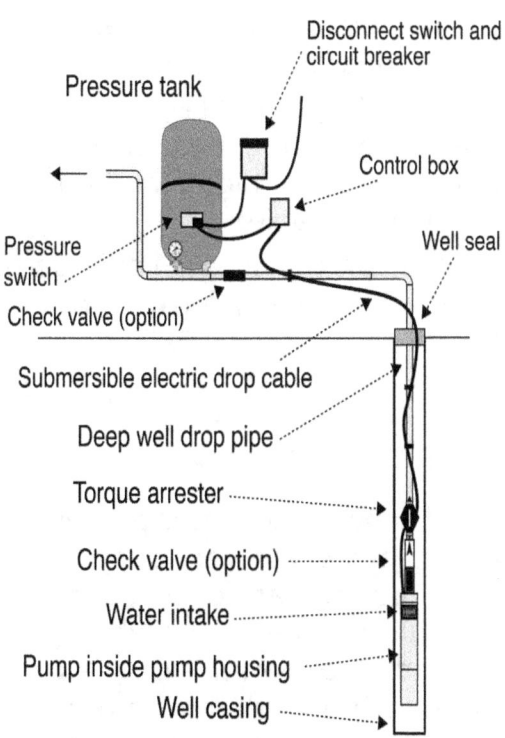

Fig. 6.4

electric installation.

- Submersible pumps should be installed 5 feet or more from the bottom of the well.
- Make sure that you use the correct pipe size and material for the well depth, pump type and water quality.
- When cables are spliced and connected, make sure that they are water tight. Besides crimping the wires together, they can also be soldered for an even more secure connection.
- Pumps usually come with a built-in check valve.

For very deep wells, you will have to use a check valve every 150 - 200 feet (45-60m).

Check valves prevent water from flowing back down into the well when the pump shuts off. They also ensure immediate water flow into the tank when the pump starts up, which prevents unnecessary physical strain on the pump.

- When selecting a submersible pump, pay special attention to the hermetic seal that protects the electric motor from water.
- The pressure switch senses the pressure coming through from the well and automatically turns the pump on or off.
- Attach a poly safety rope to your pump. You do not want to loose it down the well hole.
- Another safety feature is a torque arrestor.

This is basically a rubber support that clamps to the outside of the pipe and the top of the pump with hose clamps.

The torque arrestor protects the submerged system against the twisting force of the motor as it starts. This prevents unnecessary rotation of the pipe threads.

Absorbing the thrust of the motor also keeps the pump centered in the well casing.

Submersible pumps are great pumps, but remember that the pump must first be pulled from the casing for repairs.

These days the submersible pumps are of a high standard and very reliable.

Pumps can last for 25 years before requiring any form of maintenance.

Connecting A Solar Water Pump

Solar powered water pumps have become very popular as of late and are often used for agricultural operations where conventional power systems are absent. The modern solar energy appliances have been demonstrated to reliably produce electricity and are considered ideal

> Always disconnect the power before servicing a water pump and ensure that the motor is properly grounded.

Solar Pump Basics 1

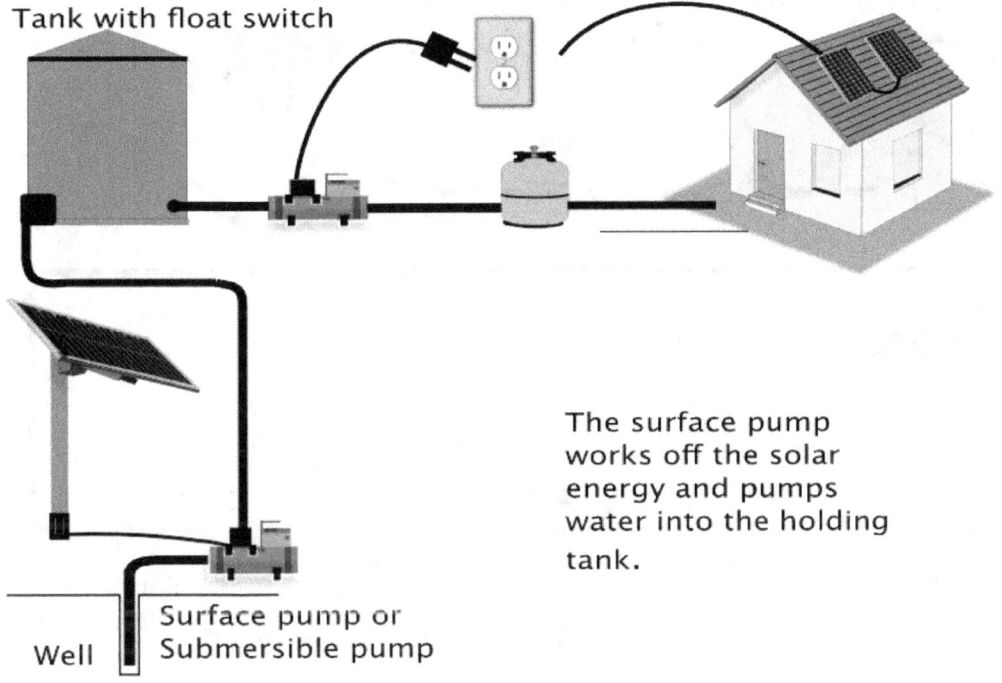

The surface pump works off the solar energy and pumps water into the holding tank.

for rural, off-grid scenarios that include irrigation, livestock and domestic use. New technology is making this a very exciting field and the future is looking bright. When buying a new solar powered water pump, make sure that you understand what you are buying, how it works and how it can be modified to suit your needs. Solar powered pumps are different from the regular AC powered pumps in that they use DC power. DC powered pumps are a very good option and they do not require large bursts of power like AC pumps do. Very much like conventional pumps, we find that solar water pumps can be divided into two categories: **Surface pumps** and **Submersible pumps**.

• Surface pumps are installed on the outside of the water source (well, spring, pond) and rely on a suction pipe to "suck" the water to the pump and then to "push" the water to the holding tank.

• Submersible pumps are installed under water (well, spring, pond) and "push" the water up to the holding tank

When installing a solar powered pump system, keep in mind that it is an integrated system that consists of various components. For best performance, buy compatible components from a supplier who is knowledgeable

Solar Pump Basics 2

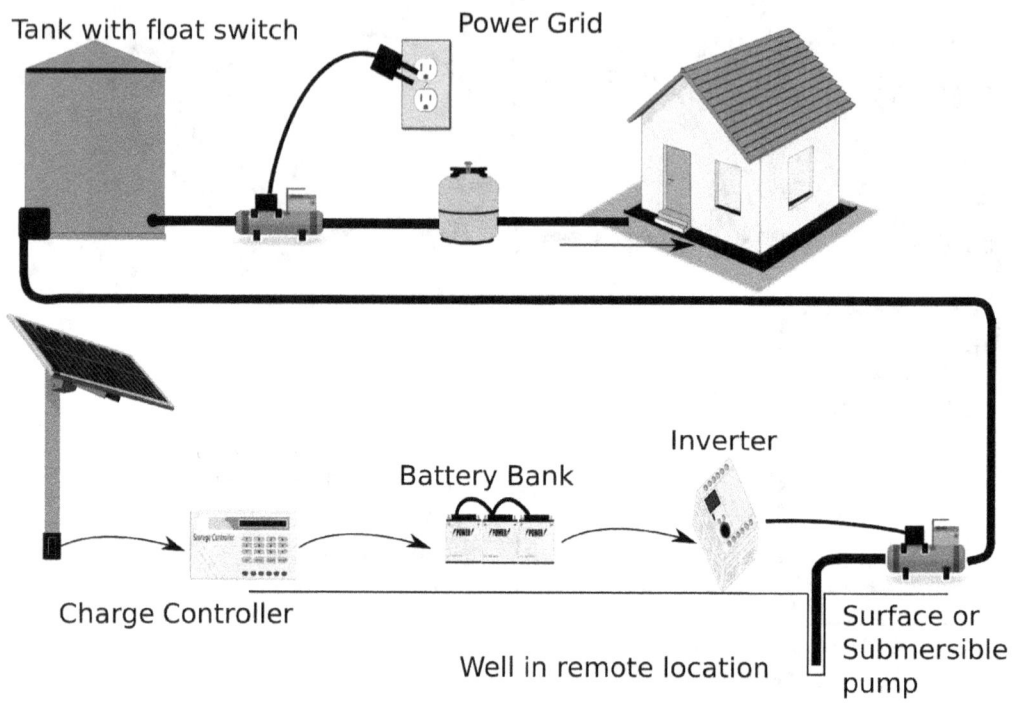

about the complete system set-up and the variables that need consideration.

In considering energy efficiency, remember that pumps are better at **pushing** than **sucking**, and always keep the distance between the pump and the water at end use as short as possible.

There are two options worth considering:

1. You can power your pump directly with energy obtained from your solar panels.
(See Solar Pump Basics 1)
This means that the pump will only work when the sun is shining. To make it a viable option, you can store your water (as you pump it) in a holding tank and use it when needed. Use a float switch in your tank to shut off the pump when full, and to turn it on when the water drops below a certain level. This is a very simple and effective option.

2. You can store the energy obtained from the solar panels in a battery bank and then rely on the batteries to power your pump.
(See Solar Pump Basics 2)
Having a battery bank allows for additional pumping during the night or when there's no sunlight available. This is a more complicated option.

If your household relies on a tank to provide water to the

Solar Pump Basics 3

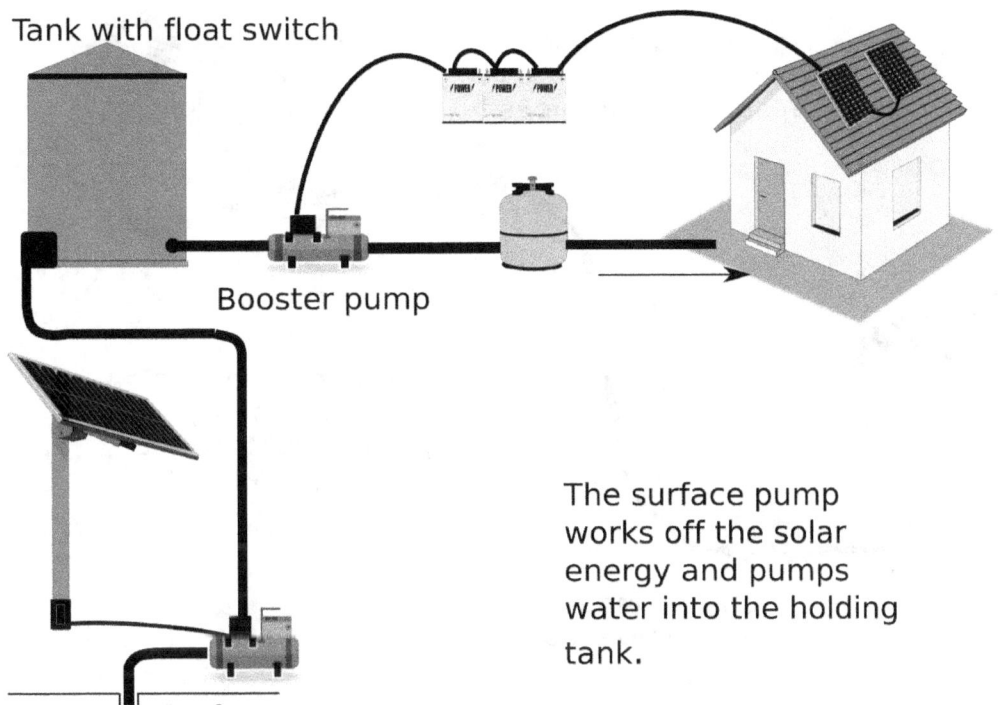

The surface pump works off the solar energy and pumps water into the holding tank.

plumbing system, then you can consider the following options:

- **Booster pump.**
(See Solar Pump Basics 3)
Solar can still provide further solutions. A DC booster pump, powered by solar panels, can assist in providing pressurized water for your home. Booster pumps are to be installed after a storage tank in the water feed line. As mentioned earlier, if your pumping installation is not properly planned, you will not receive satisfactory water service.

- **Gravity feed.** This is a system where you rely on gravity to provide water pressure in your plumbing system. Basically, the higher your tank above the outlets, the more pressure it supplies to the pipes. Gravity offers the advantage of using free energy to produce pressure, however, the tank must be elevated to a height that will deliver sufficient pressure. As a general rule, every foot of elevation provides 0.433 psi of pressure. This means that if the tank is 100 feet (30m) high, it will provide 43 psi of water pressure for your house (100' x 0.433 = 43.3 psi). That kind of elevation is not practical, but some elevation can be achieved by constructing a water tower or by placing the tank uphill of the house. Looking at the math side of things, it's

Solar Pump Basics 4

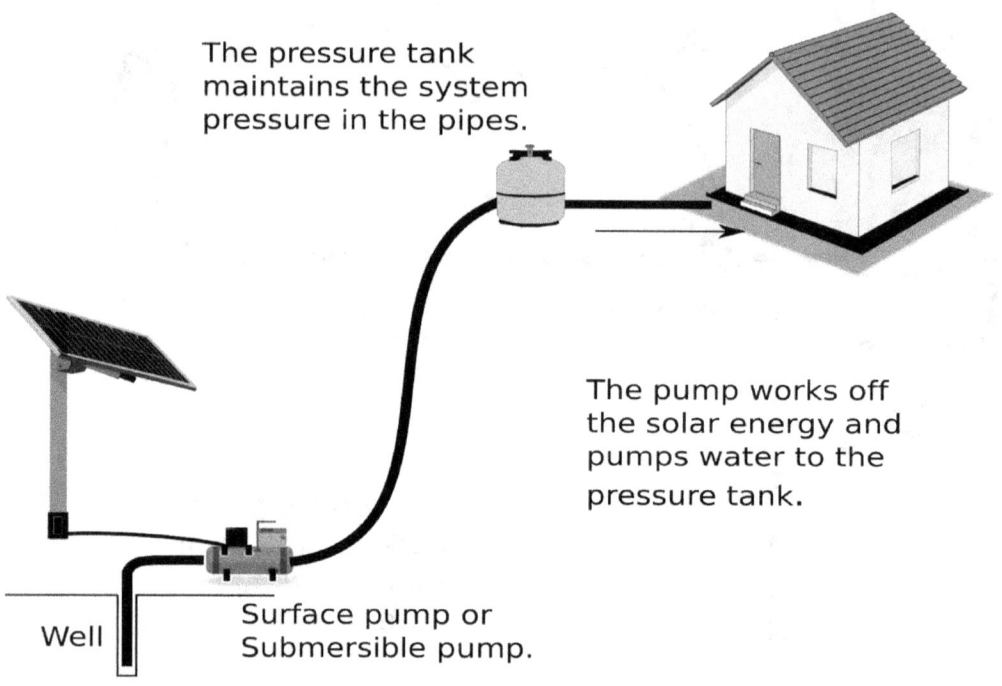

obvious that gravity feed does not offer a realistic solution for the modern home. It can still be used for irrigation and other agricultural purposes.

• **Pressure tank.**
(See Solar Pump Basics 4)
A pressure tank can also be utilized to provide water with enough pressure for your household. Water can be pumped directly from your well, with the solar powered pump, to a pressure tank where it is subsequently stored. The pressure tank is charged by the pump and this maintains pressure in the system, to the point of use in your home.

When buying your solar powered pump, you first have to calculate the lift required to get water out of your well, as well as the distance the water has to be pumped horizontally. This will determine which pump and which setup will provide the best, practical solution.

General Pump Maintenance

Always keep an eye on your well and your pump. Having a pump house, to keep your pump and well head protected from the elements, is a good option. Pay attention to the following:
• Surface pumps should be three feet above ground level to prevent contamination during

flooding and in a location with ventilation and drainage. Pumps should be allowed to breathe and not be covered in leaves or debris.
- Protect your pump and piping from freezing. If severe cold is forecast, drain your pump and remove it from the system.
- Get used to the sound of the pump motor as it pumps. A change in rhythm or unfamiliar hum may point to an issue.
- If possible, try to keep well screens clean and unclogged. This will ensure a steady flow of water and reduce friction and "laboring" in your pump.
- Make sure that your pump is securely fastened to its base. Fasten crews and make sure that there are no cracks in the surrounding cement or sealant. For submersible pumps, tether your pump to a safety rope.
- Keep all waste, chemicals and organic debris away from the system to prevent contamination.
- Be wary of any changes in water color, taste, or smell, as these and other signs may indicate a potentially dangerous issue.
- Only use galvanized steel and Schedule 40 PVC pipe and fittings for surface pumps.
- Electric pumps are not to be "experimented" with. Never tamper with the wiring of your pump. Only a licensed professional should do the work.

Conclusion

Keep things simple. Make sure that you are familiar with the operation of the pump and follow all maintenance and safety guidelines to increase the pump's longevity. A simple and efficient pump setup will include a pump that feeds a storage tank with a floating switch. A second booster pump can be used to provide pressured water from the storage tank to the household plumbing.

Before installing a new pump, check the performance rating chart to make sure it can provide you with the pumping capacity and pressure that you require. Typically, 3-4 gpm is a minimum acceptable flow rate per outlet. Note that pumping capacity and pressure are inversely related. This means that when one goes up, the other goes down.

Example:
Under normal conditions, a pump can produce 10 gpm at 20 psi. If you are using two water consuming appliances, each will receive 5 gpm of water. If we increase the pressure to 30 psi, we might end up with only 6 gpm, which means 3 gpm per appliance.

Conclusion:
An increase in pressure, caused a lower flow rate in gpm at each water using appliance. Sizing and installing a well pump can be a complex process and since pumps are not cheap it's best to consult a licensed professional who can guarantee the right

pump for your needs.

Home Water Treatment Setup

Once you have a pump installed you can consider designing a complete home water treatment system. The main aim of such a system is:
- To feed the interior plumbing with water from your holding tank.
- To filter and purify incoming water for domestic needs.
- To provide a maintenance platform for your water's treatment. Before you design such a system, you will have to determine your budget, how much energy (power) is available, and which contaminants need to be removed from your water.

Filters. You can use a system of pre- and post-filters to get rid of sediment and other contaminants. Filters come with either a mesh or a micron rating.

A wire mesh system indicates how many openings there are in one inch of screen. It can vary from inch to inch and is not very accurate.

A micron system is more accurate and represents the actual pore size of the filter; 1 micron (1μ) = 0.001 mm.

Filters calibrated in microns are intended to catch very fine particles of minute size. They are normally used at the end of the water line, before your water enters the indoor plumbing. The main thing with filters is that they slow down the flow of the water as it moves though the water line. This puts pressure on your pump and all the other treatment systems. To alleviate the influence of filters, most domestic systems place the largest filters (big pore size) first in the water line and the smallest ones last.

In a standard setup, we find that the first filter is a filter with large pores which will catch larger particles. A simple spin down filter of 100 mesh will do. A 20 micron filter will work for a system dealing with finer particles. This pre-filter will need cleaning often.

You will need a **booster pump** to provide water to the appliances in your house. It should provide between 30 and 80 psi of pressure for optimum results. An automatic pressure sensor will regulate the on and off cycle of the pump.

A DC pump is very economical and only uses electricity when in use. These on-demand pumps are very energy-efficient and you can run it off a battery if needed.

An AC pump will also work, but uses more power.

Make sure that you have stop valves on entrance and exit points of your pump to allow for maintenance when needed.

The second filter catches slighter smaller particles and although the water is not potable yet, it should be clear of

Basic Home Purifying System

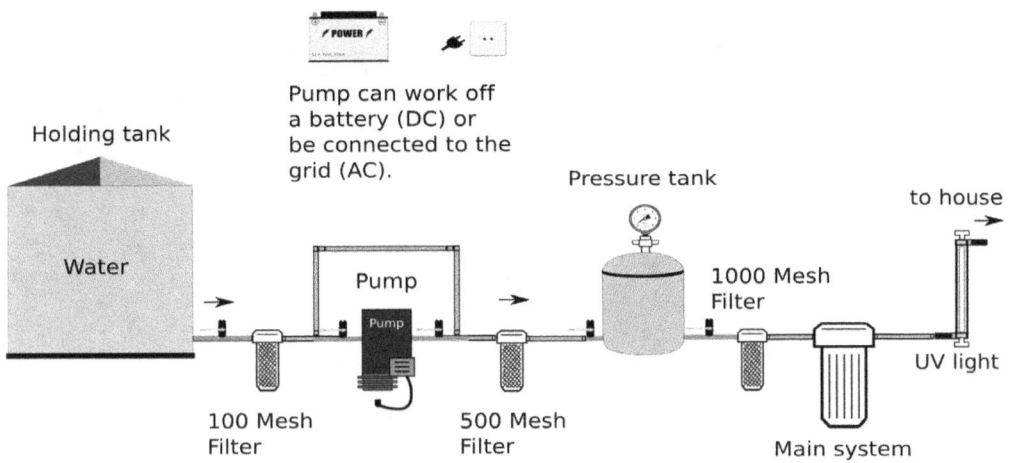

Filter options
1. Standard 100/ 500/ 1000 mesh.
2. Fine 20/ 5/ <1 micron.

Main system options:
1. Ceramic candle filter
2. Reverse osmosis sytem
3. Carbon filter

visible sediment. At this point in the feed line, the water can be diverted to showers, toilets and washing machines, except if you have known pathogens in your water that have to be removed.

Next in line is your pressure tank that will provide pressure to the indoor plumbing and minimize pump usage. Make sure you size your tank according to your pump specifications.

The third filter should be the final one that takes care of all sediment. The water should be clear of hard particles and only dissolved chemicals and pathogens will require removal from here on out.

You **main water treatment system** can be a ceramic candle filter, a reverse osmosis system or a specific filter for your individual needs. It should render your water safe to drink and give you the peace of mind to use it on a daily basis with confidence.

Finally, a UV light can also be considered to get rid of viruses. When utilizing such a light, first make sure that all sediment has been removed, for pathogens can hide behind particles and bubbles causes by turbulence. If specific impurities like dissolved chemicals need to be removed, then specialized filters can be employed in the final stage the treatment process.

See Chapter 8 for more information on the various treatment systems available.

Chapter 7
The Basics Of Managing A Well

Keep in mind that a well is an investment that can last for years if properly taken care of. From "day one" you will have to be involved in the whole process of creating a sustainable well with a healthy life-expectancy. Start with simple observations, measurements and calculations.

a.) While under construction keep notes of time of year, water table depth, soil quality and yield.

b.) Once your well is up and running keep records on productivity, seasonal fluctuations, water testing and general maintenance.

These basic observations will provide you with the information to keep contaminants out of your well and also to prevent your well from running dry due to over-pumping.

Managing your well means that you will also have to be prepared in case a flood event or an environmental disaster strikes your area.

These are unexpected events and they usually catch people off guard, but by simply being prepared, the impact on your water supply can be minimized.

The responsibility to manage a well lies squarely on the shoulders of the owner.

A well can cause problems on many levels not just for the owners, but also for members of the surrounding community.

Remember that your well uses the same local groundwater as your neighbors and that your actions will affect other groundwater users as well.

Managing Your Well

No matter which method you used to create your well, you still have to manage and maintain the structure.

Well maintenance can protect your investment, prolong the life of your well and also protect you and your family against possible contaminants.

Protect Your Well

According to the Centers for Disease Control and Prevention (CDC), here are some steps you can take to help protect your well:

- Wells should be checked and tested ANNUALLY for mechanical problems, cleanliness, and the presence of certain contaminants, such as coliform bacteria, nitrates/nitrites, and any other contaminants of local concern, (for example, arsenic and radon).
- Well water should be tested more than once a year if there are recurrent incidents of gastrointestinal illness among household members or visitors and/or a change in taste, odor, or appearance of the well water.
- All hazardous materials, such as paint, fertilizer, pesticides, and motor oil, should be kept far away from your well.
- When mixing chemicals, do not put the hose inside the mixing container, as this can siphon chemicals into a household's water system.
- Consult a professional contractor to verify that there is proper separation between your well, home, waste systems, and chemical storage facilities.
- Always check the well cover or well cap to ensure it is intact. The top of the well should be at least one foot above the ground.

Call An Expert

Along with the points previously listed, the following are indicators of when a water well systems professional should be called to evaluate the condition of your well:

1. Anytime that the well cap or seal has to be removed to look inside the well.

2. If you experience taste or odor problems, then you should contact a professional to have your water tested

3. If you experience turbidity or cloudiness of water caused by presence of suspended matter

("dirty" looking).

4. If there is a loss of capacity or pressure—the well is not producing as much water as previously produced, the pressure drops and surges, or the pump cycles on and off frequently.

5. If a test is positive for total coliforms, anaerobic bacteria, or any positive test results indicating a potential health concern. Contact a professional or your local or state regulatory agency if you experience any positive test results or believe your well has been contaminated. A water well systems professional should be hired to thoroughly clean and disinfect any well that has had a positive "anaerobic" bacterial test result, which should include removal of any pumping equipment and evacuation of the well to its bottom to be sure of maximum removal of anaerobic growth.

6. When you find defects with your wellhead, the wellhead area, or the overall water system during your routine inspections and find that you do not have the proper tools and/or knowledge to fix the issue(s). A mistake in self-servicing your well can cost you thousands of dollars, cause personal injury or death, damage personal property, and leave you and your family without water until a qualified contractor can be on-site to fix the issue(s).

Once your well has reached its serviceable life, have a licensed or certified water well driller and pump installer decommission the existing well and construct a new well. Never dispose of any chemicals or organic waste inside your old, decommissioned well.

Visual Inspections

According to the National Ground Water Association, to monitor your well's performance, you will have to carry out specific visual inspections:

- The Well Casing (pipe protruding from the ground) must be examined. Check the general condition and check if the casing extends at least 12 inches above ground. If not, hire a qualified professional to investigate remedial action.

- The Well Cap (cap on top of casing) must be present. Check the condition of the cap and any seals present, that it is securely attached, and that it will keep out insects and rodents. If not, hire a qualified professional to repair or replace the well cap.

- Survey the area above ground, surrounding the well. Check the location relative to potential sources of contamination, flooding, and physical dangers. Look for chemicals such as paint, fertilizer, pesticides, or motor oil. Remove all unwanted contaminants if possible.

- It is important to maintain at least 100 feet between the well and any kennels, pastures, feeding areas, or livestock operations. Ensure a proper distance is maintained from buildings, waste systems, or chemical storage areas (including fuel tanks)—a water well systems professional should know local codes and requirements. If there is any concern, contact your water well systems professional or your local health department.

- The ground surrounding the wellhead should slope away towards a storm-water drain. This is to divert any surface water run-off and to prevent contamination. If not, consider allowing for the ground to slope away from the well casing, while maintaining a proper height of the casing above the ground. A water well systems professional can extend the height of the casing if needed.

- You should have a concrete apron around your well casing. If there is no concrete pad surrounding the well casing, contact your local health

Annual Water Well Checkup

The National Ground Water Association suggests that at a minimum, wells should be evaluated annually by a licensed or certified water well systems professional.

An annual water well checkup should include:

- A flow test.
- A visual inspection.
- A water quality test for coliform and anaerobic bacteria, nitrates, and anything else of local concern.
- Checking of valves.
- Performing electrical testing.

A written report should be delivered to you following completion of an annual checkup. The report should include recommendations and all laboratory and other test results.

Keep this in a safe place with all other well reports.

department or regulatory agency to determine if one should be installed by a water well systems professional.

- If the well is equipped with a vented well cap, check for the presence of the vent screen in the well vent, and clear away any debris that has accumulated on the vent screen. Check the condition of the vent screen to ensure it can prevent insects and animals from entering.

- Any growth of weeds, trees, shrubs, or grasses with root systems within 10 feet of the well should be physically removed.

- Avoid the use of chemicals or herbicides near the wellhead.

- The well should not be in a roadway or driveway. If it is within close proximity to a roadway or driveway, it should be properly marked to avoid being hit by vehicles. You can protect your well by placing bollards or bump posts near the well's vulnerable position(s) to increase visibility.

- Special care should be observed if your well is located near a driveway and you live where it snows. Wells easily disappear below snow cover and are easily overlooked when pushing snow takes place.

- Be conscious of any other

You should have your well evaluated annually by a licensed professional to ensure that you get the most out of your investment.

potential threats to the wellhead—garages, ATVs, sledding hills, debris, dirt, surface water, fuels and chemicals (including fertilizers), and runoff water from kennels, pastures, or feedlots.

- Keep the well out of dog runs or animal pens.

- If your well is located in a low-lying area prone to flooding, you should consider having your water well systems professional raise the well casing to at least 12 inches above the historic record flood level, properly sealing the wiring conduit and providing casing bump protection if floating debris is a concern. Another option is to construct a new well at a location outside the flood-prone area.

Disinfecting A Well

The Centers for Disease Control and Prevention (CDC) made the following guidelines available to the public, concerning well disinfection after an emergency. Don't forget that plugging or capping your well before a disaster is a great way of preventing or reducing contamination.

If extensive flooding has occurred or you suspect that the well may be contaminated, DO NOT drink the water. Use a safe water supply like bottled or treated water.

Safety Precautions

The following safety precautions must be followed before you can disinfect your well:

1. Inspect the area around the well. Remove suspected contaminants that are visible.
2. Turn off all electricity to the well. Don't attempt to fix any machinery if not qualified.
3. Clear hazards away from the well.
4. Do not enter the well pit. Gases and vapors can build up in well pits, creating a hazardous environment.
5. Wear protective goggles or a face shield when working with chlorine solutions.
6. When mixing and handling chlorine solutions, work in well-ventilated areas and avoid breathing vapors.
7. Warn users not to drink or bathe in water until all the well disinfection steps have been completed.

How To

To disinfect your well, follow these steps:

1. If the well is equipped with an electrical pump, turn off all electricity and clear debris from around the top of the well.

Repair the electrical system and pump if needed. Contact a qualified electrician, well contractor, or pump contractor if you are not experienced with this type of work.

2. Start the pump and run the water until it is clear. Use the outside faucet nearest to the well to drain the potentially contaminated water from the well and keep the unsafe well water out of the interior household plumbing. If no pump is installed, bail water from the well with a bucket or other device until the water is clear.

If the well is connected to interior home plumbing, close valves to any water softener units.

3. Use the Tables (Imperial and Metric) on the next page to determine the amount of liquid household bleach needed to disinfect the well. Use only unscented bleach.

4. Using a 5-gallon bucket, mix the bleach from the Table with 3-5 gallons of water (12-19L).

5. Remove the vent cap. Pour the bleach water mixture into the well using a funnel. Avoid all electrical connections. Attach a clean hose to the nearest hose bib and use it to circulate water back into the well for thorough mixing.

6. Rinse the inside of the well casing with a garden hose or bucket for 5-10 minutes.

7. Open all faucets inside the home and run the water until you notice a strong odor of chlorine (bleach) at each faucet. Turn off all faucets and allow the solution to remain in the well and plumbing for a minimum of 12 hours.

8. After at least 12 hours, attach a hose to an outside faucet and drain the chlorinated water onto a non-vegetated area such as a driveway or storm water drain. Continue draining until the chlorine odor disappears. Avoid draining into open sources of water (streams, ponds, etc.).

9. Turn on all indoor faucets and run water until the chlorine odor disappears.

10. Wait at least 7-10 days after disinfection, then have the water in your well sampled. Water sampling cannot be done until all traces of chlorine have been flushed from the system.

11. Sample the water for total coliform and either E. Coli or fecal coliform bacteria to confirm that the water is safe to drink.

If the results show no presence of total coliforms or fecal coliforms, the water can be considered safe to drink from a microbial standpoint.

Follow up with two additional samples, one in the next 2 to 4 weeks and another in 3 to 4 months.

12. Check the safety of your water over the long term, continue to monitor bacterial quality at least twice per year or more often if you suspect any changes in your water quality.

If results show the presence of any coliform bacteria, repeat the well disinfection process and re-sample. If tests continue to show the presence of bacteria, contact your local health department for assistance.

Take note!

Fuel and other chemical releases and spills are common during flood events.

If your water smells like fuel, has a chemical odor or if you live in an area where the potential for a release of fuels, pesticides, or chemicals is high, contact your local health department for advice.

Water contaminated with chemicals will not be made safe by boiling or disinfection.

Until you know the water is safe, use bottled water or some other safe supply of water.

Managing your well and routinely scheduling annual check-ups, is in the best interest of your family and of the whole community.

Disinfecting A Drilled Or Driven Well (Imperial)

- Use only unscented household liquid chlorine bleach.
- Bleach concentrations are generally 5% - 8.25%.
- Quantities given in this table are approximate and are rounded to the nearest practical measurement.
- Amounts given are calculated in accordance with reaching a chlorine concentration of >100 mg/L.

Key to abbreviations:

1 cup = 8 fluid ounces = 16 tablespoons (tbsp)

1 gallon (gal) = 16 cups

Well Disinfection Table: Approximate Amount Of Bleach

Depth of water	Inside Diameter of Well casing (Inches)						
	2"	4"	6"	8"	10"	24"	36"
10 feet	¾ tbsp	3¼ tbsp	½ cup	¾ cups	1¼ cups	7 cups	1 gal
20 feet	1½ tbsp	6½ tbsp	1 cup	1½ cups	2½ cups	14 cups	2 gal
30 feet	2¼ tbsp	9¾ tbsp	1½ cups	2¼ cups	3¾ cups	1¼ gal	3 gal
40 feet	3 tbsp	13 tbsp	2 cups	3 cups	5 cups	1¾ gal	4 gal
50 feet	3¾ tbsp	1 cup	2½ cups	3¾ cups	6¼ cups	2¼ gal	5 gal
100	7½ tbsp	2 cups	5 cups	7½ cups	12½ cups	4½ gal	10 gal

Disinfecting A Drilled Or Driven Well (Metric)

- Use only unscented household liquid chlorine bleach.
- Bleach concentrations are generally 5% - 8.25%.
- Quantities given in this table are approximate and are rounded to the nearest practical measurement.
- Amounts given are calculated in accordance with reaching a chlorine concentration of >100 mg/L.

Key to abbreviations:

1 meter (m) = 100 centimeters (cm)

1 liter (L) = 1000 milliliters (ml)

Well Disinfection Table: Approximate Amount Of Bleach

Depth of water	Inside Diameter of Well casing (Centimeters)						
	5cm	10cm	15cm	20cm	25cm	60cm	90cm
3 m	12 ml	48 ml	118 ml	177 ml	296ml	1.66 L	3.78 L
6 m	24 ml	96 ml	236 ml	354 ml	592ml	3.32 L	7.56 L
9 m	36 ml	144 ml	354 ml	531 ml	888ml	4.98 L	11.34 L
12 m	48 ml	192 ml	472 ml	708 ml	1.18 L	6.64 L	15.12 L
15 m	60 ml	240 ml	590 ml	885 ml	1.48 L	8.30 L	18.90 L
30 m	120 ml	480 ml	1.18L	1.77L	2.96 L	16.60 L	37.80 L

Disinfecting A Dug Well (Imperial)

- Use only unscented household liquid chlorine bleach.
- Bleach concentrations are generally 5% - 8.25%.
- Quantities given in this table are approximate and are rounded to the nearest practical measurement.
- Amounts given are calculated in accordance with reaching a chlorine concentration of >100 mg/L.

Key to abbreviations:

1 cup = 8 fluid ounces = 16 tablespoons (tbsp)

1 gallon (gal) = 16 cups

Well Disinfection Table: Approximate Amount Of Bleach

Depth of water	Inside Diameter of Well (Feet)					
	0.5'	1'	2'	3'	4'	5'
10 feet	½ cup	1¾ cups	7 cups	1 gal	1¾ gal	2¾ gal
20 feet	1 cup	3½ cups	14 cups	2 gal	3½ gal	5½ gal
30 feet	1½ cups	5¼ cups	1¼ gal	3 gal	5¼ gal	8¼ gal
40 feet	2 cups	7 cups	1¾ gal	4 gal	7 gal	11 gal
50 feet	2½ cups	8¾ cups	2¼ gal	5 gal	8¾ gal	13¾ gal

Disinfecting A Dug Well (Metric)

- Use only unscented household liquid chlorine bleach.
- Bleach concentrations are generally 5% - 8.25%.
- Quantities given in this table are approximate and are rounded to the nearest practical measurement.
- Amounts given are calculated in accordance with reaching a chlorine concentration of >100 mg/L.

Key to abbreviations:

1 meter (m) = 100 centimeters (cm)

1 liter (L) = 1000 milliliters (ml)

Well Disinfection Table: Approximate Amount Of Bleach

Depth of water	Inside Diameter of Well (Meters)					
	0.15 m	0.3 m	0.6 m	0.9 m	1.2 m	1.5 m
3 meters	118 ml	414 ml	166 L	3.78 L	6.62 L	10.41 L
6 meters	236 ml	828 ml	3.32 L	7.56 L	13.24 L	20.82 L
9 meters	354 ml	1.24 L	4.98 L	11.34 L	19.86 L	31.23 L
12 meters	472 ml	2.49 L	6.64 L	15.12 L	26.48 L	41.64 L
15 meters	590 ml	3.72 L	8.30 L	18.9 L	33.10 L	52.05L

Chapter 8

Cleaning Water For Survival

Long gone are the days when we were able to take a swig from a bottle filled with clean water coming from a mountain stream or lake. Truth be told, these days you would actually have to travel very far to find drinkable surface water anywhere on this planet. A very large part of the earth's water supply is polluted and it's not a stretch to state that it's getting worse every year. Water contamination occurs through various means, but the result is always the same: Disease, and in more serious cases, even death. This is one aspect of self-sufficiency that you want to be very knowledgeable about.

As a general rule, when unsure about the quality of your water source, test your water and make a decision to rectify the problem. Under no circumstances should you take a chance with contaminated water.

When we start to look at our water, and the different ways to clean it, we have to make sure that we understand what type of contaminants we are dealing with. This requires clarity about what exactly needs to be removed from the water before we can drink it.

We can distinguish:
- Chemical contaminants.
- Bacteria and Protozoa.
- Viruses.

If you live in a remote, wilderness area surrounded by nature, chances are that your only issue with surface water will be biological contaminants. Keep in mind that water from a well, no matter how deep, can still contain various chemicals that occur naturally in the ground and that are harmful to human health.

Cleaning Your Water

We live in a period where, depending on your budget, technological advancements are providing us with a multitude of options to treat our water.

These days water can be purified to a degree of 99.999 percent, which pretty much guarantees potability.

If you happen to live near an urban population and near heavy industry, you will probably have to deal with biological and chemical contaminants. Sadly, most of our water supply has been contaminated with toxic chemicals found in pesticides, human waste, dumping of waste products, leaching, etc. If you are concerned that your area might have fallen victim to some of these issues, there is a possible solution at hand.

Almost every country has something that equates to the Environmental Protection Agency (EPA) or the Centers for Disease Control and Prevention (CDC). They will have information about local drinking water standards on their websites and also in print at your local water department office. They should have up-to-date **annual water qual-**

Water Glossary

Adsorption. The adhesion of ions or molecules from a liquid, or dissolved solid to a surface. This process creates a film of the adsorbate on the surface of the adsorbent.

Desalination. The removal of salts and minerals from a target substance.

Filtration. Various mechanical, physical or biological operations that separate solids from fluids by adding a medium through which only the fluid can pass.

Halogens. A group of elements that contain the chemical disinfectants iodine and chlorine.

Ions. An atom or molecule with a net electric charge due to the loss or gain of one or more electrons.

Micro-filtration. A type of physical filtration process where a contaminated fluid is passed through a special pore-sized membrane to separate microorganisms and suspended particles from process liquid.

Ultra-filtration. A type of membrane filtration in which pressure forces a liquid

ity reports for your area and this includes the following:

- The source of the drinking water. River, lake, aquifer, etc.
- A summary of the risk of contamination of the local water source.
- The regulated contaminants found in local drinking water.
- Potential health effects.
- Information on nitrate, arsenic and lead in areas of concern.
- Information on Cryptosporidium, a parasitic protozoan that causes disease.

Take note that, testing water at its source is necessary and a priority, but also testing it, as it

First determine what is in your tap water, then decide on which method is best to get rid of the contaminants.

exits your faucet, is even more important. Water is transported through water pipes, some which are remnants from days gone by. Pipes were still made from asbestos cement not too long ago. The earth expands and contracts and pipes crack. At end use people are finding harmful chemicals coming out of their faucets like lead, arsenic and asbestos. It all comes down to knowing your water source, how it's transported to

Water Glossary

against a semipermeable membrane. Solids are retained, while water and low molecular weight solutes pass through the membrane.

Nano-filtration. A membrane filtration process used with water such as surface water and fresh groundwater, with the purpose of softening and removing organic and synthetic organic matter. Nanofiltration membranes have pore sizes smaller than that used in micro-filtration and ultra-filtration, but just larger than that in reverse osmosis.

Pathogen. Bacterium, virus, or other microorganism that can cause disease.

Purification. Similar to filtration, but purified water is cleansed and purified through additional purification processes, typically reverse osmosis, distillation or deionization.

Polishing filter. Polishing filters purify pretreated water, by removing trace levels of any residual contaminants.

Cleaning Water For Survival

your house and what potential contaminants are present in the water.

Have your water tested if unsure. Your local EPA website should have information on where to get your water tested. It is a painless and simple process that will give you the answers, which will ultimately determine how to treat your water.

You can also purchase an inexpensive home test kit to test your water periodically.

The simple solution to all of the above is to educate yourself and to know what you are dealing with.

Let's take a look at some effective water treatment options.

Filtration vs. Purification

There are various water "treatment" systems on the market these days. Some are marketed as "filters" and others as "purifiers". Sometimes it causes confusion when the two terms are used interchangeably, to describe the same process. They all claim to be very effective when it comes to removing dirt and chemical particles from our drinking water. They also guarantee the removal of microbes, both viruses and bacteria. Let's look at how the two systems differ.

• Filters

Filters work on a very simple principle. They allow water to

Make Smart Decisions

• Determine what kind of contaminants need to be removed.

• Decide which system is appropriate to remove said contaminants.

• Decide which system is appropriate for your water volume needs.

• Decide which system can accommodate your family with the most suitable flow-rate.

• Check the guarantee of the system.

• Consider quality. A cracked filter/ purifier is worthless.

• Only buy the appropriate filter or purifier once you have done your research.

• Check the guarantee of the system.

• Remember that you can engineer your own system. You can add a pre-filter (that filters out larger debris beforehand), UV light, silver dust, etc.

• Are filter cartridges replaceable? Do they require cleaning to unclog holes?

travel through a tiny hole (typically measured in microns) that determines which organisms can pass through. A micron system is very accurate and represents the actual pore size of the filter; 1 micron (1μ) = 0.001 mm.

Some filters use a mesh rating. A wire mesh system indicates how many openings there are in one inch of screen. It can vary from inch to inch and is not very accurate.

Membrane filters use thin sheets with precisely sized pores that prevent objects larger than the pore size from passing through.

Depth filters use thick porous materials such as carbon or ceramic to trap particles as water flows through the material.

These days there are filters that have the ability to remove viruses, bacteria and protozoa. When you see a product with a label that reads as "absolute pore size", it means that the size given is the absolute and definitive size. A product labeled as "mean pore size" means that the size given represents an average, and that some of the pores might be bigger and some might be smaller. The Centers for Disease Control and Prevention recommends buying **absolute pore size filters**, since they give you a definitive value.

Filters can consist of elements made from the following materials:

Problem Solving

When living close to nature, always be alert to suspected surface water contamination.

Evidence of contamination will be visible in vegetation and the sudden decline in animal life in your area.

Dead vegetation and animals (especially birds and frogs) are an indication that toxins are present in the ecosystem.

Granular-activated carbon, micro-porous ceramic, carbon block resin, sand, micro-filtration, ultra-filtration and nano-filtration membranes.

Filters can also use different processes (or a combination of) to remove particles:

Sieving, absorption, adsorption, ion exchange, biological metabolite transfer, cross flow membrane filtration, etc.

When buying a filter:

• Know the filter's size and what it's rated for.

• Know the pathogens and contaminants present in your area. This way you are guaranteed to get rid of the relevant unwanted microbes.

Example: If you are worried about Cholera in your water because of a suspected cholera outbreak, you will need a micro-filter with a particle size rating of 0.2-1.0 microns.

The modern filters are quite

advanced and they can filter out almost anything. Filters are simple to operate and require no holding time. They add no unpleasant taste and may even improve the taste and appearance of water.

A known issue with filters is that they are not very effective when it comes to dealing with dissolved minerals or chemicals. If you suspect the presence of such elements, consider purifying your water.

Remember than filters should be cleaned and maintained on a regular basis. As a filter clogs, increasing pressure is required to drive the water through it; this increase in pressure can force micro-particles and organisms through the filter.

Pathogens

- **Protozoa** (Size: 1-15 micron)

  Example:
  Cryptosporidium, Giardia.

- **Bacteria** (Size: 0.1-5 micron)

  Example:
  Campylobacter, Salmonella, Shigella, E. coli, Cholera

- **Viruses** (Size: 0.004-0.1)

  Example:
  Enteric, Hepatitis A, Rotavirus, Norovirus

• Purifiers

Purifiers are similar to filters in that they also use micro-filtration, but on top of that they also utilize additional purifying processes like distillation, reverse osmosis, or deionization. Some also use chemical treatment like iodine or UV light to kill viruses.

Conclusion

As we've stated before: Buy quality and avoid products from countries that are known for their cheap prices.

Stick to a brand that is known for its quality and that provides you with the following:

a.) A safety guarantee that's approved by government agencies. A typical safety guarantee we see on filters: "Tested to exceed NSF/ ANSI Standard 53 or 58 protocol."

This should also include a list of the contaminants that can be successfully removed.

b.) Clear guidelines on how to use and maintain their product which should include factors like flow rate, how often cartridges have to be replaced, ease of use, etc.

c.) A thorough explanation on how the cleaning mechanism works.

Select a brand with a reputation for integrity and quality. Companies should know when to label a product as a filter and when to label it as a purifier. If not, they are gambling with your health.

Make **cleanliness** of your water a first priority, and **taste**, a second priority. It boggles the mind how people complain about the taste of the water, but they haven't determined what's causing it. Many companies have learned that a charcoal filter does wonders for taste and they are investing heavily in these, but this is not the most important component in the "cleaning" process.

Comparing Treatment Options

When comparing different methods and systems, we have to realize that there is no perfect method out there, since it all comes down to the needs of the consumer. Some systems use electricity, some use gravity and others use solar power. Some will suit well owners, others will not. In the end, know that you do not need to be stuck with one single option, but that you can also combine different systems and methods.

1. Boiling Water

Boiling is very effective at killing microorganisms. It is actually still one of the safest methods out there and when in doubt, boil it!

Common sense dictates that boiling does not remove harmful chemicals. These chemicals are invisible to the naked eye and more often that not, their effects are only felt years later when they have entered and damaged internal tissue and cells. Do not drink water with suspected chemical toxins.

Boiling water is one of the simplest and most effective ways of purifying biologically contaminated water for drinking.

Mechanics. When boiling water, we need to use a container like a pot, can or pan. Any canister made of heavy glass or metal will do. Remember to wash your hands and to remove debris and sediment before you boil the water.

The following applies as a standard rule of thumb:

- 30 min of heating at 70C or 160F kills all pathogens.
- 10 min of boiling at +85C or 185F kills all pathogens.
- The best solution: Boil it at 100C or 212F for one minute.

Keep in mind that the boiling point of water varies according to the atmospheric pressure. This means that water boils at lower temperatures at higher altitude. At elevations above 6,500 feet, boil your water for 3 minutes.

Upside

- Boiling does not impart additional taste or color.
- It is a single step process that

inactivates all enteric pathogens.

• Efficiency is not compromised by contaminants or particles in the water as is the case with halogenation and filtration.

Downside

• Does not improve taste, smell or appearance of water.

• Fuel sources may be scarce, expensive, or unavailable.

• Does not prevent recontamination during storage.

2. Faucet/ Sink/ Desktop/ Pitcher Filters

These filters come in all shapes and sizes. Replacing these filter cartridges has become a big industry and the emphasis is on replacing, which means revenue.

Mechanics. There are different models out there.

- Some use simple gravity filters. As the water flows through different layers of filtration material, the unwanted particles are simply left behind.

- Similar products rely on water pressure to separate contaminants from the clean water.

- The most common household water filters utilize activated carbon granules. An individual charcoal particle has a very large internal surface area. This area is unique in that the surface is filled with nooks and crannies, that attract and trap unwanted particles through a process called adsorption. This separation action provides clean water for your faucet.

These filters are often used in the city where the water coming out of your faucet has already been filtered and purified. If the water in your area has already been cleaned by the local water works, then chances are that you will not have high amounts of contaminants. Following this logic, it's very possible that these filters simply just make clean water, cleaner.

In a nutshell, these filters are effective at improving the taste of water, but are not considered a solution for someone looking at cleaning contaminated water from a well, river or rooftop. Do research before buying.

When buying any of these filters:

- Check which contaminants they remove from the water (usually lead, mercury and pesticides).

- See if they are certified by an independent water quality association like the WQA Gold Seal.

- Check how often you have to replace the cartridges.

- The predicted flow-rate.

Upside

• Filters are easy to replace.

• Provides basic filtration only.

• Water tastes better.

• Easy to store for emergencies.

Downside

• Predominantly slow when it comes to flow-rate.

- Expensive in the long run.
- Cartridges get easily clogged and have to be changed every 1-3 months.
- Only suitable for a small family of two.

3. Reverse Osmosis Filters

Reverse Osmosis Systems incorporate a 4-5 stage process of filtration and purification and is considered one of the best systems out there. This system is used world-wide by governments and industries to purify water for various purposes.

Mechanics. Pressure is applied to a contaminated solution of water. This pressure causes the water to pass through a membrane to the other side. When the contaminants hit the membrane they are held back and only the clean solution of water is allowed to pass. These systems rely on various pre- and post-filters that are used in conjunction with the reverse osmosis membrane. The filter material has a very small pore size of around 0.0001 micron. This is considered a very effective method of cleaning your drinking water.

Upside

- Effective at removing protozoa, bacteria, viruses and chemicals.
- Will reduce a greater percentage of Total Dissolved Solids (TDS), including up to 90% of fluoride. This is one benefit that sets it apart from most other systems.
- Cartridges last for months and do not easily get clogged.
- Suitable for a large family due to decent flow rate.
- Softens hard water which causes scale buildup.
- Improves taste.
- Can be easily combined with other systems (UV light, polishing filter) to remove or kill all pathogenic waterborne microbes.
- Reverse osmosis also removes radioactive plutonium or stron-

Problem Solving

When your main filtration system has a very small pore size, flow-rate can be a problem. One solution is to install a pre-filter on your line, that will get rid of sediment and larger particles. This will help with keeping your main filter unclogged and to speed up the general filtration process.

If you use well water with heavy sediment, you can install 3 pre-filters, all with different pore size ratings.

First in line should be a large 20 micron filter followed by a 10 micron filter and then a 5 micron filter. This helps with keeping the whole system balanced and reduces the need for constant filter replacements.

tium in drinking water. If you live near a nuclear power plant, this is the system to use to guarantee clean water.

Downside
- It's not cheap.
- The system wastes water.
- Removes wanted minerals from water.
- This system needs electricity.

4. Reservoirs & Removable Gravity-Fed Filters

These systems are becoming more and more popular by the day. The reservoirs are made from high grade stainless steel and come in a range of sizes. The filters are made from different media types that consist of millions of little micro-pores.

Different filter element options are available and the filters are easy to clean.

Mechanics. Several methodologies are utilized by these filters. First, the contaminated water is manually poured on top into a reservoir, from where it travels through a system of gravity-fed filters. You can use a basic set-up of one filter, or you can use two or four filter combinations to remove specific chemicals (fluoride) from your water.

Contaminants are filtered, adsorbed and absorbed. Heavy metals ions (mineral molecules) are extracted through an Ion exchange process where they are essentially electrically bonded to the media.

Finally, the water is collected in the lower reservoir where it can be accessed via a faucet.

Upside
- No electricity needed.
- Portable system and easy to keep hygienically clean.
- Very decent flow-rate. Depending on filter setup, 3.5-7 gallons per hour.
- Filters are removable and they last fairly long (3000 gallons per filter).
- Effective at removing heavy metals, protozoa, bacteria, viruses (most) and chemicals like chlorine and volatile organic compounds (VOCs).
- You can combine different kinds of filters with these systems.
- They are very handy in that the containers come in different sizes and the gravity-fed filters are purpose built to cater for your filtration needs.
- It is a very visible process that works on simple, yet efficient, technology.
- Adds no unpleasant taste and often improves taste and appearance of water.
- This system brings you closer to your water due to the fact that you can replace filters yourself and that you can decide which filters are necessary at any given period.
- Can be easily combined with

other systems (UV light, chemical disinfectants) to remove or kill all pathogenic waterborne microbes.

Downside

• You have to manually refill the reservoir.

• Eventually filters clogs from suspended particulate matter and may require some maintenance.

• Fairly expensive.

5. Ceramic Multi Candle Filter

A filter option that has to be discussed is the multi candle replaceable filters. These are very reliable filters and can be used to service the whole household.

Mechanics. Up to 6 ceramic filters are connected to a filter holder module and placed inside a housing. The filter system is designed to be fitted to the main municipal water inlet to the home, to supply large volumes of clean water. Ceramic water filters work by simply allowing the water to filter through millions of tiny pores in the water cartridge surface. This process catches the particles that are too large to pass through, typically organic and inorganic compounds, and they subsequently gather on the filter surface. As a second line of defense, the inner workings of the ceramic consists of an intricate maze with a complicated structure. This is where even smaller particles, that passed through the cartridge surface, are trapped. Certain ceramic filters also incorporate a compound made of colloidal silver into the ceramic, and this compound keeps the ceramic surface clean because it repels bacterial growth.

> Don't underestimate the toxicity of chemicals. Most chemicals are carcinogenic and have the potential to cause cancer later on.

Upside

• Removes up to 99% of pathogenic bacteria including salmonella, cholera and E. Coli, as well as cysts like cryptosporidium and giardia, sediment, and various organic chemicals.

• High flow rate. Depending on brand, these filters offer superior flow rate of up to 520 gallons (2000 liters) per hour (8 gal per minute) at 45 PSI (3 bars).

• Filters come with a 0.9 microns absolute filtration rating with 99.99% efficiency.

• Considered one of the most efficient systems for cysts removal compared to any available alternative technologies such as chlorine, ozone, UV, reverse osmosis, etc.

• Can provide whole house filtration where drinking water is desired.

• Suitable for homes drawing

Pathogen Distribution

Viruses spread primarily through animal and human fecal matter. They can survive for a long time, but can only reproduce once inside a host. Typical viruses found in water: Norovirus, Rotovirus and Hepatitis A.

Bacteria are found where nutrients and temperatures allow. There are three types of indicator bacteria to test for:

- **Total Coliform.** Commonly found in the environment and harmless if found on its own. Having this bacteria indicates that there is a gateway for bacteria to enter and the problem must be resolved by finding the source.
- **Fecal Coliform.** Commonly found in feces of animals and people. Indicates that fecal contamination is present and that the drinking water is at risk.
- **E.coli.** Most are harmless and present in the fecal matter of humans and animals. Some strains will cause serious illness. If found in drinking water, it means that there is a risk of harmful germs present. It indicates fecal contamination.

Protozoa are observed in freshwater and in marine environments. They can remain dormant as cysts in surface water for long periods of time. They are also found in animal and human waste.

water from lakes, streams or micro-biologically contaminated wells.
- Small footprint for parallel installation for large volume safe drinking water.
- Can be easily combined with other systems (UV light, gravity filter, chemical disinfectants) to remove or kill all pathogenic waterborne microbes.
- Adds no unpleasant taste and often improves taste and appearance of water.

Downside
- Requires electricity for flow of water.
- You have to manually replace the filters.
- Eventually filters clogs from suspended particulate matter and may require some maintenance.
- Fairly expensive.

6. Carbon filters

Carbon filters are very popular filters and are often seen where municipal water is used. They can be used as a stand-alone filter or as a component in a reverse osmosis system. They are very effective when it comes to removing odors and bad taste from water. We distinguish two types:

Granulated activated carbon filters and **Carbon block** filters.

Mechanics. As mentioned before, an individual charcoal particle has a very large internal surface area. This area is unique in that the surface is filled with nooks and crannies, that attract and trap unwanted particles through a process called adsorption. Carbon filters are effective at removing the contaminants that bond to carbon. This separation action provides you with clean water where you need it. Some of the more advanced, modern **carbon block filters** also utilize a combination of nanomesh technology (and sub-micron filtration) combined with the chemical adsorption process. If you are using a carbon filter in combination with a sediment filter, note that the sediment filter should be used first to get rid of the heavier particles that would otherwise clog up the carbon filter. Once the heavier particles are removed, the purer and cleaner water can move into the carbon filter where the remaining unwanted compounds will bond with the carbon and be removed as well.

Upside

• Modern, advanced carbon block filters remove most pathogens such as bacteria, viruses and parasites like Giardia and Cryptosporidium. They function not just as a filter, but also as a purifier.

• Removes chemicals like chlorine, pesticides, herbicides and heavy metals such as lead, copper, arsenic and mercury.

• Removes volatile organic compounds (VOC's).

• Very efficient at removing bad tastes and odors.

• Does not remove the healthy minerals found in water.

• An affordable option.

Downside

• Does not remove dissolved solids or inorganic compounds.

• Does not remove fluoride.

• Requires electricity.

7. Distillation Systems

Distillation systems are very popular for they can create potable water from almost any source. People use distilled water for drinking and also for maintenance of machinery.

Mechanics. First the water is heated which turns it into water vapor.

Then, as the vapor condenses, it turns back into a liquid.

Finally, it gets collected in a reservoir or collection vessel as distilled water.

This process allows only the vapor to escape, and the contaminants to remain behind.

Upside

• Distillation is a method that is effective at removing protozoa, bacteria, viruses and most chemical contaminants.

• It is also effective at removing salt from saltwater.

• It removes the "hardness" of hard water.

Downside

- Some statistics consider distilled water to be too clean and void of essential minerals and ions. There are no definitive answers surrounding this issue.
- The process of distilling water with a DIY system is a slow one.
- The yield is considered minimal.
- They do require electricity which might be an issue in an emergency.

Buying a commercially available unit is money well spent.

See Chapter 12 for a complete explanation of how homemade distillation units are made.

Water Sanitation

Viruses: E.g. enterovirus, hepatitis A, Norovirus, Rotavirus

Can cause the following: Gastrointestinal illness (for example, diarrhea, vomiting, cramps), hepatitis, meningitis.

Sources of viruses in drinking water are: Human and animal fecal waste.

Methods that may remove some or all of viruses from drinking water are:

- **Boiling.** Rolling boil for 3 minute minimum has a very high effectiveness in killing viruses;
- **Standard Filtration** is not effective in removing viruses. Modern Nanofiltration techniques can remove some viruses successfully.
- **Chemical disinfection with iodine** or chlorine has a high effectiveness in killing viruses;
- **Chemical disinfection with chlorine dioxide** has a high effectiveness in killing viruses;
- **Disinfection** has a high effectiveness in killing viruses when used with iodine, chlorine, or chlorine dioxide.

8. Ultraviolet Treatment Systems

Another very simple method to purify your water is to use a short-wave ultraviolet light. Water flows in the one end of the cylinder, is treated with Ultraviolet rays, and flows out the other end.

The wavelength range of short-wave UV is between 2000-3000 Å and this is considered the most effective range for deactivating harmful pathogens. It does not get rid of the microorganisms, it just renders them harmless.

Short wave UV has to be produced through electricity, since it does not naturally occur in the environment around us. These UV units are very convenient and are usually mounted vertically against a wall.

Mechanics. UV Water Purification systems use special lamps that emit UV light that can disrupt the DNA of micro-organisms (pathogens). Specific wave lengths of light have the ability to affect specific organisms. When organisms in clear water are exposed to UV light, their genetic code comes under attack and this eliminates the microorganism's ability to function and reproduce. A pathogen that cannot replicate, also cannot infect other organisms. This process inactivates a large percentage (up to 99%) of harmful microorganisms without adding any chemicals to the water.

Upside

• An ultraviolet light is very effective at deactivating all waterborne pathogens; protozoa, bacteria and viruses.

• Very simple installation and low maintenance. Just change the UV light bulb annually.

• Great addition to any existing filtration system. Extra doses of UV can be used for added assurance and with no side effects.

• Very reliable system that works 24/7.

• Adds no unpleasant taste.

• Does not waste any water.

• Can be easily combined with other systems (reverse osmosis, gravity filter, chemical disinfectants) to remove or kill all pathogenic waterborne microbes.

Downside

• It requires electricity, though not much.

• It only works with clear water.

• UV light degrades plastics and PVC. It's best to use non-PVC piping immediately before and after the UV system. Copper is best.

• It definitely has no effect on chemicals and toxins in your water supply.

Make sure that the water has no floating particles, since organisms can hide behind these. Turbid water will render the system useless. It is best to first use a pre-filter, to get rid of large particles, after which the water can be exposed to the ultraviolet light. A 5 micron pre-filter will do the job.

Some things to look for in a UV light.

- The length of the lamp. The longer the lamp, the longer the organisms are exposed to the UV waves.

- How often does one need to change the bulb inside?

- Does it have a protective casing and wall mount?

- Does it allow for accessories and control mechanisms like valves, monitors and alarms?

9. Ozone Purification

The use of ozone as a form of water purification has been around for more than a century. These days we find complete domestic ozone water treatment systems where people are concerned about the quality of their well or rain water. Ozone is effective as both a disinfectant and an oxidant and has become one of the most effective water

Water Sanitation

Protozoa: E.g. Cryptosporidium & Giardia

Can cause the following:
Gastrointestinal illness like diarrhea, vomiting, and cramps.

Sources of Protozoa in water:
Human and animal fecal waste.

Methods that may remove some or all of Protozoa from drinking water are:

- **Boiling.** Rolling boil for 3 minutes is very effective.
- **Filtration** has a high effectiveness in removing Protozoa when using an absolute, less than or equal to 1 micron filter (NSF Standard 53 or 58 rated "cyst reduction / removal" filter).
- **Chemical disinfection with iodine** or chlorine has no effect on Crypto. It has a low to moderate effectiveness against Giardia. Chlorine dioxide has a low to moderate effectiveness in killing Cryptosporidium and a high effectiveness in killing Giardia.
- **Combination filtration and disinfection** has a very high effectiveness in removing and killing Protozoa when used with chlorine dioxide and an absolute, less than or equal to 1 micron filter (NSF Standard 53 or 58 rated "cyst reduction / removal" filter).

treatment options available.

Mechanics. A typical ozone system has a UV ozone generator that creates tiny bubbles saturated with ozone. It is pumped into your water tank at a constant rate and circulates the contents of the tank. Ozone (O_3) is a gas that consists of typical oxygen (O_2) with one extra oxygen atom that's weakly attached. As the ozone aerates the water tank, the one extra oxygen atom attaches to any substance that can be oxidized. Once this (3rd) atom has been given up, the by-product is pure oxygen.

Upside

- Kills bacteria, viruses, spores and fungus.
- It is considered a replacement for chlorine and since it's not a chemical it leaves no residue or harmful by-products behind.
- It typically improves the odor and color of your water.
- It can oxidize metals like iron, hydrogen sulfide, manganese, and inactivates pesticides and organic material.
- It prevents the buildup of residue deposits in pipes.
- Can be easily combined with polishing filters to further remove unwanted chemical contaminants.

Downside

- Ozone will not remove nitrates, sulfates, fluoride, sodium or chlorides.

- Ozone does not last long once dissolved in water and must be pumped constantly into the water.
- These systems do require electricity (minimal).
- They are not cheap.

10. Water Softeners

We find different forms of minerals in our groundwater. These minerals are dissolved in the water, which means a standard filter will not be able to remove them. Typical symptoms of hard water is a residual buildup of scale on household appliances and also on the inside of pipes. It also prevents soap from lathering and just leaves it slimy and sticky.

Mechanics. A water softening device uses a process called "iron exchange" to remove minerals from water.

The process takes place within a tank full of negatively charged polystyrene beads. Due to the negative charge, the beads bond to the positively charged sodium ions. As the water passes past the beads, the sodium ions (+) are replaced with calcium and magnesium ions (-).

A typical water softening device will remove calcium and magnesium ions from your water and it will replace them with more desirable ions like sodium. The calcium and magnesium ions are the particles that cause the water to be "hard".

Upside
- Some advanced systems also remove heavy metals, nitrates, arsenic, chromium, selenium, sulfate, and even radioactivity.
- Water treated with a water softened system tend to be softer on appliances, laundry, hair and skin.

Downside
- Water softeners do not remove any bacteria, protozoa or viruses.
- These systems do require electricity (minimal).
- They are not cheap.

11. Land-Based Marine Filters

If you live close to the ocean and you are concerned about water availability, then you might look into installing a desalination system. They are often found on sailboats and yachts and are very effective at making seawater potable for human consumption.

These days you can install a complete setup for your house or buy a portable unit. The small, portable units are hand pumped and are for emergencies only.

Mechanics. A desalination system sucks in the seawater with a pump, and then sends it through a range of pre-filters to get rid of larger sediment.

Next, the water flows through a reverse osmosis chamber that is

Water Sanitation

Bacteria: E.g. Campylobacter, Salmonella, Shigella, E. Coli

Can cause: Gastrointestinal illness like diarrhea, vomiting, and cramps.

Sources of bacteria in drinking water are:
Human and animal fecal waste.

Methods that may remove some or all of bacteria from drinking water are:

- **Boiling.** Rolling boil for 3 minute has a very high effectiveness in killing bacteria;

- **Filtration** has a moderate effectiveness in removing bacteria when using an absolute less than or equal to 0.3 micron filter;

- **Chemical disinfection with iodine** or chlorine has a high effectiveness in killing bacteria;

- **Chemical disinfection with chlorine** dioxide has a high effectiveness in killing bacteria;

- **Combination filtration and disinfection** has a very high effectiveness in removing and killing bacteria when used with iodine, chlorine, or chlorine dioxide and an absolute less than or equal to 0.3 micron filter (NSF Standard 53 or 58 rated "cyst reduction / removal" filter).

specifically designed to remove the dissolved salt particles. This is done through high pressure filtering which sends the contaminated water through a very fine membrane at high pressure. The residual particles are left behind on the membrane and the clean water is flushed out into a collection chamber.

Upside

- Provides safe water.
- Very reliable and membrane filters last for a long time.
- Can be used for freshwater or saltwater.
- Small units available.

Downside

- Expensive and requires contractor installation.
- The output is very small. About 15% of original water volume.

12. Solar Disinfection

In an emergency situation, water can be disinfected with sunlight.

Solar disinfection (SODIS) has been proven to inactivate the waterborne microbial species (viruses, bacteria, and protozoa) that cause diarrheal diseases.

Mechanics. Users of SODIS fill 0.3-2.0 liter plastic PET soda bottles with low-turbidity water, shake them to oxygenate, and place the bottles on a roof, rack or reflective surface (such as aluminum foil) for 6 hours (if sunny) or 2 days (if cloudy).

The combined effect of thermal heating of solar light and UV radiation causes DNA damage and thermal inactivation.

These germicidal properties can inactivate disease-causing organisms.

Upside

• Proven reduction of viruses, bacteria, and protozoa in contaminated water.

• Proven reduction of diarrheal disease incidence.

• Simplicity of use and acceptability.

• No cost if using recycled plastic bottles.

• Minimal change in taste of the water.

• Recontamination is low because water is served and stored in the small narrow necked bottles.

Downside

• Need to pretreat water of higher turbidity with straining or filtration.

• Limited volume of water that can be treated all at once.

• Length of time required to treat water.

• Large supply of clean, suitable plastic bottles required.

There are a few ways that the efficiency of the system can be enhanced:

• Paint the surface underneath your bottles black to enhance solar heating.

• Place your bottles (filled with water) on a reflective surface to boost the amount of sunlight absorbed.

• You can increase the initial levels of dissolved oxygen by shaking the bottle vigorously for 30 s before sealing.

• Filter the water before you start the process. Clear water will definitely provide better results.

• Glass can be considered, but not all types of glass allow for the transmission of UV rays. Plastic bottles have the potential to leach compounds into water after exposure to strong sunlight conditions.

Take this into consideration when deciding how long and under what conditions you want to rely on this method.

Conclusion

When the time comes to make a decision about which water treatment system or method is best, you will have to consider the following:

◊ How many people will require water.

◊ The quality of the water to be treated.

◊ The availability of energy or fuel.

◊ The available space to implement the treatment.

◊ The importance of taste.

◊ Flow rate.

No method is best. Boiling is by

far the best option when worried about pathogens. Consider that filters sometimes do miss some of the viruses and that halogens (Chemical disinfectants) do not deactivate Cryptosporidium. This leads us to the conclusion that in most situations a 2- or 3-step process is preferred.

A typical solution (power source is needed) will look like this:

1. **Step one.** Look at your water quality and based on the evidence, make a decision as to what needs to be removed. If you are worried about chemicals like fluoride, chlorine or lead, then make sure that your main treatment method will take care of these.

   Alternatively, if you are using well water and you are concerned about microbes, make sure that you select the appropriate treatment method that guarantees elimination of bacteria, protozoa and viruses.

2. **Step two.** Pretreat your water to get rid of contaminants at an early stage in the feed water. This means that you have to get rid of contaminants (like sediment) that may later, down the line, affect purification equipment.

   UV lights and reverse osmosis systems should not be used if your water is contaminated with sediment.

   Pretreatment examples: sediment filters, carbon filters, and water softeners.

3. **Step three.** Use your main purification method.

   Gravity-fed filters and ceramic multi candles are very effective.

   A Reverse Osmosis system removes 90 to 99% of all the contaminants found in water and is often used as the main treatment method in homes served by well water. It effectively removes a broad range of contaminants. Don't forget to use a pressure tank for increased pressure in your plumbing system.

4. **Step four.** A polishing filter and/ or a UV light is a good option to finalize the process. The polishing filter acts as the last before water exits your faucet and it should have a small pore size that should filter out the absolute smallest of contaminants (dissolved chemicals).

Concerning filters, The Centers for Disease Control makes the following recommendations:

Protozoa

Micro-filtration has a very high effectiveness in removing protozoa. It has no effectiveness in removing bacteria, viruses or chemicals.

Examples	Required Product Type	Required Certification
• Cryptosporidium • Giardia	Micro-filtration Filter. Depending on brand, pore sizes range from: 0.1-5 micron	Tested to exceed NSF/ ANSI Standard 53 or 58 protocol.

Bacteria

Ultra-filtration has a very high effectiveness in removing protozoa and bacteria. It has a low effectiveness in removing chemicals and viruses.

Examples	Required Product Type	Required Certification
• E. Coli, • Salmonella • Cholera	Ultra-filtration Filter. Depending on brand, pore sizes range from: 0.01-0.1 micron	Tested to exceed NSF/ ANSI Standard 53 or 58 protocol.

Viruses

Nano-filtration has a very high effectiveness in removing protozoa, bacteria and some viruses. It has a moderate effectiveness in removing chemicals.

Examples	Required Product Type	Required Certification
• Rotavirus, • Norovirus, • Enteris, • Hepatitis A	Nano-filtration Filter. Depending on brand, pore sizes range from: 0.001-0.01 micron	Tested to exceed NSF/ ANSI Standard 53 or 58 protocol.

Chapter 9

Disinfecting Water With Chemicals

We can disinfect water through various chemical processes which can deactivate or kill the unwanted pathogenic microorganisms. Examples of chemical disinfectants are found in a group of elements called Halogens, of which the most common are iodine and chlorine. These chemicals are specifically used in the treatment of water for safe drinking and they come in tablet and liquid form.

In this chapter we will focus on the following chemical disinfectants available:

- Iodine.
- Unscented chlorine bleach.
- Granular calcium hypochlorite.
- Water disinfection tablets.

When using these chemicals, you have to pay close attention to the following:

- Follow the instructions provided by the manufacturer of the disinfectant.

- Pay special attention to the percentages. When choosing household bleach or iodine, make sure that you know the percentages of the active ingredients which will determine the ratio for mixing.

- Clear water and murky, brown water differ in treatment process. Disinfection does not work as well when water is cloudy or colored. It's best to filter the water before you start the disinfection process.

- Pay special attention to the temperature. Specific temperatures are required for these chemical disinfectants to be effective.

The Chemical Treatment Of Water

In an emergency situation, we should pay special attention to the quality of our drinking water. Floods, hurricanes and earthquakes are known to disrupt water supply lines and in these situations it's best to stick to bottled, boiled or disinfected water. Bottled and boiled water is by far the best option and disinfecting your water with chemicals should be a last resort.

Water Disinfection

Take note. Chemicals, are toxic in large doses and can still have an adverse affect on your health when taken in small doses, no matter which authority tells you it's safe. It is this writer's opinion that chemicals should only be used as an absolute last resort, when boiling or filtration is not an option; when you are faced with a life or death situation.

Iodine

Iodine is a strong chemical that can penetrated the cell walls of microorganisms. Iodine can kill many of the most common pathogens in natural fresh water sources, but it provides no protection against Cryptosporidium. For iodine to be effective, we have to pay special attention to the following:

1. This is a strong chemical that is light sensitive and must be stored in a dark bottle.
2. Iodine works best for water that are over 66 F (21°C). In colder water the iodine will take longer to activate and it will also be less effective.
3. Note that some people are allergic to iodine. Pregnant women, women over 50 or people with thyroid problems should also consult their doctor before ingesting water with iodine.
4. Iodine should not be used to disinfect water over long periods of time as prolonged use can cause thyroid problems.

When talking about iodine, we often see the word tincture. Tincture defined: a medicine made by dissolving a drug in alcohol: a bottle containing tincture of iodine. **Iodine comes in tincture, tablet and crystal forms.**

When using iodine tablets or crystals, you will have to follow the instructions provided by the manufacturer.

For common household iodine (or "tincture of iodine")

you can follow the following steps:

1. Locate your iodine in your medicine cabinet or first aid kit.

2. Prepare a container that is clean and that has been disinfected. A one liter or quarter gallon container in perfect.

3. Fill the container with the contaminated water. Make sure that the water is settled and clear. You can filter it beforehand with a cotton cloth or coffee filter.

4. Add 5 drops of 2% tincture of iodine to each quart or liter of water that you are disinfecting. If the water is cloudy or colored, add twice the amount (10 drops) of iodine.

5. Stir and let the water stand for at least 30 minutes before use.

Some additional information:

• Whenever possible use warm water (66 F/ 21°C) and let it stand a minimum of 20 minutes after mixing and before drinking.

• When using cold water around 40-59F (5 - 15°C) you can increase the waiting time after mixing to 40 minutes.

Chlorine Bleach

If you don't have safe bottled water and if boiling is not possible, you can make small quantities of filtered and settled water safer to drink by using a chemical disinfectant such as unscented household chlorine bleach. According to the Centers for Disease Control and Prevention (CDC), chlorinating water will kill most micro-organisms like bacteria and viruses. Unfortunately it **provides low protection against protozoa.** Some rules for purifying water with chlorine bleach:

• Do NOT use scented bleaches, color-safe bleaches, or bleaches with "added cleaners" as these will contaminate your water.

• Check the expiration date as the effectiveness of chlorine bleach is just six months.

• Do NOT use pool chlorine. It's a much stronger constitute than laundry bleach!

• The disinfection action of bleach depends as much on the waiting time after mixing as to the amount used. The longer the water is left to stand after adding bleach, the more effective the disinfection process will be.

Disinfectants can kill most harmful or disease-causing viruses and bacteria, but are not as effective in controlling more resistant organisms, such as the parasites Cryptosporidium and Giardia. (Chlorine dioxide tablets can be effective against Cryptosporidium if the manufacturer's instructions are followed correctly.)

To disinfect water with unscented household liquid chlorine bleach you will first have to assess the quality of the water. If the water is cloudy and colored you will have to filter it through a clean cloth, paper towel, or coffee filter OR allow it to settle. Be patient, for clear water allows for a faster and more efficient disinfection process. Once the water has settled, you can draw off the clear water to be used for disinfecting.

Consult the table provided and follow these steps:

1. Follow the instructions for disinfecting drinking water that are written on the label of the bleach product container.

2. If the necessary instructions are not given, check the "Active Ingredient" part of the label to find the sodium hypochlorite percentage, and use the information in the table provided as a guide. Typically, unscented household liquid chlorine bleach will be 8.25% sodium hypochlorite, though concentrations can differ.

3. Using the table provided at the bottom of the page, add the appropriate amount of bleach using a medicine dropper, teaspoon, or metric measure (milliliters).

4. Stir the mixture well.

5. Let it stand for at least 30 minutes before you use it.

6. Store the disinfected water in clean, sanitized containers with tight covers.

If you don't have liquid bleach, consider the disinfection method described on the next page.

Disinfect water using household bleach

Volume of Container	Concentration Percentage of Sodium Hypochlorite present in bleach			
	1% concentration	4-6% concentration	8.25% concentration	unknown concentration
1 Quart/ 1 Liter	10 drops/ 0.5 ml/ ⅛ teaspoon	2 drops/ 0.1 ml	2 drops	10 drops/ 0.5 ml/ ⅛ teaspoon
1 Gallon/ 3.8 Liter	40 drops/ 2.5 ml/ ½ teaspoon	8 drops/ 0.5 ml	6 drops/ 0.5 ml	40 drops/ 2.5 ml/ ½ teaspoon
5 Gallons/ 18.9 Liter	200 drops/ 12.5 ml/ 2½ teaspoons	40 drops/ 2.5 ml/ ½ teaspoon	30 drops/ 2 ml/ $1/_3$ teaspoon	200 drops/ 12.5 ml/ 2½ teaspoons

Granular Calcium Hypochlorite

Water can also be disinfected with Granular Calcium Hypochlorite. You can buy it online under the name Shock Chlorination. Make sure that the percentage of Calcium Hypochlorite to inert ingredients is 68% or higher.

Many people prefer not to use bleach. Some of the reasons include:

a.) It takes up more storage space.

b.) It only has a shelf life of 6 months.

c.) It's hard to get in the correct concentration.

If you prefer to use Calcium Hypochlorite, then pay attention to the following steps:

1. You first have to prepare the chlorine solution that you will use to disinfect your water. For your safety, do it in a ventilated area and wear eye protection. Gather clean, sanitized containers and prepare a measuring aid.

2. Add one heaping teaspoon (approximately ¼ ounce or 7 grams) of high-test granular calcium hypochlorite to 2 gallons (7.5 L) of water and stir until the particles have dissolved. The mixture will produce a chlorine solution of approximately 500 milligrams per liter. This is your chlorine solution that you will use to treat the contaminated water.

3. To disinfect the contaminated water, add one part of the chlorine solution to each 100 parts of water you are treating. Let's do the math:

This is a ratio of 1:100.

This gives us 10 ml added to 1 liter of water, or ½ liter to 50 liters of water.

This is about the same as adding 1 pint (16 ounces) of the chlorine solution to 12.5 gallons of water.

4. Let the mixture sit for at least 40 minutes.

If the chlorine taste is too strong, pour the water from one clean container to another and let it stand for a few hours before use.

CAUTION: HTH is a very powerful oxidant. Follow the instructions on the label for safe handling and storage of this chemical. Make sure that you label the storage container as DANGEROUS and POISON. You do not want children to come into contact with it. Store the granules dry for long-term storage. When buying calcium hypochlorite, make sure that it is the plain product and that it does NOT contain any water clarifiers or anti-algae agents.

Disinfection Tablets

You can disinfect water with light-weight tablets that contain chlorine, iodine, chlorine dioxide, or other disinfecting agents. These tablets (or crystals, drops) are available online or at pharmacies and sporting goods stores. Follow the instructions on the product label as each product may have a different strength. Pay attention to the clarity of the water, since chemical disinfectants are more effective when used with clear water. If the water is cloudy and colored you will have to filter it through a clean cloth, paper towel, or coffee filter OR allow it to settle. Once the water has settled, you can draw off the clear water. This is the water that should be used for disinfecting. After treating the water, make sure not to wait too long before consuming it. The water can become active with pathogens after a period of time (30-40 hours) and it's advisable to drink the treated water within 24 hours.

• *Iodine Based Tablets*

Although not a perfect solution, iodine tablets can come in pretty handy out in the field when you need a light-weight solution to take with. Usually these tablets come with an additional Vitamin C (ascorbic acid) tablet to improve the taste of the iodine disinfected water. Add the Vitamin C only after the iodine has done its job of disinfecting the water. Follow the instructions of the manufacturer and pay special attention to the following:

- **Warning.** Not for persons with iodine allergies or restrictions like persons with thyroid problems, are on lithium, are allergic to shellfish (they may also be allergic to iodine), women over fifty and pregnant women should consult their physician prior to using iodine for purification.

- **Effectiveness.** Iodine kills bacteria and viruses before they mutate.

- **Treatment Time.** The waiting time prescribed is 30 minutes.

- **The shelf life.** The guaranteed shelf life is usually 4 years for an unopened bottle. Around one year for an opened container with exposure to air.

- **Storage instructions.** Product should be stored in a cool, dry place.

• *Iodine Based Crystals*

Iodine in crystal form is very easy to use and is a handy option to have. It's advisable to heat the water slightly before you add the crystals. Warm water aids in dissolving the crystals and thus increases its efficiency.

Follow the instructions of the manufacturer and pay special attention to the following:

- **Warning.** Keep away from children. Non-dissolved crystals can be deadly if consumed. Not for persons with iodine allergies or restrictions like persons with thyroid problems, are on lithium, are allergic to shellfish (they may also be allergic to iodine), women over fifty and pregnant women should consult their physician prior to using iodine for purification.

- **Effectiveness.** Iodine kills bacteria and viruses before they mutate.
- **Treatment Time.** The waiting time prescribed is usually 20-30 minutes. Warmer temperatures will increase the effectiveness of the crystals.
- **Shelf life.** Indefinite shelf life when bottle is kept tightly capped.
- **Storage instructions.** They should be properly sealed to avoid exposure to air.
- **Extra.** Limits radiation intake in case of a nuclear disaster.

- ***Chlorine Dioxide Based Tablets (Or Drops)***

This is a very popular option that consists of a concentrated oxidant that makes it hard for protozoa, bacteria and viruses to survive. This is considered more powerful than standard solutions (e.g. iodine and chlorine) as it removes colorations, odors and also unpleasant taste from the treated water.

- **Warning.** Keep away from children. Volatile and sensitive to sunlight: do not expose tablets to air, and use generated solutions rapidly.
- **Effectiveness.** Effective against viruses, bacteria, giardia and cryptosporidium. It's important to note that it will take approximately four hours to treat the water for Cryptosporidium and Giardia, so plan ahead!
- **Treatment Time.** The waiting time prescribed is 30 minutes to 4 hours.
- **Shelf life.** It is guaranteed 4 years, but can be much longer if sealed.
- **Storage instructions.** Usually each tablet comes sealed in its own compartment.
- **Extra.** No nasty taste or odors.

- ***Chlorine Based Tablets***

Sodium dichloroisocyanurate or Troclosene Sodium (also shortened as NaDCC), is a form of chlorine used for disinfection. They are effervescent tablets which kill micro- organisms in water to prevent cholera, typhoid, dysentery and other water borne diseases. It is approved by the EPA for long term drinking water treatment.

- **Warning.** Chlorine should be used for persons with iodine allergies.
- **Effectiveness.** It is used for killing bacteria, cysts, algae, fungi, viruses and protozoa.
- **Treatment Time.** 30 Minutes.
- **Shelf life.** Guaranteed for 5 years. If opened they should be discarded after 3 months.
- **Storage instructions.** Usually each tablet comes sealed in its own compartment which is designed to keep out UV light and moisture.
- **Extra.** These capsules are effervescent, meaning they self-dissolve.

Chapter 10

Designing Your Own Filtration System

Years ago, having a traditional homemade filtration system was considered a normal component of a healthy homestead and people relied on it to provide clean drinking water. It's safe to assume that the environment was in a better state back then and that the biggest concern was probably dealing with water-borne pathogens that caused diarrhea.

These days we have volatile compounds in our homes and infrastructures that can create all kinds of problems. Dissolved chemicals are very hard to remove from your water and therefore I recommend that a system based on the traditional model only be used when you are sure that your water source has not been contaminated with chemicals or radioactive material. You can build a simple gravity filter that relies on gravity and layers of natural filtering material. You do not need electricity and you do not need to buy any commercial filters. Note that the thicker the layers of filtering material, the longer it will take for the water to pass through. In theory this also means the cleaner the water will be. You can experiment with different layers of filtering material: fine sand, coarse sand, fine gravel, coarse gravel, activated charcoal granules and even clay.

You also have the option of assembling a specialized gravity filtration system.

These filters remove almost everything from your water and do not require electricity. The only downside is that they are expensive and have to be replaced.

How To Design And Build A Filtration System

During a rain event, the water falls to the earth and with the aid of gravity, it is filtered through layers of sand, clay, gravel and rock. The result is clean groundwater fit for consumption. Applying the same logic, we can mimic the earth's perfect filtration system by creating layers of filtering material that removes contaminants as the water passes through. This is all possible with the aid of gravity.

A Homemade Filtration System

If you find yourself in an emergency situation, where water and electricity are no longer readily available, then the following method of water filtration might come in very handy. This is obviously not a preferred method for water treatment, but when you run out of options, this application might just save your life.

A traditional filtration system consists of a large container, made of natural material, filled with different layers of filter media:

Fine sand, coarse sand, fine gravel, coarse gravel, activated charcoal granules and even clay. The media all differ in particle size and application. They have been proven to clear water of turbidity, certain pathogens and iron. The larger gravel-like pieces will filter out the thicker sediment and debris. The smaller, compacted sand-like particles will filter out the microscopic contaminants.

Activated charcoal has microscopic pores that trap contaminants. It also attracts negative ions to its surface and thus removes them from the flowing water.

Besides plastic, glass and steel, you can also experiment with clay containers. In water-starved regions of the world, porous clay pots are constructed by mixing and pressing indigenous clay with sawdust, which leaves behind a porous structure. The inside is painted with silver particles and it is allowed to seep into the pore structure. Water simply filters through and is purified as it passes through the tine pores and over the silver.

First let's look at a modern, yet basic, version of this filtering system. This method is based off a traditional method used in the early 1900's by land owners to clean their drinking water. Important to note that you should be familiar with the quality of your water source. This means that if your water comes from a well, you should have it tested before and after filtration. This will give you a clear picture of what kind of contaminants are present and how effective your filtration system is.

CAUTION: Once filtered you should still boil your water. Boiling is still the best method to get rid of pathogens.

How To

Making a filtration system is very simple. Just keep the concept of filtering in mind. Your water must flow through various layers that will catch particles and contaminants. You have the freedom to experiment with different media layers made up of different filtering material. Traditionally, sand, gravel and activated charcoal are used. You can also use a combination of fine sand and a layer of coarse sand. The same goes for the gravel. Make sure that you wash and rinse the sand, gravel and charcoal before you use it. Separate the media and swish each around in a bucket filled with clean water. Repeat till all floating particles have been removed.

In the end, your water must pass through all the layers of filtration and be collected in a collection vessel. Make sure it's safe to drink by testing a number of samples. If the tests still indicate the presence of pathogens, then you should boil the water for at least three minutes. If the tests indicate the presence of dissolved chemicals, discard the water and look at another treatment option.

Material Needed For A Homemade Filtration System

- 3 x large food grade (BPA free) plastic buckets with lids. The bigger the better. You can still get 5 gallons/19 liters from stores like bakeries. Only use plastic containers with the following numbers and letters:

1-PETE, 2-HDPE, 3-LDPE, 5-PP.

- Large bags of fine gravel and coarse gravel. Enough to fill two containers.

- Large bags of fine sand and coarse sand. Pool filter sand works well.

- Activated charcoal. You can buy activated charcoal from pet stores that sell aquariums.

- 1 x Spigot

- Large drill bit or hole saw.

- Threaded Plumber's fittings with O-rings.

- Fine screen material to keep the material from flowing into the next bucket.

Let's look at the basic setup.
• **Step 1. Planning.** Prepare the material that you will need. Place the three containers on top of one another.

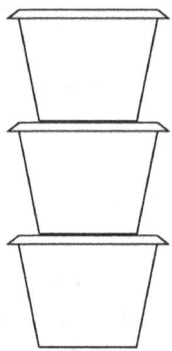

Take a marker and mark where you want to drill the holes.
• **Step 2.** Drill the holes. Drill holes in the bottoms of buckets A and B. The plumber's fittings will connect the buckets through these holes and should fit snugly. Do not drill a hole in the bottom of bucket C.

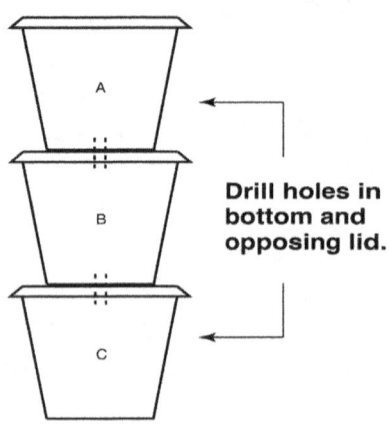

Drill holes in bottom and opposing lid.

You should also drill a hole in both the lids of buckets B and C. Make sure they line up nicely with the holes of the buckets sitting on top of each.
Bucket A's lid does not need a hole, except if you wish for it to serve as an inlet for an exterior water source.
• **Step 3.** The hole in bucket A connects to the hole in the lid of bucket B. Likewise, the hole in the bottom of bucket B, connects to the lid of bucket C.

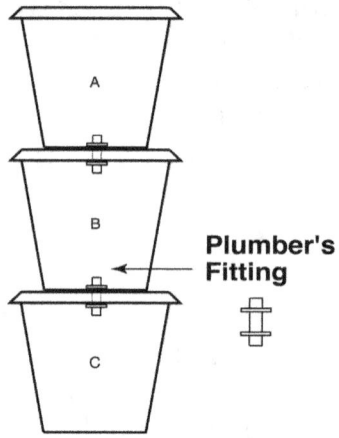

Plumber's Fitting

The plumber's fitting that connects the buckets should be watertight.
• **Step 4.** The screen filter should be placed on the top part of the threaded plumber's fitting opening. This is the part that is exposed to the layer of sand or charcoal. It will prevent sediment from flowing into the next container. It's very important to pay attention to fittings and con-

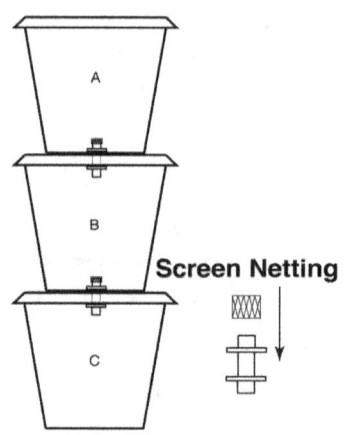

Screen Netting

nections for this is where water leaks will ruin the project. Stretch it nice and tight over the opening and use a rubber band or O-ring fitting to hold it in place.

• **Step 5.** For the last bucket at the bottom you can drill a hole at the side near the bottom.

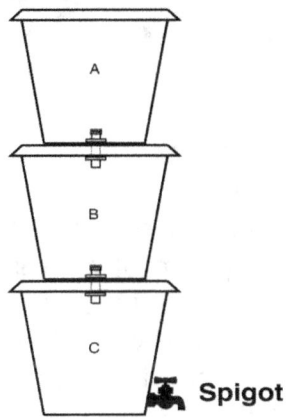
Spigot

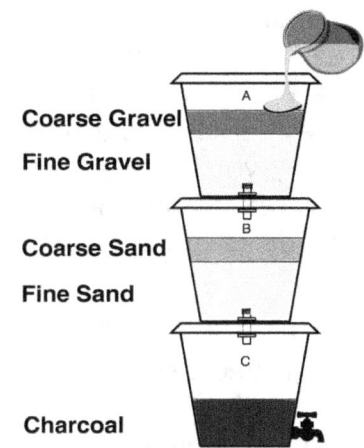
Coarse Gravel
Fine Gravel
Coarse Sand
Fine Sand
Charcoal

Get a threaded spigot, or flow valve, and screw it into the side of the bucket. This functions as your outlet. From here you should have a container ready to fill when needed.

• **Step 6.** Fill your buckets. One option is to fill bucket A with coarse gravel on top, and fine gravel at the bottom. Fill bucket B with a thin layer of coarse sand on top and the thick layer of fine sand at the bottom.
Put the charcoal in bucket C. Use charcoal that's been crushed into small pieces to be effective. If your water has visible particles floating at the surface, strain it through a thick cotton cloth or coffee filter before pouring it into your filtration system.

• **Step 7.** Test your water. Filtering is a basic method that will remove some microbes, but it will definitely not remove dissolved chemicals. Clean and rotate the system every 6 months and keep it away from household cleaners or other contaminants.

CAUTION. No matter which filtration method you use, remember to still boil your water before consumption.

Alternative: Bio-filter

The above-mentioned system introduces us to the basics of a traditional filtration system. These days many people avoid using plastic products when it comes to food and water needs. If you want a truly traditional homestead solution, you can consider substituting the plastic containers with one large oak barrel/ tub or even a large clay or concrete urn with a lid. You have to make sure that the lid closes properly to prevent contamination.

The aim will be to design a bio-filter (fig. 10.1) which consists of a filtration system with various layers of media filter (created by you) and a layer of bio-film (created by nature), to deal with disease causing pathogens.

The filter media needed:
- Fine, graded sand, 0.15 mm effective size.
- Coarse, graded sand, 0.35 mm effective size.
- Fine, graded gravel.
- Coarse, graded gravel.
- Activated charcoal, crushed (Optional).
- Use a thick cotton strainer at the top to catch the obvious particles when dealing with especially turbid water.

All media should be as sterile as possible. You do not want any organic matter in your filter, for incoming microbes will feed on it. Buy sand from a supplier that adheres to NSF 061 and/or AWWA B100-01 standards.

Mechanics: Water is poured into the reservoir at the top where small perforations allow the water to gently trickle down to the first layer of filter media (sand). It will form a little pool of standing water (3-5cm). It's important that the sand remains covered with a layer of water which provides oxygen to the system, so don't fret over the water at the top.

After a period of 3-4 weeks a bio-film (*Schmutzdecke*) will develop in the standing water,

Fig. 10.1 Contaminated water

Standing water
Biofilm
Fine sand
Coarse sand
Fine gravel
Coarse gravel

Outlet
5cm
60cm
3cm
5cm
6cm

PVC pipes, side by side, with holes. Use plumber's fittings, Schedule 40.

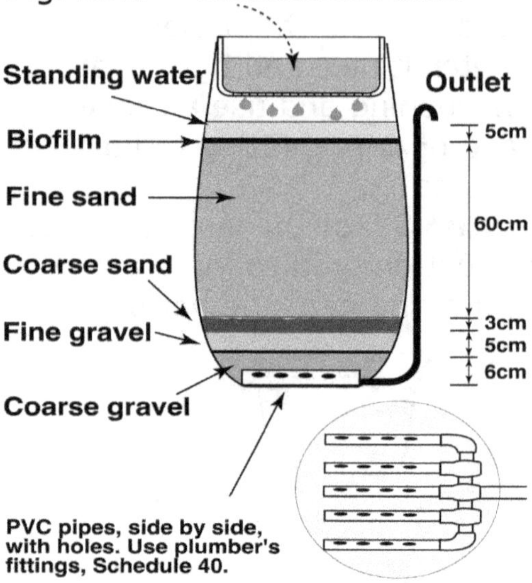

on the top layer of sand. This is a thin, slimy film of bacteria that adheres to the surface of the sand. A bio-film will only form if your water used is contaminated with microbes. This might sound counter intuitive, but healthy microbes will gather at the top of the sand layer and they will aid in fighting incoming disease-causing pathogens. Start using the filter once the bio-film is present.

Caution! When you pour the water in at the top, try not to disturb the bio-film. To achieve this, design a container that will gently sprinkle water over your bio-film. This is basically a reservoir with numerous holes punched in the bottom. You can use a plastic bucket, stainless steel or clay pot with holes drilled in the bottom. Once water is poured in, the holes will sprinkle the water like a shower head and it will not displace the bio-film clusters.

Once the water moves through the bio-film layer, it continues to percolate through the different layers of filter media. This addresses turbidity, some pathogens and general contamination like iron. Important that you don't want any air pockets in the filter media. These pockets cause pathogen development and foul odors.

Once it reaches the bottom, it is siphoned through the PVC outlet pipes to the outside where it should be collected in a collection vessel. Thanks to air pressure, the position of the outlet nozzle will determine the standing water level inside the container.

You have the option of filtering the water through an activated charcoal filter, to send it through a UV light or simply to boil it. Your system will require maintenance when it becomes ineffective at filtering water. Clean your system when the flow rate drops down to a trickle (could take years). Replace all filter media with new material.

It's necessary to use coarse gravel at the bottom of this system. The PVC outlet pipes are located at the bottom of your system and a layer of sand will clog the outlet in no time. The last layer of media is coarse gravel, which cannot clog or obstruct the outlet pipe.

CAUTION. No matter which natural filtration method you use, remember to still boil your water before consumption.

Gravity-Fed Filtration System

Using filters made from natural material is an inexpensive option that gives you access to fairly clean water, for as long as you want.

Following the same principles, we can also combine this method by substituting the filtering media with modern commercial filters. There are many options available and some of the modern ceramic filter candles are

Material Needed For A Gravity-Fed System

- 2 x large food grade (BPA free) plastic buckets with lids. The bigger the better (5gallons/ 19liters). You can still get these from bakeries. Only use plastic containers with the following numbers and letters: 1-PETE, 2-HDPE, 3-LDPE, 5-PP
- Your gravity filter system.

Select the brand that suits your needs. Remember to read the instructions beforehand. Many filters have to be "primed" before use.

- 1 x Spigot kit or Faucet system
- Drill bit

actually classified as purifiers that take out all biological and chemical contaminants. This means that they can remove almost all of the contaminants, that we expect to find in polluted or contaminated water. These filters are very effective and we find that some are specifically designed to rid water of chemical elements such as arsenic and fluoride. Such filters should be used in areas with documented chemical contamination of this kind.

It is best to have your water tested before you buy a specific gravity-fed filter. Knowing exactly what needs to be removed from your water source will guarantee cleaner water. Although expensive, these gravity-fed candle filters are good for around 3000 gallons (11000 liters) before they have to be replaced.

How To

A great option for any home, whether for daily use or in case of an emergency, is to incorporate standard gravity-fed purification elements (ceramic filters) into your treatment system. Although there's a wide selection of manufacturers to choose from, you will be looking for something that filters bacteria, chemicals such as chlorine, herbicides, pesticides, pharmaceuticals, VOC's, and heavy metals. Look for a product that relies on micro-filtration tech-nology with a good flow rate and that's been tested to exceed NSF/ ANSI Standard 53 or 58 protocol. Avoid the cheaper products from unknown manufacturers.

It's very easy to assemble this system. You can use one, two or even three filters. Utilizing multiple filters means an increase in flow rate and also guarantees longer usage before replacement is needed.

• **Step 1.** Drill holes in the bottom of the top bucket for the filters you will use.

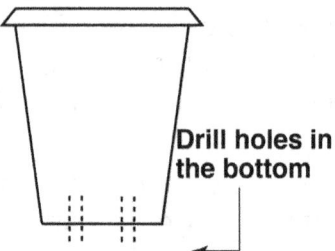

The holes must be the same diameter as the filter threads to allow for a snug fit.

• **Step 2.** Follow the instructions provided with the filter and place seals and O-rings where needed. You will definitely need a seal between the filter and the bucket.

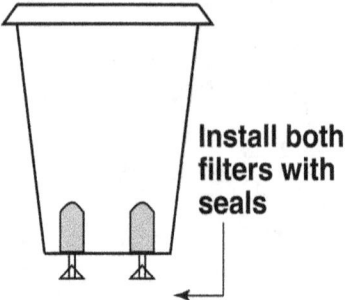

Make sure that the filters are sitting tight and flush against the bottom.

- **Step 3.** Drill holes in the lid of the bottom bucket and make sure that they line up with the holes in the top bucket. This allows the filter shafts to stick into the lid of the bottom bucket.

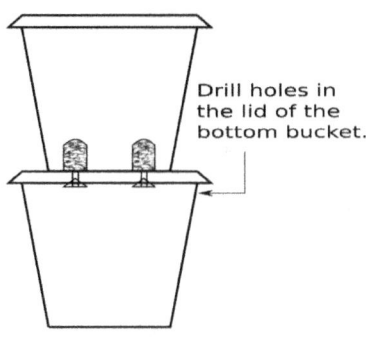

Drill holes in the lid of the bottom bucket.

- **Step 4.** Install a spigot or faucet about 2" (5cm) from the floor of the bottom bucket.

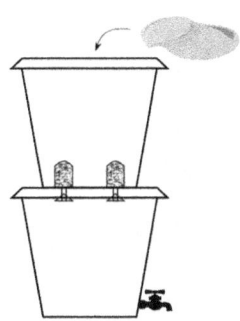

- **Step 5.** Fill the top bucket with water and watch as the drinkable, filtered water fill the bottom container.

Note that different companies will make different guarantees as to the safety of the water once it's been purified. If unsure about the water quality in your area, have it tested before you use a purifying system.

How To Make Activated Charcoal

It is very hard to produce activated charcoal without a furnace and chemicals. In a dire emergency, consider the following method developed at Santa Clara University. Use coconut shells as your carbon source.

1. **Put your material** in a 55 gal steel drum and place it on a large bonfire. The container should have a hole at the top to allow air to escape. The wood will burn inside the container and as it carbonizes, gases will escape with noises and sparks. Let it burn for around 3-5 hours. Wait till the fire has completely burnt out, before touching the container. **It will be very hot.** Inspect the charcoal and select the pieces that have been fully carbonized. Avoid ash and hard pieces.
2. **Next**, take out your charcoal and mash it into a powder or into small granules.
3. **In a ventilated area**, place it in a solution made of water and 25% *table salt, for 24 hours. The salt helps in opening the pores of the charcoal.
4. **Take it out** and rinse thoroughly to get the salt off.
5. **Finally**, let it dry in the sun or in an oven for several hours.

Never use charcoal from the grill. It's saturated with toxins!

*Optional, instead of salt, you can also use Calcium Chloride ($CaCl_2$) and natural hickory or oak as carbon source.

Designing Your Own Filtration System

Chapter 11

Designing A Distillation System

In an emergency situation you will be very grateful to have a distillation system in your home.

These days complete distillation machines for domestic use are readily available and you can buy them at select consumer outlets. Such a system is not cheap, but it can pretty much purify any water as it mimics nature's very own hydrological cycle that we discussed earlier in this book.

A typical distillation system is highly effective at removing bacteria, protozoa and viruses. It also removes chemical contaminants like arsenic, barium, cadmium, chromium, lead, nitrate, sodium, sulfate and various organic compounds. Such a system will of course also remove salt from seawater. There are claims from very credible authors in the scientific community that it can also clean radioactive water, but I have no way of testing that claim.

The critics of distilled water as a potable water source argue that it's too clean and that all minerals are removed, which has a negative impact on human health. The counter argument, used by proponents of distillation systems, is that we actually get most (95%) of our trace minerals from our food. They further suggest adding a pinch of salt (actual organic rock salt) to your distilled water to compensate for the loss in minerals. Whatever the case, having water will keep you alive longer than not having water.

Let's take a look at the various distillation options available.

How To Design And Build A Distillation System

A simple distillation system can be very helpful in an emergency situation. It is a simple system that needs heat for evaporation, lower temperature for condensation and a vessel to collect the water droplets in. The yield can be greatly improved by using large quantities of contaminated water (or brine) and to make the system vapor-tight. You will need an energy source to provide the desired amount of heat.

Distillation

A commercial distillation system is a good investment and a welcome addition to any home, but remember that it is still a machine and that it requires electricity. If you choose to purchase one of these systems, know that you should always have a backup generator at the ready, to supply you with electricity when the grid is down.

The knowledge of how to make your own system can be priceless, so let's look at two practical methods to distill water at home:

- One simple and cheap option involves filling a container with water, connecting it to a heat source and once steam is produced the droplets are collected in a collection vessel. This is a practical system you can assemble in your kitchen.
- If you want to be even more prepared, then you can look at constructing your very own outdoor solar still. This method requires more work, but is also a very efficient method that does not require a conventional heat source, since it uses energy

---
Distillation Glossary

Brine. Water impregnated with salt.

Condensation. Water that collects as droplets on a cold surface when humid air is in contact with it.

Evaporation. The process of a substance in a liquid state changing to a gaseous state due to an increase in temperature.

PVC board. Also known as Chevron board and Andy board, a board made of light weight, foamed PVC, which is lightweight, moisture and corrosion resistant.

Solar still. A solar still distills water, using the heat of the Sun to evaporate, cool then collect the water.

Tempered glass. A type of safety glass processed by

from the sun.

Note that we are dealing with water vapor and tiny droplets. This is a slow process and the output is especially slow. If you have a large family, it might be best to have one system available for every family member.

Distillation With A Heat Source

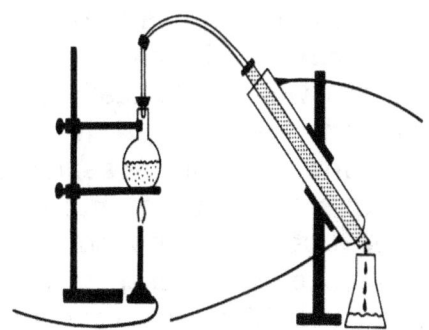

Basic distillation of water is a concept that we were all introduced to as students, in science class.

The process:
1. Heat water in a container and eventually vapor will start to rise from its surface. This is evaporation. The contaminants remain behind in the container.
2. Condensation occurs when we catch the vapor droplets against a cold surface.
3. Wait for the droplets to run down and to collect in a collection vessel.

The key ingredient in this experiment is heat. Heat speeds up the process and is the catalyst for evaporation to take place.

Preparation

The concept of water distillation is a simple one. Keep the basics in mind and you'll be able to scale down or up, according to your needs. Avoid plastic or any material that can leach chemicals into your system. The main components are:

• **You need a heat source.**
This can cause some problems. Electricity or generators can provide ample energy for this task, but what if you do not have a readily available power source? You will have to invest in a simple wood stove, rocket

---

Distillation Glossary

controlled thermal or chemical treatments to increase its strength compared with normal glass.

Trough. A channel used to convey a liquid.

VOCs. Volatile organic compounds, (also called VOCs), are organic compounds that can easily become vapors or gases. Along with carbon, they can also contain elements like hydrogen, oxygen, fluorine, chlorine, bromine, sulfur, or nitrogen.

Vapor. The gaseous state of a substance that is in a liquid or solid state at room temperature.

Yield. The amount or quantity that's obtained from a specific process or reaction.

stove (wood fuel), solar powered water heater or a gas burner.

- **A container.** Stainless steel or copper will do. Basically a clean container that produces steam and that can be sealed off to direct the steam into your copper tube. You can also use a pressure cooker. Pressure cookers work well because they seal the steam in the container and are vapor-tight.
- **Copper tubing.** Buy some 10-15 feet of copper tubing at your hardware store. The thickness will depend on the fittings you want to use (¼ or ½ inch).

If you want to use a pressure cooker, the hole in the top of the lid will be of a small diameter and you will have to decide what size fitting will work best. (A ¼ inch fitting seems to work well.) If you are using a copper boiler or kettle, then you will need a wider diameter tubing. Only buy the copper tubing once you know the corresponding fitting size.

You will also have to shape the tubing into a coil. Before you bend the tubing, fill it with water and then let it freeze in your freezer overnight. Get a long cylinder or PVC pipe (4-5") and then wrap the copper tubing around it. This method helps prevent cracks and kinks in your

Pressure Cooker Fittings

Assess the thickness of the metal lid of the pressure cooker and select the correct compression fitting (brass).

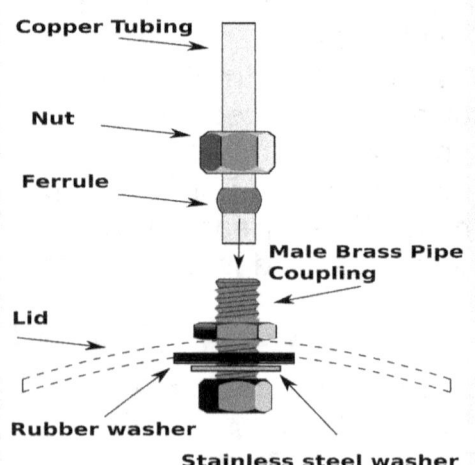

- If the metal is thick enough for a screw thread, then use a screw "tap" to make a thread in the metal. Just screw the fitting that you bought into the threaded hole.
- If the lid is too thin, then drill a hole through the lid (over the original hole).
- Use a Male brass compression coupling and screw it through the hole. Remember to add a rubber washer and also a stainless steel washer to make your cooker vapor-tight.
- You can also use a Tee- or Elbow fitting if you need the tubing to point in a specific direction.

tube. Shaping the copper tubing into a coil increases the efficiency of the system. When steam meets resistance, it condensates back to water. If the coil is straight, the steam will simply escape out the back end.

- **Fittings.** You will need brass or copper compression fittings. A male fitting will work well if you are using a pressure cooker. Ask for advice at your hardware store to see which fitting will do the job for the system you have planned. If you are not familiar with fittings and soldering, get some "push connect" fittings. They are easy to use and will work just fine.

A flexible coupling will also come in handy for large diameter pipes where an exact fitting is not available.

- **Corrugated copper tubing** (Optional). Having a flexible section of tubing at one end (or both) just makes life easier. It provides you with options in that you can fit or place your coil in different positions. This is to prevent kinks in your main coil. Use it if you feel it will improve the efficiency of your system.

- **A container or material to cool down the coil and to speed up condensation.** This can be a tricky one, because most of us want a system that does not use electricity. Let's look at the possibilities:

a) If you are connected to the grid, then you have electricity to cool water or to make ice. Simply fill a large container with

> Bring the water to a rolling boil before you put the lid on for distillation. This kills off all pathogens that might enter the tube.

ice or ice water. You use the electricity to make the ice or ice water.

b) If you want a system that doesn't use electricity, and you live in an area with a cold climate, then you can use snow or cold water to cool down the coil. It can be any contaminated water, from a river, pond or lake, since its only function is to cool down the outside of the coil.

c) If you don't have electricity, then things become more complicated. Making ice or cold water without electricity is near impossible. To cool the coil to some degree, wrap a wet towel around it and place your whole system outside where it can receive some air flow. A solar powered fan will help as well. Not having a way to cool down the coil will slow down the process considerably. Remember, the reason we cool the coil is to speed up the condensation process.

d) If none of the above are available, just leave the coil as it is. It will just take much longer for the water droplets to condensate.

- **A collection vessel.** Any container will do. Make sure that it is sterile and keep it as close as possible to the tubing opening, where the water flows out.

How To

Before you start, lay all the pieces out in front of you. All parts should be sterile and clean. Make sure that all the fittings are the right size and that the steam will be contained without any leaks. You want a vapor-tight system!

A system that leaks is a faulty system and will have a negative impact of the distillation process. This is a very important issue that deserves your full attention.

This system follows a very simple process.

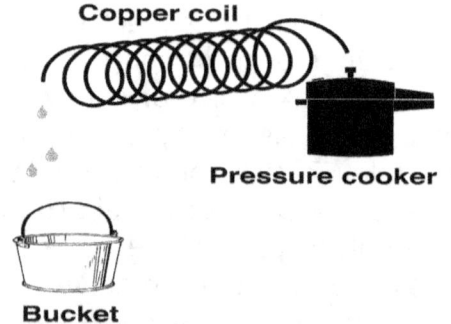

1. Connect your tubing (with the correct fitting) to the container you selected. Make sure that your fittings are vapor-tight.
2. Submerge the tubing in the cold (ice) water at a downward angle. The droplets caused by the condensation inside the tubing will run downhill into your collection vessel.
3. Fill the container with the contaminated water. Start to heat the container.
4. When the water starts to boil, let the first few clouds of vapor out before you close the lid. This will get rid of some Volatile Organic Compounds (VOCs).
5. Constantly check for leaks during the distillation process.
6. Eventually the water will start to trickle out of the tubing into the collection vessel.

Discard the first 100 ml if the contaminated water had obvious impurities like acetone or gasoline.

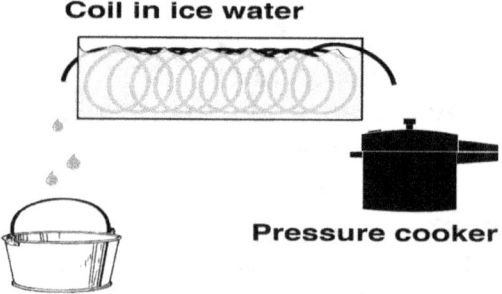

You have the option of pouring the water through a charcoal filter at the end, which will get rid of unwanted odors.

Solar Still Distillation

These days many people prefer to rely on mother nature and to get their energy from the sun. If you are also leaning in this direction, then you might consider building a passive solar powered water distillation system. Note that although you require sunlight for maximum efficiency, you can still use a solar still in cloudy weather, but the yield will be about half of

normal.
This method works on a simple principle:
1. Heat the water with the aid of the sun, this causes evaporation.
2. Allow it to condense, against a pane of glass.
3. Let it flow down and out into a collection vessel.

Preparation

Solar stills use very simple technology and depending on your budget and skill-set, you can ramp up the size as much as you desire.
CAUTION! Before you start this project, note that lifting and moving large panes of glass is very dangerous. Use suction cup lifters, gloves and protective eye wear. This is a two-man job.
For a basic still, you will need the following material:
- 1 x Tempered glass pane 30½" x 60". A thickness of ⅛ inch (3.175 mm) will do. This is placed on top of the still. You want a solid piece of glass that can rest safely on your still under all conditions. Make sure the surface is clean, smooth and even. This is important! Tempered glass can crack and chip easily. Keep metal tools away from the glass. (If looking for recycled, a sliding door's glass will do.)
- 1 x Tempered glass pane 28½" x 58½". The thickness of this pane is not crucial, since it is placed in the bottom of the still. These measurements are correct if used with ¾ inch PVC side-boards. Dark, tinted glass is best for heat absorption, but clear glass with black silicone underneath will also suffice.
- PVC board as needed for the construction of the still. A ¾ inch width should be thick enough.
- Silicone caulk as needed.

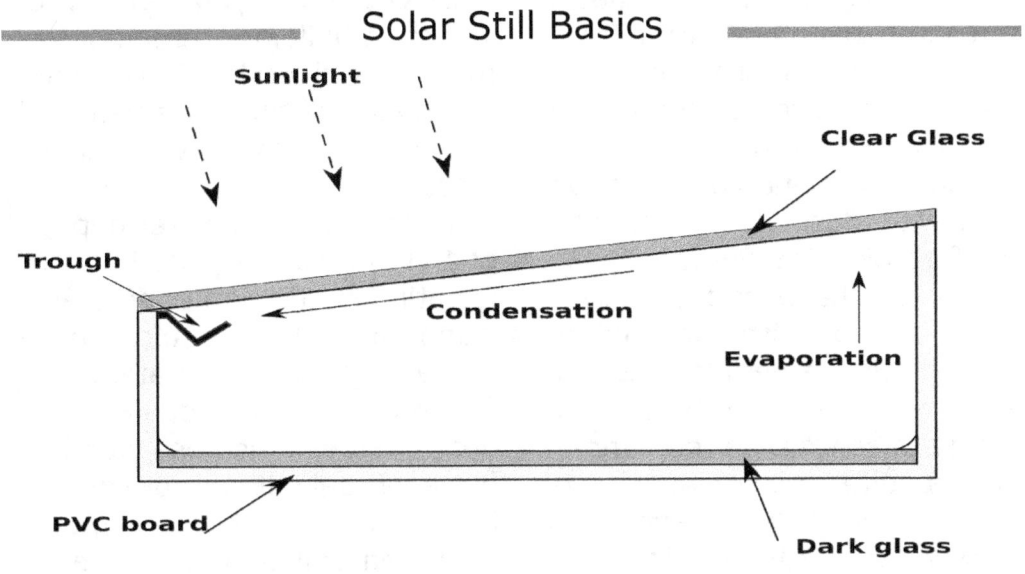

Solar Still Basics

- PVC fittings. These are used for the inlet and outlet.
- PVC cement or a substitute to join the PVC boards.
- Wall anchors and screws to join the PVC boards.
- Aluminum angle strips. Use these strips at the bottom of the still to seal the pane of glass in place with silicone. Use ¾ - 1 inch angles.
- Trough material. Copper, stainless steel, aluminum or PVC pipe.

A couple of things to keep in mind before you start construction:

1. A bigger surface area will increase the yield. The bigger the better. Remember that you are dealing with glass. Too big and things get complicated. A 30" x 60" pane of glass is already a very large still. You will need an extra pair of hands (and suction cup lifters) to help with construction.
2. The angle of the Cover glass. You can experiment with different slope angles for your glass that will be placed on top of the still. Anything from 10°- 40° will do. In the end you will have to select a design that you are comfortable with. Personally, I think it's cheapest and easiest to start off with a low angle around 10°. Remember that the angle will determine the width of the glass as well. A steep angle means quite an increase in cover glass size. At a 10° angle, the glass width is 30½ inches. At a 30° angle, the width increases to 34½ inches.
3. For maximum efficiency, the system must be vapor tight. This is crucial. Loss of water vapor causes less yield.
4. Use thin, tempered glass for the cover of the still. The sun will eventually crack the normal glass due to the constant fluctuation in temperature. Too thick cover glass will have a negative affect on the energy efficiency of the system.
5. You can use heat resistant, food grade silicone caulk. This material is safe to use and very little of it will be exposed to the water in the system. If you are absolutely against silicone, use the aluminum angle strips for the glass at the bottom of the sill.
6. The glass pane should be as close to the water as possible. This makes for a more efficient system and less vapor is lost during evaporation. Have the glass cut before you start. When you start building, remember that you will work off the measurements of the glass pane. It's always easier to cut wood than glass.
7. Water depth. A water depth of 1 cm of contaminated water is optimum. It is an energy efficient depth that requires minimal energy for evaporation.
8. Alternatives. Besides PVC board you can also use wood, but eventually it will rot from the exposure to heat and moisture. Remember that we are dealing with potable water for

human consumption so avoid treated (sealant or oil) wood. If you do not have glass available to place at the bottom of the still, then simply cover and seal the whole inside of the still with black silicone caulk. Make sure that you get it as smooth and even as possible.

If you prefer not to use silicone, and you do not have glass available, then you can consider using a large, flat, black tray as the holding vessel for the contaminated water.

How To

Prepare the material needed to construct your still. Think carefully about the assembly and how complicated a design you want to create.

1. You should have the two precut panes of glass prepared and ready. Take the smaller of the two (tinted), the one that you want to put at the bottom of the still, and place it on a large piece of PVC board. We will call this the base-board. A board of 30" x 60" will do, but you can go smaller to suit your budget. Trace the outline of the glass on the base-board. Leave space all around the glass for the side-boards to be placed upright. If you are using ¾ inch thick boards, then leave a ¾" space all around the glass.

Remember when cutting or buying the clear cover glass pane, that it is slightly wider (½") than the PVC base-board. The cover

Comparison Of Inclination Angles

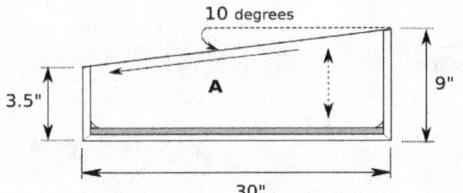

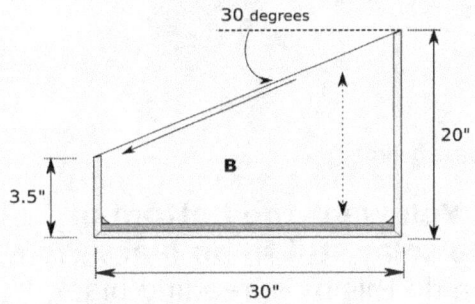

Model A (10°) has a short distance between the bottom and top of the still. This allows for improved evaporation output, but due to the low inclination angle it causes slower speed of distilled water output.

Model B (30°) has a larger distance between the bottom and top of the still. This affects evaporation due to increase in distance, but the increase in inclination angle allows for faster distillation output. This model is more expensive due to the increase in size of material used. In the end, the difference in yield between the two models is not huge. Design a still that you find practical in design and make sure that it's vapor-tight.

Designing A Distillation System

glass pane sits at an angle on the still, and is therefore wider. Example: If you used a 30" x 60" base-board, the top glass cover should measure 30.5" x 60".

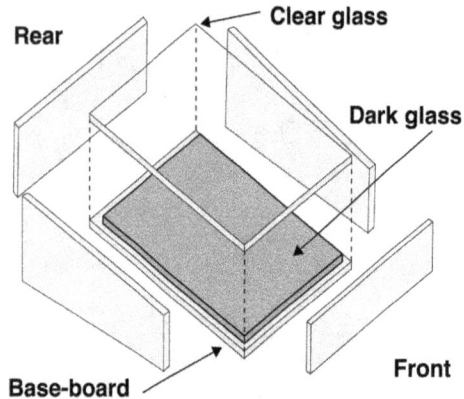

2. You want the bottom of the solar still to be black.
You can do this by spreading black silicone caulk over the base board. This will function as the bottom of your still.

Use a putty knife and spread it evenly. Place the pane of tempered glass on top.

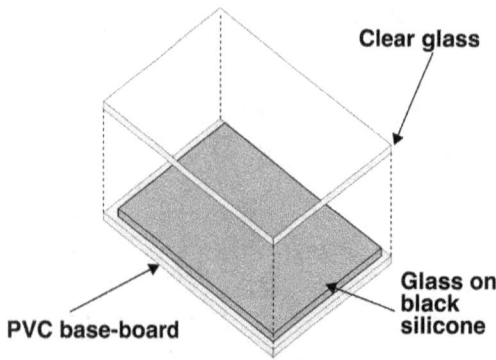

Clean the excess silicone and let it dry.

3. You will also need PVC boards for the sides of the structure.
We will call them the *side-boards*.

Cut your boards according to your measurements, but remember to cut the front and back boards first. The back board must be higher than the front board. Remember to cut both the front and back boards at an angle to accommodate the glass that will rest on top (*See p. 349 for a comparison of inclination angles*).

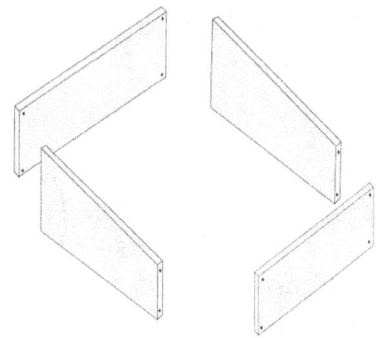

- If you prefer a low angle of 10°, then a back board of 9" will do. The front board is 3½" high.
- For an angle of 30°, a back board of 20" will do. The front board is 3½" high.

Cut PVC board at sloped angles to fit the glass cover

You should use wall anchors, screws and PVC cement to fasten all boards and joints securely. Once assembled, the four side-boards should fit perfectly around the glass pane on top of the base board.

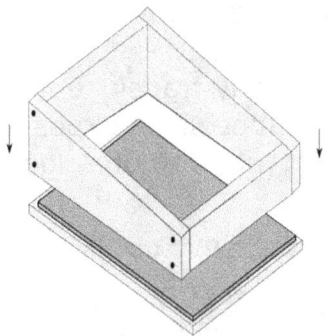

4. Once dry, take the assembled side-board frame and fit it on the base-board with the glass. The bottom will be quite heavy on account of the glass.

You now have a structure that resembles an open box with a glass covered bottom.

Seal the glass with aluminum angle strips or just use silicone caulk.

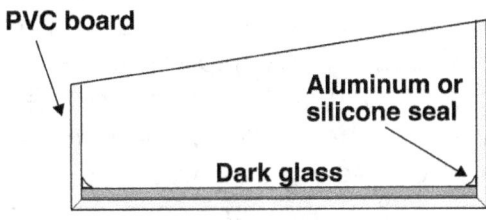

5. Next comes the trough. To make the through you can use a piece of stainless steel, aluminum, copper or even PVC pipe (Eventually the PVC pipe will sag due to the heat, so best to put little supports underneath to prop it up. In general, you want to avoid anything that can leach chemicals into your water.).

The trough should be long enough to cover the inside width of the still.

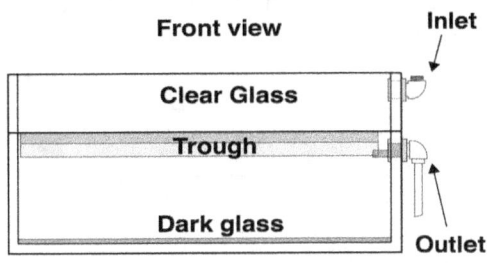

Bend the trough material in a shape that will allow it to rest on top of the front PVC board and also to form a tray for the water to run down. Use silicone and screws to fasten it to the front PVC board.

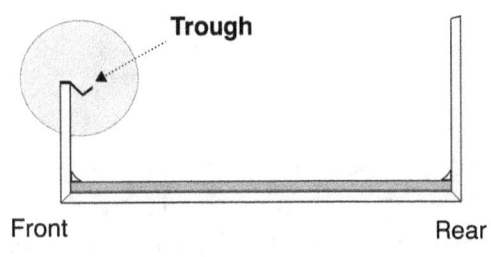

The trough should be slightly tilted to run downhill. To achieve this, make sure that you bend the trough, where the water flows, on a crease that runs at a slight angle.

6. Once the water runs down the trough, it runs into a small diameter PVC pipe that forms the outlet.

Both the inlet and outlet are on

the same side of the still.

Drill a hole through the PVC side-board where you want to place your PVC fittings. You should have about 1½" pipe that protrudes from out of the side board.

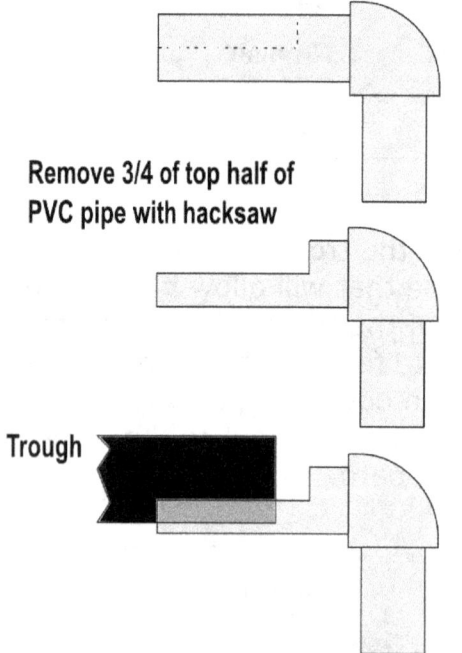

Remove 3/4 of top half of PVC pipe with hacksaw

Trough

Use a hacksaw and remove three quarters of the top half section of the PVC pipe. Use PVC cement and connect it to the joint that sits on the outside. This section of pipe basically just provides space for the trough to rest on.

Alternative: For a simpler trough solution, just cut a section out of a piece of PVC pipe. This pipe should be long enough to run across the width of the still. One end connects to the PVC joint that flows into the collection vessel. The other end can go through the PVC side-board. Make sure it's closed up and sealed with silicone.

From the trough and the PVC joint, the water runs down into a collection vessel.

Do not use any translucent piping or tubes since exposure to the sun will promote algal growth.

Using a PVC trough

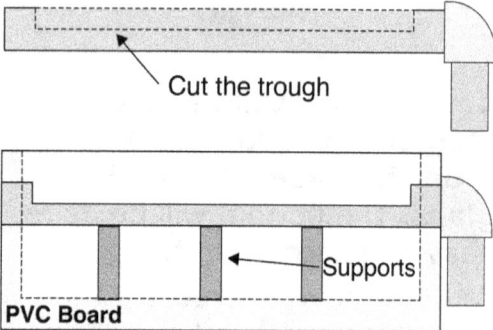

7. In a similar fashion you can create a filling inlet, to pour the contaminated water into your system.

You will have to drill a hole in the side-board, the same side as the outlet. Use plumber's fittings and add a down-pipe that almost touches the bottom of the still. On the outside, you can use a PVC pipe with a threaded end. This will allow you to place a cap on it when not in use, or to use

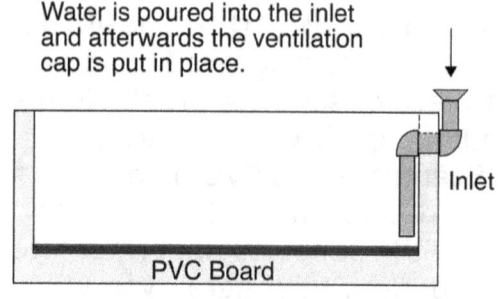

Water is poured into the inlet and afterwards the ventilation cap is put in place.

a funnel when needed.

8. If your system is properly constructed, it will be almost vapor-tight. This means that when you fill the unit with water, air pressure will build up inside the still (from displaced air) and this will put stress on your joints. The solution is a ventilation outlet, that is basically just a cap with a tiny hole drilled in the middle, which will allow the air to escape. Glue a small piece of screen material to the inside of the cap, to keep ants out. Just place the cap on the inlet pipe when the still is in use.

All these fittings can be bought at your local hardware store. They consist of PVC pipes, elbows and caps of appropriate dimensions that fit snugly together. Remember to use PVC cement where needed.

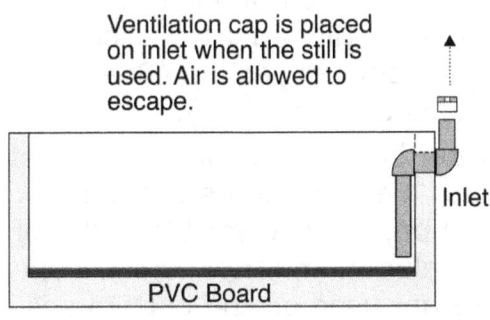

collect in the trough. If necessary, you can glue a thin strip of PVC onto the glass, just above the trough. This strip will **guide** the droplets to fall straight onto the trough.

There are no guarantees as to how much your yield will be. It all depends on your workmanship and how vapor-tight your still is. If you have a contaminated water solution with expected chemicals like hydrocarbons (acetone and gasoline) note that they will rise to the top, which means that you can discard the first 100ml of distilled water. This is considered unfit for drinking.

Depending on weather conditions and quality of construction, a still that's 30" x 60" should be able to provide you with more than a gallon (3.785 L) of water a day.

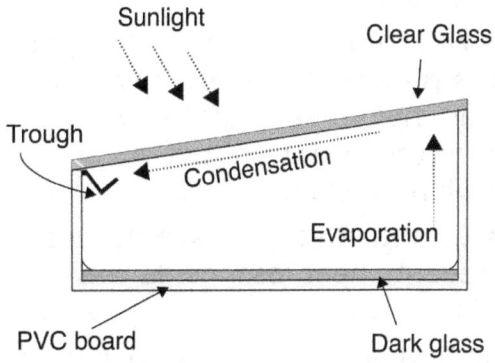

9. Once the inside of the still is done, and the trough is in place, you can place the glass cover on top of the still. This is the final piece of the puzzle as the glass cover will effectively seal the whole unit. Make sure that the water droplets will

Options

A basic solar still will come in real handy when you run out of water. Depending on your budget, you have quite a few options on how to improve on this design.

- **Pre-filter your water if possible.** A simple cotton cloth or coffee filter will do. Try to eliminate larger particles before you pour the contaminated water into the still. Once the water has been distilled, you can also use a home-made charcoal filter to improve the taste of the water.

- **Use waterproof aluminum duct tape.** This is optional, but it will definitely help with keeping heat in and also to make the whole system vapor-tight. Seal every joint and opening for added efficiency.

- **You can also add reflective material to your sill to concentrate the energy from the sun.** A reflective panel, like a mirror, can be placed above the back-board of the still. Angle it to reflect the sun's rays onto the glass top.

- **You can add a drain plug to drain excess water.** Having a drain hole means that you can flush the system to get rid of unwanted sludge.

- **Insulate your system.** This prevents heat loss inside the still.

Go to a store that sells roof insulation and get the thinnest insulation material available. You should be able to find sheets around 1-1½ inch thick.

The insulation material should be placed at the bottom and sides of the still. Insulating the whole system means that you will have to buy extra PVC board to "wrap" the insulation material in.

Use a 2 x 4" timber wood frame with cross sections as a base frame to place your bottom section on. This will accommodate the combined weight of the structure.

- **Having access to the inside of your still might be good for maintenance and cleaning.** The big issue here is how the new design will affect the overall effectiveness of the still. Every little hole, gap or crack will cause precious water vapor to escape. This will have a huge effect on the still's water production capability. Adding hinges might seem like a good idea, but they will most definitely cause vapor loss.

A practical solution is to separate the still into a top section and a bottom section (See diagrams on next page).

Note that you will have to adjust your measurements of the top section, to allow for the width of the aluminum strips and silicone sealant used on the bottom section. There is no need to connect the two sections with screws, since the weight of the top section will be enough to hold it in place. It should be a snug fit, and once the bottom of the still is filled, the actual water will seal the openings in between the two sections and this will prevent vapor loss.

Place handles on the sides of the top section to make it easier to lift it up. Use pop-rivets.

As mentioned earlier with the insulation discussion, you can use a 2 x 4" timber wood frame with cross sections as a base frame to place your bottom section on.

This information should provide you with the know-how to build a solar still that will suit your pocket and water needs.

Keep in mind why we use solar stills; they can purify any water and they do not require a conventional energy source. This could be a resource worth investing in.

I would like to thank a gentleman named Robert Howell, who lives on the coast of North Carolina. An expert at solar stills, his knowledge and creativity proved very useful in contributing to the design of this still.

In his own words, "I have built and used this type of water distiller for close to 20 years. In my opinion a more pure water for drinking can not be found. I have purified water from rivers, creeks, the ocean and more sources."

Creating A Top And Bottom

- You can make your still more versatile with an option of separating it into a top and bottom section.

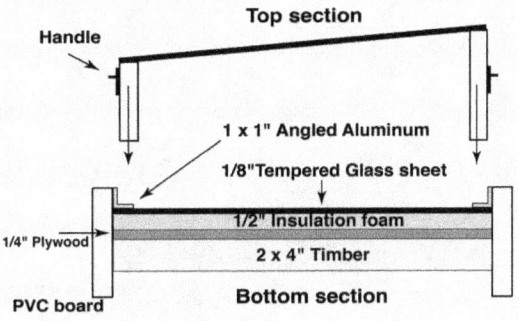

- The top section consists of the frame, the cover glass and handles on the sides.

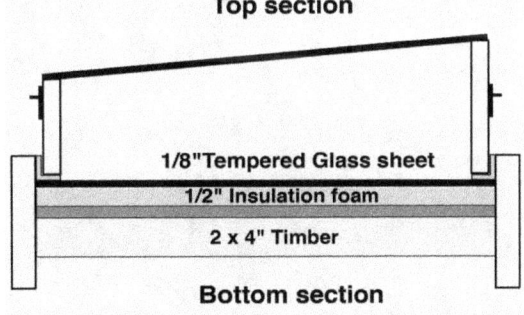

- Once the bottom of the still is filled with water, it will be vapor-tight.

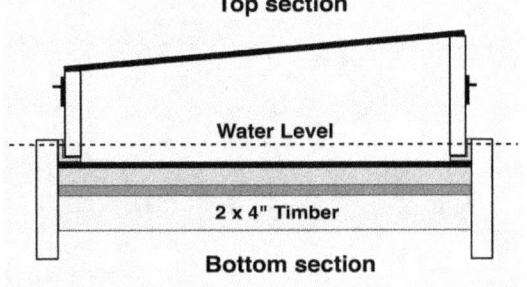

Designing A Distillation System

Chapter 12

Testing Water: Disease and Contamination

Testing for pathogens and contaminants are essential. The United States Environmental Protection Agency (EPA) recommends that, as the owner of an individual water system, it is up to you to make sure that your water is safe to drink.

As mentioned earlier, we have reached a point in human history where a very large part of our fresh water supply is contaminated with microbes, pollutants and toxins. The ever increasing need for population growth to feed the economy will do that to our environment. This already gives us reason to test our water, but if still not convinced, read on.

About four years after the Fukushima disaster in Japan, I found myself visiting the island nation of Taiwan, to the south of Japan. During this trip, I had the opportunity to speak to many Japanese citizens who had actually decided to leave Japan and to immigrate to Taiwan. In short, they felt that at the time of the incident, they had not been warned of the affect that the nuclear disaster would have on the environment and on humans; nobody had warned them that the radiation would eventually spread into the water and into the food chain, which it did. They had to take the initiative themselves, to do their own research and to make a decision to leave the area.

This is something we are starting to see more often. Large companies are reluctant to face responsibilities for their actions and major ecological disasters are almost always ignored and even dismissed as having "minor consequences".

Getting Your Water Tested

It's important to be aware of the potential dangers that your drinking water may pose for you and your family. Be informed about:
- Why test?
- What to test for?
- When to test?
- Where to test your water?
- Water related diseases and contaminants in private wells.

Well owners have to pay special attention for a private well uses groundwater as its water source. There are many reasons why our groundwater can become contaminated. Here is a list of the most common sources of contaminants:
• Naturally occurring chemicals and minerals (for example, arsenic, radon, uranium)
• Land use practices in your area (fertilizers, pesticides, livestock, animal feeding operations, biosolids application)
• Manufacturing processes
• The overflow of sewers
• Faulty wastewater treatment systems (for example, nearby septic systems). This is an important one. Watch out for septic tanks!
You are responsible for testing your well water and making sure it is safe.

Visualize a province, state or county with surface water like rivers, lakes and dams. These water sources have a network of pipes that flow out of them and that connect with cities, suburbs and households all over the area. Any ecological disaster, like a sewerage spill or chemical disaster, near the source, will affect the millions of households down the line. The issue with large populations is, that the news of the spill will only reach us once the damage has been done. On top of that, damage might last for decades and for various reasons, the powers that be, might not even have reported it in the first case.

What To Test For?

According to the Centers for Disease Control and Prevention, if you are getting your water from a privately-owned well or from local surface water sources like streams, rivers or lakes, you should test for water quality indicators:

• **Total Coliforms**

Coliform bacteria are microbes found in the digestive systems of warm-blooded animals, in soil, on plants, and in surface water. These microbes typically do not make you sick; however, because microbes that do cause disease are hard to test for in the water, "total coliforms" are tested instead. If the total coliform count is high, then it is very possible that harmful germs like viruses, bacteria, and parasites might also be found in the water.

• Fecal Coliforms (E. Coli)

Fecal coliform bacteria are a specific kind of total coliform. The feces (or stool) and digestive systems of humans and warm-blooded animals contain millions of fecal coliforms. E. Coli is part of the fecal coliform group and may be tested for by itself. Fecal coliforms and E. Coli are usually harmless. However, a positive test may mean that feces and harmful germs have found their way into your water system. These harmful germs can cause diarrhea, dysentery, and hepatitis. It is important not to confuse the test for the common and usually harmless WQI E. Coli, with a test for the more dangerous germ E. Coli O157:H7.

• pH Level

The pH level tells you how acidic or basic your water is. The pH level of the water can change how your water looks and tastes.

If the pH of your water is too low or too high, it could damage your pipes, cause heavy metals like lead to leak out of the pipes into the water, and eventually make you sick. It's advisable to test pH levels often, since they can be affected by various conditions.

When To Test?

The CDC recommends the following:

Wells

At a minimum, check your well every spring to make sure there are no mechanical problems; test it once each year for total coliform bacteria, nitrates, total dissolved solids, and pH levels. If you suspect other contaminants, you should test for those as well. However, spend time identifying potential problems as these tests can be expensive. The best way to start is to consult a local expert, such as the local health department, about local contaminants of concern. You should also have your well tested if:
• It is a new well and you want to document the water quality.
• There are known problems with well water in your area.
• You have experienced problems near your well (i.e., flooding, land disturbances, gas leakage, chemical spill, or discovered nearby waste disposal sites).
• You replace or repair any part of your well system.
• You notice a change in water quality (i.e., taste, color, odor).
• If someone in your household is pregnant, nursing an infant, or if there are unexplained illnesses in your family.
• As the well owner, you suspect

When To Test Your Water (EPA Guideline)	
Conditions or nearby activities	Recommended Test
Recurrent gastro-intestinal illness.	Coliform bacteria.
Household plumbing contains lead.	pH, lead, copper.
Radon in indoor air or region is radon rich.	Radon.
Scaly residues, soaps don't lather.	Hardness.
Water softener needed to treat hardness.	Manganese, iron.
Stained plumbing fixtures, laundry.	Iron, copper, manganese.
Objectionable taste or smell.	Hydrogen sulfide, corrosion, metals.
Water appears cloudy, frothy or colored.	Color, detergents.
Corrosion of pipes, plumbing.	Corrosion, pH, lead.
Rapid wear of water treatment equipment.	pH, corrosion.
Nearby areas of intensive agriculture.	Nitrate, pesticides, coliform bacteria.
Coal or other mining operation nearby.	Metals, pH, corrosion.
Gas drilling operation nearby.	Chloride, sodium, barium, strontium.
Odor of gasoline or fuel oil, and near gas station or buried fuel tanks.	Volatile organic compounds (VOC).
Dump, junkyard, landfill, factory or dry-cleaning operation nearby.	VOC, Total disolved solids (TDS), pH, sulfate, chloride, metals.
Salty taste and seawater, or a heavily salted roadway nearby.	Chloride, TDS, sodium.

there is something wrong with your water quality.

Public Water

If you do not own a well, you can still have your water tested. If you are purchasing water from a public water system, then your water is monitored, tested and the results reported to the federal, state or tribal drinking water agencies responsible for making sure it meets the National Primary Drinking Water Standards. Your water company must notify you when there are contaminants in the water they provide, that may cause illness or other problems.

You can also have your water tested if:
- If you suspect lead (or other chemicals) may be in some of your household plumbing materials and water service lines.
- There are known problems with water in your area.
- You notice a change in water quality (i.e., taste, color, odor).
- If someone in your household is pregnant, nursing an infant, or if there are unexplained illnesses in your family.

Where To Test?

Local health or environmental departments often test for nitrates, total coliforms, fecal coliform, volatile organic compounds, and pH (see above). Health or environmental departments, or county governments should have a list of the certified (licensed) laboratories in your area that test for a variety of substances. They cover both city water and well water and will require you to send them a sample. They test for hundreds of target contaminants and should have you covered. Typically, the process looks like this:

1. Find a lab and call them.
2. Tell the lab the water tests you need based on the EPA guidelines.
3. Ask if they're certified to do all the tests or use another certified lab.
4. Wait for the water test kits to arrive in the mail. The kits include empty bottles, directions, and forms to fill out.
5. Read the directions, fill the bottles with tap water, and mail them back to the lab.
6. Watch for test results in the mail and compare whether your results fall within the safe limits on well testing.
7. Make adjustments to your system if needed.

Self-monitoring test kits can also be bought online, at large hardware stores (plumbing) and select home retail outlets. These test kits are easy to use. A basic test kit will test for bacteria, lead, copper, pesticides, nitrites/nitrates, chlorine, hardness, and pH.

Kits are also available that test specifically for contaminants like arsenic, lead, radon, coliform bacteria, etc. Kits include all test materials and no laboratory testing is necessary.

Testing Odors Or Colors In Your Water

Certain odors or colors can indicate the presence of contaminants in your water system. As a simple rule of thumb, follow these guidelines to determine the cause of the odor:

- **Odor in all faucets** -> Problem with main water supply, whether public or from a private well.
- **Odor in some faucets** -> Problem in plumbing system (pipes) nearest to the faucet with odor.
- **If the odor disappears after a few minutes** -> Problem in plumbing system.
- **If the odor lingers indefinitely** -> Problem with main water supply or plumbing system or combination.

For odors in the kitchen around the sink, fill a glass with water and move away from the sink. If your water does not smell, then the problem lies with the drain system.

Often we find that plumbing systems and water heaters are the main culprits when it comes to odors and colors. Public water is treated with chlorine and when organic matter gets trapped in your plumbing system, the reaction with chlorine can cause sulfurous odors.

When dealing with wells we find that surface water contamination and septic tank seepage are the main causes of a change in color or odor.

Colors with a residue (usually metals) will of course stain linen and clothes when washed with the contaminated water.

Test immediately if unsure about the source of the odor or color:

- **Sulfurous odor**

This is the rotten egg smell we sometimes find in drain pipes. This odor can be caused by a variety of problems and it's best to call in an expert plumber to do an assessment.

Cause: The smell may be due to hydrogen sulfide gas forming in the water heater itself, sulfur in the building water supply, bacteria in the drain pipes or in the water system, deteriorating water heater electrodes or dangerous sewer gas leaks. It can also be caused by decaying organic material underground, which causes hydrogen sulfide gas. The water picks up this gas underground and carries it to the surface where it is released again. To test for hydrogen sulfide gas, open both the hot and cold water faucets, to decide which faucets have the rotten egg odor.

- If the rotten egg odor is coming only from the hot water faucet, it means that the problem is probably in the water heater.
- If the odor is strongest when both the hot and cold faucets is first turned on, and then becomes weaker (or disap-

pears) after the water has run for a while, the problem likely is sulfur bacteria in the well or plumbing system.
- If the smell is strongest when you first turn on both the hot and cold faucets and it stays constant and persists with use, the problem likely is hydrogen sulfide in the groundwater that supplies the well.
Solution: Have your water tested and call in an expert plumber to do an assessment.

- **Chemical, chlorine odor**

If you smell bleach or similar chemicals check for chlorine in your water system. Chlorine is not good for you, especially when you inhale it (showers).
Cause: The water treatment plants use chlorine for water disinfection.
Solution: If it persists, contact your water utility office or have it tested. Activated carbon filters, shower filters and reverse osmosis systems can remove chlorine.

- **Petroleum odor**

It's not often that you smell gasoline or petroleum in your water.
Cause: Anything that smells like gasoline, petroleum or turpentine indicates an external contaminant. This indicates that the source can be a chemical plant, gasoline tanks or runoff from neighboring areas. A known chemical found in gasoline is MTBE. Although not explicitly harmful, it's always best to avoid chemicals (especially used in gasoline) and to have your water tested.
Solution: Stop drinking the water and call in the help of an expert plumber. Have your water tested and check with your local water utility office to locate the source.

- **Fishy, earthy or moldy odors**

This smell can be caused by harmless bacteria or the presence of metal.
Cause: This can indicate decaying organic matter (bacteria) or pollution of well water from surface drainage. Another cause could be the presence of Barium or Cadmium. Barium is a metal that occurs naturally in nature. Cadmium can be caused by deteriorating galvanized plumbing.
Solution: Locate and test for the source of contamination.
To remove Barium use: an Ion Exchange system, Reverse Osmosis, or a Water Softener system.
And for removing Cadmium: Filtration, Ion Exchange, Water Softener system, and/or Reverse Osmosis.

- **Musty, unnatural odor**

Cause: This usually indicates organic matter in the water supply. If you smell other familiar chemicals, remember that certain pesticides have a very distinct smell that can be easily identified.
Solution: Try to locate the source of the contamination. It

can also indicate the presence of pesticides in your water. If you suspect the presence of pesticides, stop drinking the water and have it tested immediately.

- **Methane gas**

Cause: This can be caused by mining or oil drilling activities in the area. It is also found in wells located near swamps and marshes.

Solution: It's best to contact a certified well contractor in your area to see if a well vent or similar can solve the problem.

- **Foam in your water**

The presence of foam in your water is never a good sign, especially when accompanied by any odors.

Cause: This could be caused by a leaking septic tank close to a well. This is a very serious problem and the cause of the leakage will have to be investigated.

Solution: Stop drinking the water and call in the help of an expert plumber to shock chlorinate your well.

- **A blue or green color**

This is a situation you sometimes find in homes with old plumbing systems.

Cause: This color indicates the presence of copper. There is copper in drinking water, but in small amounts.

Solution: If you suspect higher amounts in your water (usually well water), then stop drinking the water and have it tested. Anything higher that 30 ppm should be considered as problematic.

- **A red or brown color**

Sometimes you will see red stains in your sink and/ or an odor like sewerage, fuel, or cucumber.

Cause: These colors are indicative of iron or manganese. Ion is a common substance found inside the earth. It can dissolve in water and easily enter a home's water supply system. The red stains you see around drains and basins are deposits of either ionic ferrous iron, ferric iron, or "iron bacteria". Iron-bacteria are more commonly found in water from dug wells or shallow wells.

Solution: Ion or manganese filters and/ or water softeners can aid in removing the unwanted substances and calling an expert to disinfect your well (chlorine) can help in getting rid of the iron bacteria.

- **Cloudy, milky white water**

Sometimes water can appear milky white as it flows out the faucet.

Cause: One likely cause is tiny air bubbles in the water. A more serious problem can be surface water contamination.

Solution: Fill a clear glass with water and set it on the counter. If the water starts to clear at the bottom of the glass first, the cloudy, white appearance is trapped air. This is not a serious issue and might be caused by your well's pumping system. If it does not settle, then investigate for surface water contamination.

Water Related Diseases And Contaminants In Private Wells

As the owner of a private well, you are responsible for testing your water. Make sure that your house has a water treatment system installed if you are using your well as a source for drinking water. Certain unwanted pollutants can be routinely removed by a water treatment system. Depending on which contaminant is present, you can install sediment filters, a reverse osmosis system, ozone purifying system, etc.

If temporarily unable to find a treatment solution, then use the emergency solutions provided by the Centers for Disease Control and Prevention.

- **Hepatitis A**

Hepatitis A is a virus that can be found anywhere that human feces is present. This means that it can enter your well through the flow of storm-water runoff, septic tanks, sewerage overflow and contaminated waste water.

Immediate solutions:

1. Boil your water for 1 minute.
2. Disinfect your well. Chlorination kills the virus.
3. Remove the source of contamination.
4. Get advice from your local health department for recommended procedures.

- **Giardia**

Giardia is a parasite that causes diarrhea. It is found worldwide where humans live and defecate. This parasite can enter your well through the flow of storm-water runoff, septic tanks, sewerage overflow and contaminated waste water.

Immediate solutions:

1. Boil your water for 1 minute.
2. Install a reverse osmosis filtration system. It must have a pore size smaller than 1 micron. Tested and certified NSF Standard 53 for cyst removal and reduction.

- **Campylobacter**

Campylobacter is a bacteria we often see in the summer months. It causes diarrhea. It is found worldwide in human and animal feces. This bacteria can enter your well through the flow of storm-water runoff, septic tanks, sewerage overflow and contaminated wastewater.

Immediate solutions:

1. Boil your water for 1 minute.
2. Remove the source of contamination.
3. Get advice from your local health department for recommended procedures.

- **Shigella**

Shigella is a bacteria we often see in shallow wells after an occurrence of flooding. It is a bacteria that causes a very contagious type of diarrhea. It is found worldwide in human and animal feces. This bacteria can enter your well through the flow of storm-water runoff, septic tanks, sewerage overflow and contaminated waste water.

Immediate solutions:

1. Boil your water for 1 minute.
2. An ultraviolet light device can be used.
3. In emergencies, use iodine tablets.
4. Get advice from your local health department for recommended procedures.

• **Noroviruses**

Noroviruses cause intestinal illness, or gastroenteritis, and have been associated with outbreaks on cruise ships and in communities, restaurants, camps, schools, institutions, and families. Noroviruses may be found in water sources, such as private wells, that have been contaminated with the feces of infected humans. Waste can enter the water through various ways, including sewage overflows, sewage systems that are not working properly, and polluted storm-water runoff. Wells may be more vulnerable to such contamination after flooding, particularly if the wells are shallow, have been dug or bored, or have been submerged by floodwater for long periods of time.

Solution:

1. Bring your water to a rolling boil for one minute (at elevations above 6,500 feet, boil for three minutes).
2. Water should then be allowed to cool, stored in a clean sanitized container with a tight cover, and refrigerated.

Because of the small size of the virus, using a traditional point-of-use filter will not remove it from your water.

• **Rotaviruses**

Rotaviruses are wheel-shaped ("rota-") viruses that cause intestinal illnesses and are the most common cause of severe diarrhea among infants and children. The virus may be found in water sources such as private wells that have been contaminated with the feces from infected humans. Waste can enter the water through different ways, including sewage overflows, sewage systems that are not working properly, and polluted storm-water runoff. Wells may be more vulnerable to such contamination after flooding, particularly if the wells are shallow, have been dug or bored, or have been submerged by floodwater for long periods of time.

Solution:

1. Bring your water to a rolling boil for one minute (at elevations above 6,500 feet, boil for three minutes) Water should

then be allowed to cool, stored in a clean sanitized container with a tight cover, and refrigerated. Because of the small size of the virus, using a traditional point-of-use filter will probably not be effective at removing it from your water.

• E.Coli

E. Coli is a bacteria that can cause severe illness. One particular strain produces a very powerful toxin that needs to be avoided at all costs. This bacteria can enter your well through the flow of storm-water runoff, septic tanks, sewerage overflow and contaminated waste water.

Solution:

1. Boil your water for 1 minute.
2. Disinfect your well. Chlorination kills the virus.
3. Remove the source of contamination.
4. Get advice from your local health department for recommended procedures.

• Cryptosporidium and Salmonella

"Crypto" is a parasite that causes diarrhea. It is found worldwide where humans live and defecate. This parasite can enter your well through the flow of storm-water runoff, septic tanks, sewerage overflow and contaminated waste water.

Immediate solutions:

1. Boil your water for 1 minute.
2. Install a reverse osmosis filtration system. It must have a pore size smaller than 1 micron. Tested and certified NSF Standard 53 for cyst removal and reduction.
3. Consider an ultra-violet light source or ozone as alternative disinfection process.
4. Get advice from your local health department for recommended procedures.

• Arsenic

Arsenic can enter a well through industrial waste released into the environment, the use of arsenic containing fertilizers and through the natural occurring ground deposits found in the earth.

Solutions:

1. Filter your water through reverse osmosis.
2. Ultra-filtration
3. Distillation
4. Get advice from your local health department for recommended procedures.

• Nitrate

Nitrate can occur naturally in surface and groundwater at a level that does not generally cause health problems. High levels of nitrate in well water often result from improper well construction, well location, overuse of chemical fertilizers, or improper disposal of human and animal waste. Sources of nitrate that can enter your well include fertilizers, septic systems, animal feedlots, industrial waste, and food processing waste. Wells may be more vulnerable to such contamination after flooding, particularly if the wells are shallow, have been dug or bored, or have been submerged by floodwater for long periods of time.

Solutions:

1. Treatment processes such as ion exchange,

2. Distillation

3. Reverse osmosis

Contact your local health department for recommended procedures.

Heating or boiling your water will not remove nitrate. Because some of the water will evaporate during the boiling process, the nitrate levels of water can actually increase slightly in concentration if the water is boiled. Mechanical filters or chemical disinfection, such as chlorination, DO NOT remove nitrate from water.

- **Lead**

Lead rarely occurs naturally in water; it usually gets into the water from the delivery system. Lead pipes are the main contributor to high lead levels in tap water. Other sources include parts of the water delivery system such as lead solder used to join copper pipes, brass in faucets, coolers, and valves. Although brass usually contains low lead levels, the lead can still dissolve into the water, especially when the fixtures are new. Private wells more than 20 years old may contain lead in the "packer" element that is used to help seal the well above the well screen. Some brands of older submersible pumps used in wells may also contain leaded-brass components. Corrosion of pipes and fixture parts can cause the lead to get into tap water.

Solutions:

1. First, try to identify and remove the lead source. If you have a private well, check both the well and the pump for potential lead sources.

2. Heating or boiling your water will not remove lead. Because some of the water evaporates during the boiling process, the lead concentration of the water can actually increase slightly as the water is boiled.

3. If it is not possible or cost-effective to remove the lead source, flushing the water system before using the water for drinking or cooking may be an option. Any time a particular faucet has not been used for several hours (approximately 6 or more), you can flush the system by running the water for about 1-2 minutes or until the water becomes as cold as it will get. Flush each faucet individually before using the water for drinking or cooking. You can use the water flushed from the tap to water plants, wash dishes or clothing, or clean. Avoid cooking

with or drinking hot tap water because hot water dissolves lead more readily than cold water does. Do not use hot tap water to make cereals, drinks or mix baby formula. You may draw cold water after flushing the tap and then heat it if needed.

4. You may also wish to consider water treatment methods such as reverse osmosis, distillation, and carbon filters specially designed to remove lead. Typically these methods are used to treat water at only one faucet.

Contact your local health department for recommended procedures.

- **Radon**

Radon is a colorless, tasteless, odorless, radioactive gas. It occurs naturally and is produced by the breakdown of uranium in soil, rock, and water. It can also dissolve into our water supply. High levels of dissolved radon are found in the groundwater in some areas flowing through granite or granitic sand and gravel formations. If you live in an area with high radon in groundwater it can get into your private well. Showering, washing dishes, and laundering can disturb the water and release radon gas into the air you breathe.

Solutions:

Radon can be removed from water by using one of two methods:

1. Aeration treatment - spraying water or mixing it with air and then venting the air from the water before use, or

2. GAC treatment - filtering water through granular activated carbon. Radon attaches to the carbon and leaves the water free of radon. Disposing of the carbon may require special handling if it is used at a high radon level or if it has been used for a long time.

In either treatment, it is important to treat the water where it enters your home (point-of-entry device) so that all the water will be treated. Point-of-use devices such as those installed on a tap or under the sink will only treat a small portion of your water and are not effective in reducing radon in your water. It is important to maintain home water treatment units properly because failure to do so can lead to other water contamination problems. Some homeowners use a service contract from the installer to provide carbon replacement and general system maintenance.

Remember to have your well water tested regularly, at least once a year, after installing a treatment system to make sure the problem is controlled.

- **Fluoride**

Fluoride is a mineral that occurs naturally and is released from rocks into the soil, water, and air. Various studies (some controversial) have been done to advocate for or against the fluoridation of drinking water.

For some reason, fluoride is always linked to tooth decay and there seems to be an obsession with the believe that it can and will prevent tooth decay.

It's interesting to note that only about 5% of the world population is fluoridated and more than 50% of these people live in North America. Fluoridation is rejected in most advanced societies including Germany, Austria, France, Denmark, Norway, Switzerland, Sweden, Japan and various others.

Decisions about adding fluoride to drinking water are made at the state or local level. The best way to find the fluoride level of your local public water system is to contact your water utility provider.

If you rely on well water you should have your water tested for fluoride. It is unusual to have the fluoride content of water be 4 mg/L or higher. If a laboratory report indicates that you have such a high fluoride content, you may want to consider alternate sources of water for drinking and cooking, or installing a device to remove the fluoride from your home water source. There are a variety of gravity-fed filters available that cater specifically for this need.

Conclusion

When reading through this list, it is obvious that the one issue that affects almost all wells is storm-water drains and run-offs. Failure to secure an effective storm-water drain runoff can be considered one of the main causes of well contamination. Storm-water runoff picks up all kinds of pollutants like pesticides, bacteria, and chemicals. It travels downhill at a rate that is too fast to be absorbed into the ground. When faced with floods, we are at mother nature's mercy, but with storm-water runoff we have options:

- Make sure that you storm-water drain is located away from your well, and also pay attention to the run-off direction of your water.

- Harvest the rain from your roof and make sure all gutters empty into downspouts that lead to tanks or barrels.

- Make sure that there are no potential contaminants in the vicinity of your well. Pay special attention to pesticides, compost manure piles, livestock areas and septic tanks.

- Consider permeable surfaces around your well to allow for water absorption. This includes investigating driveway paving, decks and lawns (bad).

- Get involved in the community. Make fellow community members aware of the importance of storm-water management.

How To Manage Waste

When disaster strikes and you run out of water, you will have to implement a water conservation plan. Water is responsible for our sanitation needs and without proper sanitation facilities, waste from infected individuals can contaminate a community's land and water, increasing the risk of infection for other individuals.

The absence of basic sanitation facilities can:

— Result in an unhealthy environment contaminated by human waste. Proper waste disposal can slow the infection cycle of many disease-causing agents.

— Contribute to the spread of many diseases and conditions that can cause widespread illness and death. Without proper sanitation facilities, people often have no choice but to live in, and drink water from, an environment contaminated with waste from infected individuals, thereby putting themselves at risk for future infection. Inadequate waste disposal drives the infection cycle of many agents that can be spread through contaminated soil, food, water, and insects such as flies.

Sanitation is an aspect of preparedness that the whole household should be aware of.

Basic hygiene when handling waste

• Wash hands with soap and water immediately after handling human waste or sewage.

• Avoid touching face, mouth, eyes, nose, or open sores and cuts while handling human waste or sewage.

• After handling human waste or sewage, wash your hands with soap and water before eating or drinking.

• After handling human waste or sewage, wash your hands with soap and water before and after using the toilet.

• Before eating, removed soiled work clothes and eat in designated areas away from human waste and sewage-handling activities.

• Do not smoke or chew tobacco or gum while handling human waste or sewage.

• Keep open sores, cuts, and wounds covered with clean, dry bandages.

• Gently flush eyes with safe water if human waste or sewage contacts eyes.

• Use waterproof gloves to prevent cuts and contact with human waste or sewage.

- Wear rubber boots at the work site and during transport of human waste or sewage.

- Remove rubber boots and work clothes before leaving work site.

The Need For Latrines And Toilets

Proper sanitation facilities (for example, toilets and latrines) promote health because they allow people to dispose of their waste appropriately. Make sure that you place your toilet away from area where food gets prepared or where people eat.

Keep toilets at least 100 feet away from water sources such as lakes, rivers, streams, and wells.

Make sure that people can wash their hands with soap, running water, and paper towels.

The emergency preparedness industry is rapidly expanding and these days emergency sanitation devices are readily available.

In an emergency, you can consider the following options:

- Emergency biodegradable bags
- Heavy-duty home refuse bags.
- Buckets
- A latrine trench
- An outhouse

1. Bags

Bags can be used on their own, inside an empty toilet or in a container (bucket or portable toilet. We can use two kinds of bags:

1.) Special biodegradable waste bags (expensive).

2.) Heavy-duty home refuse bags (cheap).

1. Biodegradable bags

You can buy portable toilet waste bags for an easy solution to your toilet needs. They consist of an inner waste bag and a leak-proof outer carry bag.

A zip locking mechanism makes for a secure seal.

These bags come with a chemical disinfectant in powder or gel form.

These bags can be used with empty toilets, portable toilets or with buckets.

2. Heavy duty home refuse bags

An immediate (and very practical) solution for every individual in the household is to use heavy-duty home refuse bags. These bags typically are placed inside reusable buckets or empty toilets and when done, have to be buried or properly disposed of according to local health regulations.

You will need carbon material to cover the poo: A supply of sawdust, coffee husk chaff, finely shredded paper, kitty litter or coconut fiber.

Bags are very handy and can be a lifesaver in an emergency situation.

Upside:
- No infrastructure required.
- Lightweight and easy to transport
- May be used where space is severely limited or in flooded areas.

Downside:
- Active supply chain needed to provide approximately 1 bag, per person, per day.
- High costs of some types of bags.
- Need for disposal site and possibly collection services.
- Potential for bags to be discarded in open areas or areas posing risks to others.
- Requires intense hygiene campaign to educate the household on handling and disposal of bags.

2. Buckets

Bags work very well with 5 gallon (19L) plastic buckets. Buy two 5 gallon buckets with lids. One bucket will be the "pee" bucket, and the other will be the "poo" bucket. You can buy a snap-on toilet seat for both 5 gallon buckets and each unit will function as a toilet.

When dealing with waste we have to separate the fecal matter from the urine. Urine (of a healthy individual) is actually sterile and that's why it's better to separate it from the fecal matter. The harmful pathogens resides in the bucket with the fecal matter. Urine can later (wait a few days) be sprinkled on land as a fertilizer.

Prepare:
- Two 5 gallon plastic buckets.
- These days snap-on toilet seats are available for 5 gallon buckets.
- Carbon material to cover the poo: A supply of sawdust, coffee husk chaff, finely shredded paper, kitty litter, fireplace ash, or coconut fiber.

Pay Attention To The Following:

1. Make sure that you mark the two buckets as "pee" and "poo" (or similar).

2. Prepare a space that will provide some privacy and place the bag with carbon covering material nearby, along with a scoop or tongs.

3. Once it's time to clean the buckets, try not to mix the pee and poo. Once the pee and poo are mixed, the smell will be quite unpleasant.

4. After using the pee bucket, the user should put the toilet paper in the poo bucket. Then, remove the seat from the pee bucket and cover it with a lid that closes well.

5. After using the poo bucket, sprinkle as much carbon material as needed to completely cover the surface of the poo. This eliminates odors and keeps flies away.

6. Put the toilet seat back down on the poo bucket so it doesn't invite pests.

7. Sanitize your hands thoroughly!

3. Digging An Emergency Trench Latrine

You can always buy a portable camp toilet to be used in an emergency. Eventually, if the disaster persists and you run out of bags and ways to dispose of them, you will have to consider digging a trench.

1. Choose a convenient site the minimum safe distance from water sources. Locate latrines and portable toilets at least 100 feet away from surface water bodies such as lakes, rivers, streams, and at least 150 feet downhill or away from any drinking water source (well or spring), home, apartment, or campsite.

2. Dig a trench 1 foot (0.3 m) wide by 3 feet (0.75 m) deep. The length of the trench will depend on the number of users ; about 3 feet (0.75 m) per user is enough.

3. Place boards along each side of the trench for people to stand on. Privacy can be improved by building a screen around the latrine. Use simple material.

4. Leave the dirt, obtained from digging the pit, in a pile near the trench with a shovel. Once used, each person covers over their feces with a shovelful of earth to keep away flies and reduce odors. Additionally, throw lime, mulch, sawdust, coffee husk chaff, finely shredded paper, kitty litter, or ash in the latrine to minimize odors and to keep flies, mosquitoes, and rodents away.

5. Store stacks of the traditional anal cleansing material used in the area close by, and top them up regularly. Old telephone books and Yellow Pages make excellent emergency toilet paper.

6. You can cover the latrine between use with plywood, or another material.

Close the trench when it is within 1 foot (0.3 m) of ground level. Cover it over with earth and pack it down tightly. The area should not be disturbed for two years.

4. Digging An Outhouse

Make sure that you follow your local health regulations and the restrictions that apply to outhouses. Pay special attention to restrictions that apply to water sources, animal enclosures and domestic living areas. Design your outhouse to suit the weather and climate in your area. Give special attention to ventilation (vents), humidity (mold resistant paint), snow (insulation) and insects (screens). Important to note that you definitely want the soil around your outhouse to drain easily.

How to:

1. Dig a hole 4 ft deep, and 3.5 x 3.5 ft square. Make sure that the hole is square and that the walls are even and straight down.

2. You want to line the inside of the hole with wood. Keep the bottom of the hole open. This is where the waste will seep into the ground. Construct a "box" out of wood (2 x 4's with cross pieces) and wrap it in waterproof material like tar paper that construction workers use.

The tar paper will keep moisture out of the pit. The "box" should fit snug in the hole and the bottom and top sections should be left open.

3. Level out the area around the hole and create a square foundation for your outhouse with some 4 x 4's. You should have a square pit in front of you with the ground visible at the bottom and a nice frame of wood around the hole.

4. For the actual outhouse, you have two options:

a.) One option is to build a solid structure with four posts that

Keep small children away from the outhouse! Small kids should always be accompanied by an adult when using the out-house!

can be set in concrete. Use a post hole digger to dig your post holes. This is a permanent fixture on your property and you can make it as sturdy as you desire.

b.) Another option is to make it a temporary structure that can be mobile when needed. Build the structure, but keep in mind that you will relocate it at a later stage.

For both construction options you can use plywood and timber posts. The corner posts should be timber (4 x 4's) and you can use ½" plywood for the floor, roof, and seat. Use ¼" plywood for the walls. The roof should be sturdy enough to keep rain and/ or snow out. Depending on climate, you can use shingles, roof panels or sheet metal.

You will also need a vent pipe. The vent pipe should be venting from as high as possible in the chamber, since hot air rises. One option is to use 4" PVC pipe (with holes) that runs from the seat all the way up through the roof. You can place a pipe cap on top to keep water out.

Another option is to run the vent pipe on the outside of the

Testing Water: Disease and Contamination

building, all the way up to the top next to the roof. Use black pipe and place it on the sunny side so that the sun-heated black pipe will draw the air by convection. This avoids having to put a hole in the roof as well.

Some Tips To Keep In Mind:

- Use brass screws or coated screws to avoid rust.

- Waterproof the area around and underneath the toilet seat. You can use a plastic sealant. You can do this anywhere you think moisture might be an issue.

- If you live in a cold area you should definitely try to insulate the building.

- Paint the inside of the structure with mold resistant paint.

- Put screens on all pipe holes, vents and window openings to keep the bugs and critters out.

- Set your door so that it opens inwards.

You will need to keep carbon material in your outhouse to sprinkle over the waste when done. Store stacks of the traditional anal cleansing material used in the outhouse, and top them up regularly. Old telephone books and Yellow Pages make excellent emergency toilet paper.

Important! Do not put bleach, chemicals, pesticides, trash, cigarettes, or anything non-biodegradable down the hole. Keep things simple and try to live with the odors that will escape from time to time.

Small children should never be allowed near an outhouse. If you have small children in the house, consider locking the outhouse door with a simple combination lock.

What To Do With A Filled Up Outhouse?

This is something that should only happen when there is a sudden increase in users. For a normal family, a standard outhouse should last for decades. If, for whatever reason, your outhouse has reached the end of its holding capacity, you can consider moving it to a new location.

How to:

1. Prepare a new location. Dig a new hole.

2. Move the outhouse to the new location.

3. Fill up the old hole with saw dust, ash and natural organic matter. Use the dirt from the new hole to completely fill up the old hole.

4. Plant some trees, shrubs or flowers over the hole. This indicates where the old hole was. Do not disturb the area for at least two years.

Discussing Waterborne Diseases

As the name suggests, waterborne diseases facilitate their journey of contamination through the medium of water. When we have unnatural high rainfall, we see an increase in floods, overflowing rivers, stagnant pools and also rotten organic matter. This creates a perfect environment for these diseases to spread and ultimately, to infect human populations. Waterborne diseases are caused by pathogenic microbes that can be directly spread through contaminated water. Most waterborne diseases cause diarrheal illness. Eighty-eight percent of diarrhea cases worldwide are linked to unsafe water, inadequate sanitation or insufficient hygiene. These cases result in 1.5 million deaths each year, mostly in young children. The usual cause of death is dehydration.

The best way to avoid a waterborne disease is to be aware of proper hygiene and to avoid coming into contact with contaminated water. As living organisms ourselves, we can rely on our immune systems and supportive care to recover from a waterborne disease. In some cases, depending on the pathogen and also the infected person's ability to fight an infection, some people may require antibiotics or other forms of treatment. In most cases, restricting the patient to clear fluids and avoiding solid food for a period of time will do the trick. However, other symptoms in combination with diarrhea can be a sign of something more serious lurking underneath. **Keep an eye open for the following additional symptoms:**

- Fever greater than 101 F or 38.3 C
- Blood in stool
- Severe vomiting
- Major intestinal pain
- Severe dehydration
- Diarrhea lasting longer than 3 days

When we leave diarrheal illnesses untreated, we eventually end up with a state of dehydration. Since the human body consists of mostly (75%) water, it is no surprise that a loss of fluids causes severe discomfort and can lead eventually to death.

There are three stages of dehydration:

1. Mild dehydration. Loss of 0-5% of body water (weight).

Symptoms:

- Anxiety
- Loss of appetite
- Dry, sticky mouth
- Tiredness, and feeling less active than usual
- Thirst

- Darker, concentrated urine
- Dry skin
- Headache
- Dizziness or lightheadedness

2. Moderate dehydration.
Loss of 5-10% of body water (weight).

Symptoms:
- Nausea and vomiting
- Dizziness
- Fatigue
- Mood swings

3. Severe dehydration. Loss of 10% or more of body water (weight).

Symptoms:
- Patient becomes delirious and
- confused
- Loss of coordination
- Little or no urine output

The best and easiest way to treat dehydration is by preventing it altogether.

Re-hydration

Oral re-hydration is the first step in managing diarrheal dehydration. Special care should be taken with the young and the old.

Consider the following methods:

- Over the counter re-hydration packets are available and contain vital minerals that need to be replaced in the patient.

- Clear fluids are readily absorbed by the body and can be considered when symptoms are less severe.

- You can also manufacture your own home-made re-hydration drink.
Take one liter of water and add:
- 1 teaspoon of salt
- 6 teaspoons of sugar
- ½ teaspoon of baking soda

The patient should drink as much as possible.
As the patient improves, they can progress to thin cereals, congee (watery rice), broths and juice.

- For children the BRAT diet works well to treat dehydration:
- Bananas
- Rice
- Applesauce
- Toast or crackers

This diet is effective at slowing down the movements of the intestines and thus it gives the body more time to absorb fluids and it slows down water loss.

- Alternatively, consider natural remedies like ginger tea, hot water with raw honey, a teaspoon of apple-cider vinegar in a glass of water, bananas and yogurt.

- As a last resort, consider antibiotic medication to treat your diarrhea. This will hopefully provide your body with an opportunity to rehydrate. Your personal physician should be able to give you some advice in this regard.

Caution! Check all expiry dates before using any medication.

Treating Waterborne Diseases

The following section is to be used as a guide to be consulted when needed. The information is provided by the Centers For Disease Control And Prevention (CDC) and is aimed at informing citizens of the dangers present in untreated water.

In an emergency, people are often forced to consume food and water from unknown sources. The disruption of power and water lines can cause sanitation problems and neglecting the treatment and proper disposal of human waste can have catastrophic consequences. As mentioned earlier, most waterborne diseases cause diarrheal illness. The majority of diarrhea cases are linked to unsafe water, inadequate sanitation or insufficient personal hygiene. If untreated, diarrhea leads to dehydration and in severe cases, to death.

Pathogens are extremely versatile and can survive in a variety of environments, but they especially flourish in areas with contaminated water. These pathogens that we find in water, all pose a potential threat to human health.

When it comes to waterborne diseases, the CDC recommends paying special attention to the following list:

- Cholera
- Giardiasis
- Shigellosis
- Amebiasis
- E. Coli
- Typhoid
- Hepatitis A
- Leptospirosis

These are serious pathogens that can cause severe illness.

Make sure that you are familiar with the symptoms of each.

CAUTION. Although most waterborne diseases cause symptoms that resemble food-poisoning, it's also important to be aware of the fact that anti-diarrheal medication is not to be administered to treat the symptoms. Only administer medication once the illness has been positively diagnosed by a healthcare professional.

A big problem with diarrhea is the waste that it creates. Waste, especially fecal matter, is highly contagious and should be disposed of according to local health regulations.

This is a serious issue that should never be neglected.

(See p.371 for advice on waste management.)

Cholera

Description

Cholera is an acute, diarrheal illness caused by infection of the intestine with the bacterium Vibrio cholerae. An estimated 3-5 million cases and over 100,000 deaths occur each year around the world. The infection is often mild or without symptoms, but can sometimes be severe. Approximately one in 10 (5-10%) infected persons will have severe disease characterized by profuse watery diarrhea, vomiting, and leg cramps. In these people, rapid loss of body fluids leads to dehydration and shock. Without treatment, death can occur within hours. The cholera bacterium is usually found in water or food sources that have been contaminated by feces (poop) from a person infected with cholera. Cholera is usually found in the aftermath of natural disasters like floods and hurricanes.

The cholera bacterium may also live in the environment in brackish rivers and coastal waters.

Symptoms

Cholera infection is often mild or without symptoms, but can sometimes be severe. Approximately one in ten (5-10%) infected persons will have severe disease characterized by
- Profuse watery diarrhea
- Vomiting
- Leg cramps

In these people, rapid loss of body fluids leads to dehydration, shock and coma. Without treatment, death can occur within hours.

Treatment

Cholera can be simply and successfully treated by immediate replacement of the fluid and salts lost through diarrhea. Patients can be treated with oral re-hydration solution, a prepackaged mixture of sugar and salts to be mixed with water and drunk in large amounts. This solution is used throughout the world to treat diarrhea. Severe cases also require intravenous fluid replacement. With prompt re-hydration, fewer than 1% of cholera patients die.

Antibiotics shorten the course and diminish the severity of the illness, but they are not as important as receiving re-hydration. Persons who develop severe diarrhea and vomiting in countries where cholera occurs should seek medical attention promptly.

Giardiasis

Description

Giardia is a microscopic parasite that causes the diarrheal illness known as giardiasis. Giardia (also known as Giardia intestinalis, Giardia lamblia, or Giardia duodenalis) is found on surfaces or in soil, food, or water that has been contaminated with feces (poop) from infected humans or animals. Giardia is protected by an outer shell that allows it to survive outside the body for long periods of time and makes it tolerant to chlorine disinfection. While the parasite can be spread in different ways, water (drinking water and recreational water) is the most common mode of transmission.

Symptoms

Giardiasis is the most frequently diagnosed intestinal parasitic disease in the United States and among travelers with chronic diarrhea. Signs and symptoms may vary and can last for 1 to 2 weeks or longer. In some cases, people infected with Giardia have no symptoms.
Acute symptoms include:
- Diarrhea
- Gas
- Greasy stools that tend to float
- Stomach or abdominal cramps
- Upset stomach or nausea/vomiting
- Dehydration (loss of fluids)
- Other, less common symptoms include itchy skin, hives, and swelling of the eye and joints. Sometimes, the symptoms of giardiasis might seem to resolve, only to come back again after several days or weeks. Giardiasis can cause weight loss and failure to absorb fat, lactose, vitamin A and vitamin B12 [2,4,9-12].

Treatment

Several drugs can be used to treat Giardia infection. Effective treatments include metronidazole, tinidazole, and nitazoxanide. Alternatives to these medications include paromomycin, quinacrine, and furazolidone. Some of these drugs may not be routinely available. Different factors may shape how effective a drug regimen will be, including medical history, nutritional status, and condition of the immune system. Therefore, it is important to discuss treatment options with a healthcare provider.

Shigellosis (Bacillary dysentery)

Description

Shigellosis is a diarrheal disease caused by a group of bacteria called Shigella. Shigella causes about 500,000 cases of diarrhea in the United States annually. Shigella germs are present in the stools of infected persons while they have diarrhea and for up to a week or two after the diarrhea has gone away. Shigella is very contagious; exposure to even a tiny amount of contaminated fecal matter, too small to see, can cause infection. Transmission of Shigella occurs when people put something in their mouths or swallow something that has come into contact with stool of a person infected with Shigella. To prevent Shigella, wash your hands with soap and water for a minimum of 20 seconds. If people do not have soap and water, they can use alcohol-based hand sanitizer with at least 70 percent alcohol.

Symptoms

Symptoms of shigellosis typically start 1–2 days after exposure and include:
- Diarrhea (sometimes bloody)
- Fever
- Abdominal pain
- Tenesmus (a painful sensation of needing to pass stools even when bowels are empty)

Treatment

Diarrhea caused by Shigella usually resolves without antibiotic treatment in 5 to 7 days. People with mild shigellosis may need only fluids and rest. Bismuth subsalicylate (e.g., Pepto-Bismol) may be helpful, but medications that cause the gut to slow down, such as loperamide (e.g., Imodium) or diphenoxylate with atropine (e.g., Lomotil), should be avoided. Antibiotics are useful for severe cases of Shigellosis because they can reduce the duration of symptoms. However, Shigella is often resistant to antibiotics. If you require antibiotic treatment for shigellosis, your healthcare provider can culture your stool and determine which antibiotics are likely to work. Tell your healthcare provider if you do not get better within a couple of days after starting antibiotics. He or she can do additional tests to learn whether your strain of Shigella is resistant to the antibiotic you are taking.

Amebiasis (Amoebic dysentery)

Description

Amebiasis is a disease caused by a one-celled protozoan parasite called Entamoeba histolytica. Although anyone can have this disease, it is more common in people who live in tropical areas with poor sanitary conditions. E. histolytica infection can occur when a person:

Puts anything into their mouth that has touched the feces (poop) of a person who is infected with E. histolytica.

- Swallows something, such as water or food, that is contaminated with E. histolytica.
- Swallows E. histolytica cysts (eggs) picked up from contaminated surfaces or fingers.

Symptoms

Only about 10% to 20% of people who are infected with E. histolytica become sick from the infection. The symptoms are often quite mild and can include:
- Loose feces (poop)
- Stomach pain
- Stomach cramping
- Weight loss

Amebic dysentery is a severe form of amebiasis associated with:
- Stomach pain
- Bloody stools (poop)
- Fever

Rarely, E. histolytica invades the liver and forms an abscess (a collection of pus).

In a small number of instances, it has been shown to spread to other parts of the body, such as the lungs or brain, but this is very uncommon.

Treatment

For symptomatic intestinal infection and extra-intestinal disease, treatment with metronidazole or tinidazole should be followed by treatment with iodoquinol or paromomycin. Asymptomatic patients infected with E. histolytica should also be treated with iodoquinol or paromomycin, because they can infect others and because 4%–10% develop disease within a year if left untreated.

E. Coli

Description

Escherichia coli (E. Coli) bacteria normally live in the intestines of people and animals. Most E. Coli are harmless. However, some E. Coli are pathogenic, meaning they can cause illness, either diarrhea or illness outside of the intestinal tract. The types of E. Coli that can cause diarrhea can be transmitted through contaminated water, food, or through contact with animals or persons. Six pathotypes are associated with diarrhea and collectively are referred to as diarrheagenic E. Coli.

- Shiga toxin-producing E. Coli (STEC) also referred to as Verocytotoxin-producing E. Coli (VTEC) or enterohemorrhagic E. Coli (EHEC). This pathotype is the one most commonly heard about in the news in association with food-borne outbreaks.
- Enterotoxigenic E. Coli (ETEC)
- Enteropathogenic E. Coli (EPEC)
- Enteroaggregative E. Coli (EAEC)
- Enteroinvasive E. Coli (EIEC)
- Diffusely adherent E. Coli (DAEC)

Symptoms

The symptoms of STEC infections vary for each person but include severe stomach cramps, diarrhea (often bloody), and vomiting. If there is fever, it usually is not very high (less than 101°F/less than 38.5°C). Most people get better within 5–7 days. Some infections are very mild, but others are severe. Around 5–10% of those who are diagnosed with STEC infection develop a potentially life-threatening complication known as hemolytic uremic syndrome (HUS). Clues that a person is developing HUS include decreased frequency of urination, feeling very tired, and losing pink color in cheeks and inside the lower eyelids. Persons with HUS should be hospitalized because their kidneys may stop working. Most persons with HUS recover within a few weeks, but some suffer permanent damage or die.

Treatment

Non-specific supportive therapy, including hydration, is important. Antibiotics should not be used to treat this infection. There is no evidence that treatment with antibiotics is helpful, and taking antibiotics may increase the risk of HUS. Anti-diarrheal agents like Imodium may also increase that risk.

Typhoid

Description

Typhoid fever is a life-threatening illness caused by the bacterium Salmonella Typhi. Salmonella Typhi lives only in humans. Persons with typhoid fever carry the bacteria in their bloodstream and intestinal tract. In addition, a small number of persons, called carriers, recover from typhoid fever but continue to carry the bacteria. Both ill persons and carriers shed Salmonella Typhi in their feces (stool). You can get typhoid fever if you eat food or drink beverages that have been handled by a person who is shedding Salmonella Typhi or if sewage contaminated with Salmonella Typhi bacteria gets into the water you use for drinking or washing food. Therefore, typhoid fever is more common in areas of the world where hand washing is less frequent and water is likely to be contaminated with sewage.

If you drink water, buy it bottled or bring it to a rolling boil for 1 minute before you drink it.

Eat foods that have been thoroughly cooked and that are still hot and steaming.

Avoid raw vegetables and fruits that cannot be peeled.

Symptoms

Persons with typhoid fever usually have a sustained fever as high as 103° to 104° F (39° to 40° C). They may also feel weak, or have stomach pains, headache, or loss of appetite. In some cases, patients have a rash of flat, rose-colored spots. The only way to know for sure if an illness is typhoid fever is to have samples of stool or blood tested for the presence of Salmonella Typhi.

Treatment

Typhoid fever is treated with antibiotics. Resistance to multiple antibiotics is increasing among Salmonella that cause typhoid fever. Reduced susceptibility to fluoroquinolones (e.g., ciprofloxacin) and the emergence of multidrug-resistance has complicated treatment of infections, especially those acquired in South Asia. Antibiotic susceptibility testing may help guide appropriate therapy. Choices for antibiotic therapy include fluoroquinolones (for susceptible infections), ceftriaxone, and azithromycin. Persons who do not get treatment may continue to have fever for weeks or months, and as many as 20% may die from complications of the infection.

Hepatitis A

Description

Hepatitis A is a contagious liver disease that results from infection with the Hepatitis A virus. It can range in severity from a mild illness lasting a few weeks to a severe illness lasting several months. Hepatitis A is usually spread when a person ingests the virus from contact with objects, food, WATER or drinks contaminated by feces or stool from an infected person. A person can get Hepatitis A through:

Person to person contact when an infected person does not wash his or her hands properly after going to the bathroom and touches other objects or food

Hepatitis A can be spread by eating or drinking food or water contaminated with the virus. (This can include frozen or undercooked food.) The food and drinks most likely to be contaminated are fruits, vegetables, shellfish, ice, and WATER. In the United States, chlorination of water kills Hepatitis A virus that enters the water supply.

Symptoms

Some people with Hepatitis A do not have any symptoms. If you do have symptoms, they may include the following:
- Fever
- Fatigue
- Loss of appetite
- Nausea
- Vomiting
- Abdominal pain
- Dark urine
- Clay-colored bowel movements
- Joint pain
- Jaundice (a yellowing of the skin or eyes)

Treatment

There are no special treatments for Hepatitis A. Most people with Hepatitis A will feel sick for a few months before they begin to feel better. A few people will need to be hospitalized. During this time, doctors usually recommend rest, adequate nutrition, and fluids. People with Hepatitis A should check with a health professional before taking any prescription pills, supplements, or over-the-counter medications, which can potentially damage the liver. Alcohol should be avoided.

Leptospirosis

Description

Leptospirosis is a bacterial disease that affects humans and animals. It is caused by bacteria of the genus Leptospira and are spread through the urine of infected animals, which can get into water or soil and can survive there for weeks to months. Many different kinds of animals carry the bacterium. These can include, but are not limited to: Cattle, pigs, horses, dogs, rodents, and wild animals. When these animals are infected, they may have no symptoms of the disease. Humans can become infected through:
- Contact with urine from infected animals.
- Contact with water, soil, or food contaminated with the urine of infected animals.

Symptoms

In humans, Leptospirosis can cause these symptoms:

- High fever
- Headache
- Chills
- Muscle aches
- Vomiting
- Jaundice (yellow skin or eyes)
- Red eyes
- Abdominal pain
- Diarrhea
- Rash

The time between a person's exposure to a contaminated source and becoming sick is 2 days to 4 weeks. Illness usually begins abruptly with fever and other symptoms. Leptospirosis may occur in two phases:

- After the first phase (with fever, chills, headache, muscle aches, vomiting, or diarrhea) the patient may recover for a time but become ill again.

- If a second phase occurs, it is more severe; the person may have kidney or liver failure or meningitis. This phase is also called Weil's disease.

The illness lasts from a few days to 3 weeks or longer. Without treatment, recovery may take several months.

Treatment

Leptospirosis is treated with antibiotics, such as doxycycline or penicillin, which should be given early in the course of the disease.

Intravenous antibiotics may be required for persons with more severe symptoms. Persons with symptoms suggestive of leptospirosis should contact a health care provider.

Chapter 13

Weather Patterns And Systems

Rain in the dry season is a sight to behold. When you live close to the land, you very quickly become aware of nature, its inner-workings and how it affects your homestead. Land owners instinctively know when to expect the winds to change and when the rains will come. It is obvious, that both people and animals, look to rain events as the prime provider of water and the giver of life. The rain comes and goes with the seasons and historically we have always been able to predict these patterns. This is the very thing that allows us to plant our seeds well before the rain starts to fall. Long periods of stable weather is optimum for landowners with livestock and vegetation. Stable weather means predictable weather patterns and that means a happy homestead that provides with abundance.

Unfortunately, we are currently living in a period of human history where stable weather patterns have become a thing of the past. We are constantly reminded of the abnormal weather conditions that are causing problems all over the planet. We see spring blossoms in the middle of the winter, snow in Egypt, record highs in the arctic and cold fronts in the tropics. We call these extreme weather events and speculation abounds as to what's causing it all. We can speculate all we want, but that doesn't solve anything, because the fact of the matter is that these systems are here to stay and there is nothing we can do about them. The safest thing to do is to be prepared and to be aware of what every weather pattern represents and how it will influence local rainfall expectations in your area.

The Two Main Patterns

Let's look at the facts and what we do know about weather patterns that contribute to abnormal weather conditions. Take note that these are abnormal conditions that do not come around every year.

Currently, our civilization is confronted with two specific fronts. They have been identified as El Nino and La Nina. Both El Nino and La Nina are generated in the Pacific ocean and both conditions affect the weather all over the world.

El Nino

When taking a closer look at the El Nino phenomenon we know that it causes the tradewinds to weaken and warm water is carried to the east (the west coast of the Americas). This is a "warm phase" that, on average, comes around every five years. It results in the following conditions:

North America

This area experiences warmer-than-normal winters in the Northern US and Canada. Expect above average rainfall in the Southern United States. It is very wet between Jan. and April in California and also very wet in the Southeast to the Atlantic coast between Nov. and the following April. Expect drier than normal conditions from Dec. to the following March in the Ohio and Tennessee valleys and From Nov. to the following March in the Pacific Northwest.

= El Nino =

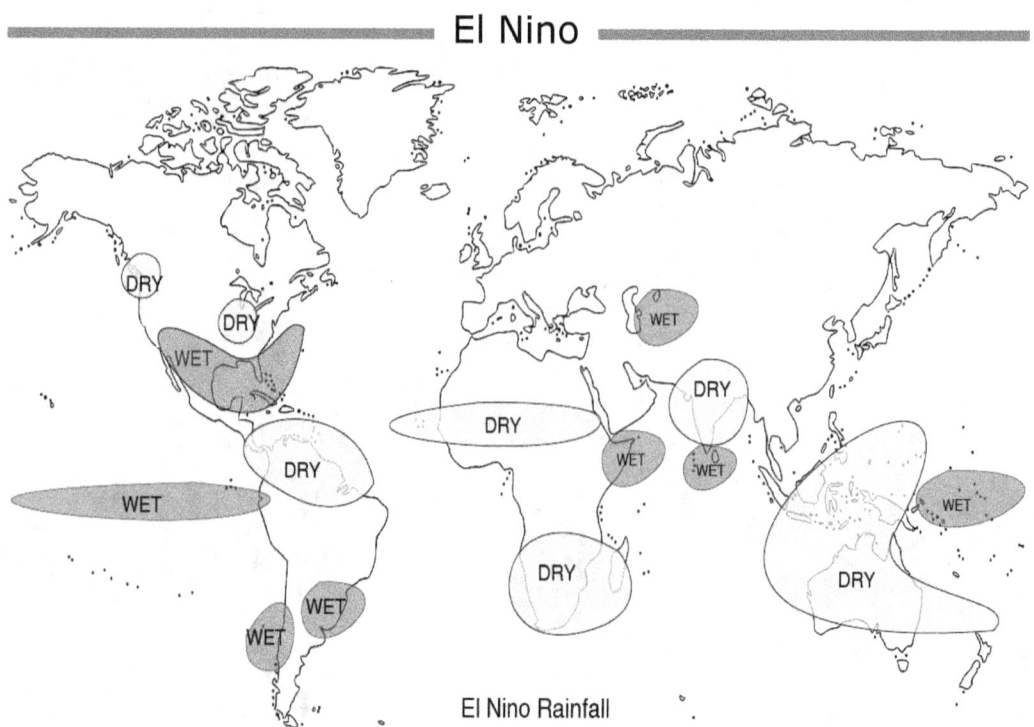

El Nino Rainfall

South America
Heavy rains along the West coast from June to Sept. and on the East coast from Sept. and into the following Jan. On islands nearby in the Pacific Ocean, expect above average rainfall from Jan. to May.

Indonesia and the Philippines
Very dry conditions in Indonesia, and the Philippines from June to the following Jan.

Australia
This area truly feels the effects of his system. Very dry conditions in central Australia from July to Nov. Very dry conditions in Eastern Australia from July to following Jan. Very dry conditions in Northern Australia from July to following March.

Africa
Expect droughts in Northern Africa from July-Sept. and further droughts in Southern Africa from Nov. to following March. Heavy rain is expected in East Africa from Oct. to the following Jan.

India
India will experience droughts From June to Sept. Heavy rains is expected in the far South from Oct. to Dec.

Northern Middle East
The areas Northwest of India and covering the Northern Middle-east region, is wet from Jan. to April.

Conclusion

Start to become more aware of the change in weather conditions. When an El Nino pattern is predicted, analyze how it will affect your local area. If you expect higher precipitation numbers than usual, take measures to harvest more water for times of need. Drier weather will require action as well. Get involved with water management in your community and make sure the majority of landowners are aware of the dangers posed by abnormal weather patterns that cause droughts. Share and discuss the El Nino related information with your local governmnet office to contribute to an emergency response system.

El Nino Field Notes
My Area And Rain Events

Expected Weather:

Preparedness:

La Nina

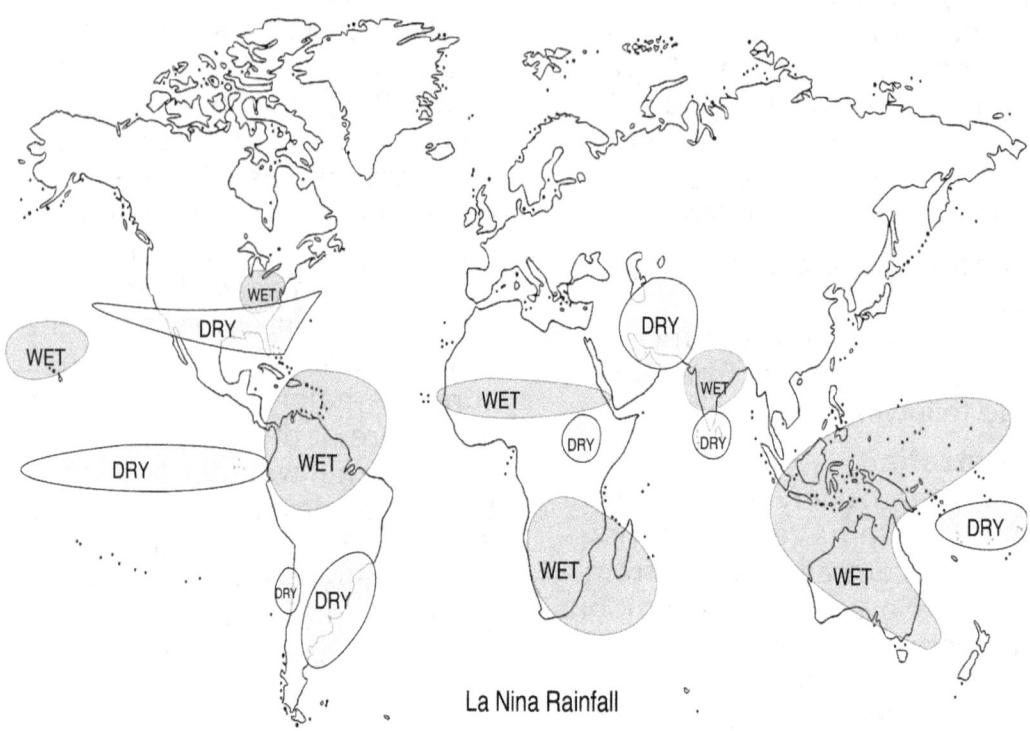

La Nina Rainfall

La Nina

La Niña is considered the "cool phase". During this event the trade winds are stronger that normal and more warm water is carried to Asia and upwelling increases. It usually follows immediately after an El Nino event and can last anything from 9-12 months. The El Nino event can be strong or weak, just depends on mother nature. This pattern is also characterized by a shift in rainfall patterns in many parts of the world:

South America

Drier than normal conditions can be expected along the West coast of South America from June to Sept. The East coast will also be drier than normal between Aug. and Dec. The nearby islands of the Pacific Ocean will be drier than normal between June and the following April.

Australia

Wetter than normal conditions over Northern Australia from Sept. to the following Jan. Southern Australia will receive more rain from May to the following Feb. Central Australia will experience above average rainfall between Sept. and the following Jan. Tasmania will have more rain from Aug. to Dec.

Indonesia, Malaysia and Philippines

Above-average precipitation to be expected between June and Dec. More tropical cyclones in the Western Pacific ocean.

USA

Expect above-average precipitation across the Northern Midwest USA from Dec. to the following March. Generally, expect below average rainfall to the South, southern and central parts of the Rockies and Great Plains and Florida. More specifically: In the Southeast from Oct. to the following April, and in the Southwest from Jan. to April.

Africa

Wetter than normal conditions are also observed over Southeastern Africa during Nov. to the following April. Northern Sub-Saharan Africa will be experiencing more rainfall from July to Sept. East Africa to the interior will be drier than normal between Nov. to the following March.

India

From June to Sept. expect the Indian monsoon rainfall to be greater than normal, especially in Northwest India. The South will be drier than usual from Oct. to Dec.

Northern Middle-East

Expect drier than average conditions from Jan. to May.

Conclusion

Be aware and expect a possible change in weather conditions when a La Nina event is predicted. Do your own research for your geographical area and analyze how much such an event will affect your homestead (if at all). If you expect higher precipitation numbers than usual, take measures to harvest more water for times of need and to arrange storage. Do this out of principle and be prepared to help other people in your community. Drier weather will require action as well and here you might need government assistance. As mentioned before, the consciousness that arises from being prepared should be contagious and should affect all members of the community all the way up to your local government office. This means a reliable infrastructure of trust and dependency that can save lives in the long run.

La Nina Field Notes
My Area And Rain Events

Expected Weather:

Preparedness:

Chapter 14

Plumbing Basics

Knowing plumbing basics is essential for DIY enthusiasts for several reasons:

- **Cost Savings:** Understanding plumbing basics allows DIY enthusiasts to handle simple plumbing repairs and installations themselves, saving them from the expense of hiring a professional plumber for minor tasks.
- **Emergency Situations:** Plumbing emergencies can occur unexpectedly, such as a burst pipe or a clogged outlet. Having basic plumbing knowledge enables DIY enthusiasts to take immediate action, minimizing potential damage and addressing the issue before professional help arrives.
- **Maintenance:** Regular maintenance is crucial for plumbing systems to function properly.

    DIY enthusiasts who are familiar with plumbing basics can perform routine tasks like checking for leaks, inspecting pipes, or cleaning filters, ensuring the plumbing system operates efficiently and preventing major problems from developing.
- **Confidence and Independence:** Knowing plumbing basics provides a sense of confidence and independence to the DIY enthusiast.

    There is great satisfaction in identifying the source of a plumbing issue, and then successfully resolving it single-handedly.

How To Apply PVC Cement

First, make sure the PVC pipe and fitting are clean and dry. Use a clean cloth or rag to wipe away any dirt or debris.

Use a PVC primer to clean and prepare the surface of the PVC pipe and fitting. Apply the primer according to the manufacturer's instructions, and let it dry completely.

Apply a thin layer of PVC cement to the inside of the fitting, and a thin layer to the outside of the pipe. Be sure to use the correct type of PVC cement for your specific application.

Quickly join the pipe and fitting together, making sure they are aligned correctly. Twist the pipe a quarter-turn to spread the cement evenly and create a strong bond.

Hold the pipe and fitting together for a few seconds to allow the cement to set. Avoid moving or disturbing the joint for at least 15 minutes to allow the cement to fully cure.

Wipe away any excess cement with a clean cloth or rag.

Repeat the process for any additional joints.

Pro Tip:

When adding PVC cement to PVC fittings, do the female side first, then you can set it down and not get debris on the glued surface while you are doing the male end.

The brass male by male nipples can cause problems when used with PVC fittings. You should never thread metal MALE fittings, into FEMALE PVC fittings, as metal expands and contracts (due to temperature) significantly more than plastic, and will crack the plastic female fittings. When you make a conversion from metal to PVC, the PVC should ALWAYS BE male, and the metal ALWAYS female.

Ideally you should spray paint all the PVC parts after finishing, as UV light weakens PVC over time, making it brittle.

How To Thread Plumber's (Teflon) Tape:

Use a wire brush or sandpaper to clean the threads on both the male and female ends of the pipes. This will remove any debris and ensure a tight seal.

Begin by holding the pipe or fitting in your left hand with the end facing you, and with the tape in your right hand.

Hold the roll of tape so it looks like a snail, with the tape unrolling off the bottom edge.

Lay the tape across the bottom edge of the pipe or fitting, then wind it on clockwise, moving upwards, in the direction of the threads and towards the larger part of the fitting, with a little bit of tension, for around 7 turns. Make sure to overlap the tape by about half its width with each rotation and also remember to wrap the tape away from you.

Wrapping in the direction of the threads is important, as wrapping the wrong way may result in the tape coming unwound as the fittings are tightened. Also be careful not to wrap over the hole.

You wind the tape clockwise so it stays on when you screw the fitting in, if you wind it the other way it will unwind when you screw it in. You hold it "snail-wise" with the tape feeding from the bottom so you can tension the tape as you wind it. If you hold it the other way the tape loops out faster than you can wind it and it's impossible to tighten it as you wind it!

Once the tape is wrapped around the male threads, screw the pipes together. Tighten the connection by hand first, then use a wrench to tighten it further if necessary.

How To Install PVC Union Couplings For Maintenance

PVC unions are a useful plumbing component that allows for easy disassembly and maintenance of a plumbing system. Here's a step-by-step guide on how to install a PVC union for maintenance:

Turn off the water supply to the section of the plumbing system that you will be working on.

Cut the PVC pipe at the location where you want to install the union, using a PVC pipe cutter or saw.

Dry fit the PVC union onto the two cut ends of the PVC pipe to ensure that the union is the correct size and that the two ends of the pipe are aligned properly.

Apply PVC primer to the outside of both pipe ends and the inside of the union sockets. This will help to prepare the surfaces for the PVC cement.

Apply a liberal amount of PVC cement to the outside of both pipe ends and the inside of the union sockets.

Quickly push the union onto the two pipe ends and twist slightly to ensure a good bond. Hold the union in place for a few seconds to allow the cement to set.

Once the cement has dried, turn the water supply back on and check for leaks. If there are no leaks, the

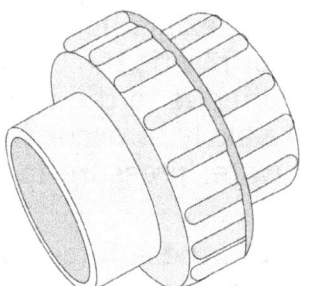

PVC Union Coupling

PVC union has been successfully installed.

When it comes time for maintenance, the PVC union can be unscrewed to separate the two pipe ends, allowing for easy access and maintenance.

Spigot vs. Tap

A spigot and a tap are similar in function as they both allow water to flow out of a container or pipe, but they are not exactly the same thing.

A spigot is a device used to control the flow of liquid from a container, such as a water tank or barrel. It typically consists of a valve or faucet that is mounted on the container, which can be opened or closed to regulate the flow of water. A spigot is often used for outdoor water sources, like for gardening or camping, and may be made of metal, plastic, or wood.

A tap, on the other hand, is a more general term that can refer to any device used to control the flow of liquid from a pipe. Taps are commonly used in homes and buildings to control the flow of water from a plumbing system. Taps can come in various shapes and sizes, and they may be operated by a handle, lever, or knob.

So, while a spigot and a tap are similar in function, they are typically used in different contexts and may have different designs and features.

When To Use Reinforced Flexible Hosing

Flexible housing for rainwater tank joints can be used in situations where there is a need to accommodate some movement or settling of the tank, where there is bulging of the tank, or where the tank is subject to significant changes in temperature.

Rainwater tanks are often exposed to various environmental conditions, such as changes in temperature, soil movement, and ground settling, which can cause stress and strain on the tank and its connections.

Also, the type of tank can influence on whether flexible housing is necessary or not. Poly tanks, like other types of plastic tanks, can be subject to bulging or distortion at the bottom due to the weight of the water inside the tank. This can be more likely to occur when the tank is only partially full, as there is less water pressure to counteract the weight of the tank itself.

Flexible housing can help to reduce the impact of these forces and can help to prevent damage to the tank or its fittings.

Flexible housing is typically made from materials such as rubber or

PVC and can be used to create a flexible connection between the tank and the piping or other components. This can help to absorb the forces that are created by movement or settling, and can help to prevent leaks or other damage to the tank or its connections. However, it's important to ensure that any flexible housing that is used is compatible with the specific type of tank and fittings that are being used, and that it is installed properly to prevent leaks or other issues.

Metric To Imperial Pipe Sizes

20mm – ½"
25mm – ¾"
32mm – 1"
40mm – 1-¼"
50mm – 1-½"
63mm – 2"

Metric Conversions

Metric Conversion Chart		
to convert	to	multiply by
Inches	Centimeters	2.54
Centimeters	Inches	0.4
Feet	Centimeters	30.5
Centimeters	Feet	0.03
Yards	Meters	0.9
Meters	Yards	1.1

Types Of PVC

The most common schedule numbers for PVC pipes used in rainwater tank installation are Schedule 40 and Schedule 80.

Schedule 40 PVC pipes are thinner and lighter than Schedule 80 pipes, which are thicker and heavier and are typically used for higher pressure applications.

Pipe Thread Acronyms

NPT - National Pipe Thread (USA)
BSP - British Standard Pipe (Commonwealth)
MIP(T) - Male Iron Pipe (interchangeable with NPT)
FIP(T) - Female Iron Pipe (interchangeable with NPT)

MPT - Male Pipe Thread (interchangeable with NPT)
FPT - Female Pipe Thread (interchangeable with NPT)
DWV - Drainage, Waste, Vent Fitting

Example:

When perusing female-threaded fittings, you may come across the abbreviations "FPT" and "FIPT," which respectively stand for "female pipe thread" and "female iron pipe thread." Likewise, for male-threaded fittings, you will encounter the same abbreviations but with an "M" instead of an "F," where the "M" represents "male." Historically, pipes were primarily composed of iron or steel, hence the inclusion of "iron" in the abbreviation. However, it's important to note that this doesn't imply that the component itself is made of iron. PVC pipes are simply designed to be compatible with metal pipelines.

Slip Socket

A slip socket is a connection method used for PVC pipes or fittings without any threads. It involves inserting a PVC pipe or fitting into a socket or hub without the need for threads. The slip socket is designed with a smooth and slightly tapered interior to facilitate

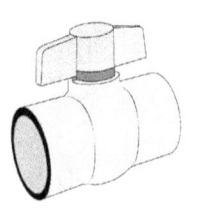

effortless insertion of the pipe or fitting. To ensure a secure and watertight connection, PVC solvent cement is applied to the inside of the socket and the outside of the pipe or fitting. This solvent cement creates a chemical bond between the two surfaces, resulting in a robust and long-lasting connection.

Threaded Socket

A threaded socket, also known as a female threaded end, is designed with internal threads to facilitate the connection of a male threaded PVC pipe or fitting. It shares a similar shape with a slip socket but features spiral grooves on its interior surface. These grooves align with the threads on the male component, enabling them to be securely screwed together. Threaded connections are commonly employed in

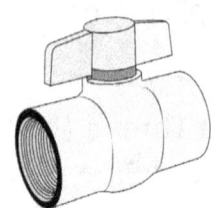 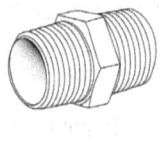

situations where easy assembly and disassembly of pipes or fittings are required.

The choice between slip sockets and threaded sockets depends on the specific requirements of your plumbing system. Slip sockets are commonly used in applications where a permanent, leak-proof connection is desired, such as pressurized water lines. Threaded sockets are often used in scenarios where frequent assembly and disassembly are required, such as in irrigation systems or certain types of industrial applications.

Sockets And Spigots (No Threads)

Sockets and spigots (not a tap!) are both slip fitting end types (no threads), but the term "slip" usually refers to sockets. A socket is a fitting that simply goes over the end of a pipe. A 1" socket end will fit on a 1" pipe. Many PVC couplings have socketed ends.

A spigot fits inside a socket, so it is the same size around as regular pipe. When you need to attach a fitting to another fitting, you use a spigot end. In their function, sockets are the female and spigots are the male (just without threads). We use PVC cement with sockets and spigots.

Female And Male Threaded

Female and male threads are straightforward to understand. In a female-threaded fitting, the threads are located on the inside, while in a male-threaded fitting, the threads are on the outside. This allows a male-threaded fitting to be screwed into a female-threaded fitting.

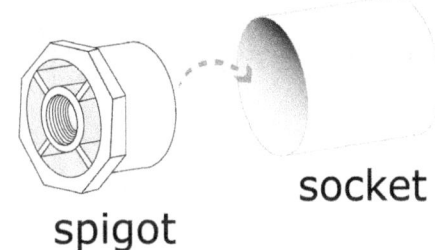

Unlike other connections, such as slip fittings, threaded connections do not require glue. Instead, using Teflon tape is sufficient to ensure a tight seal. It's worth noting that many threaded fittings feature slip fittings on the opposite end, which simplifies the process of connecting them to pipes.

How To Add A PVC Ball Valve and Garden Hose To A PVC Bulkhead

Let's use a ¾ inch PVC Ball valve as an example:

1. Connect a ¾ inch PVC hex nipple to your bulkhead fitting. (A PVC bulkhead fitting adapter looks like a PVC hex nipple and can also be used.)
2. Connect the ¾ inch PVC Ball valve to the ¾ inch PVC Hex nipple.
3. Connect the ¾ inch PVC Hose barb to the ¾ inch PVC Ball valve.
4. Fit the ¾ inch hose with hose clamps to the ¾ inch PVC Hose barb.

Remember to use Plumber's tape on all threads.

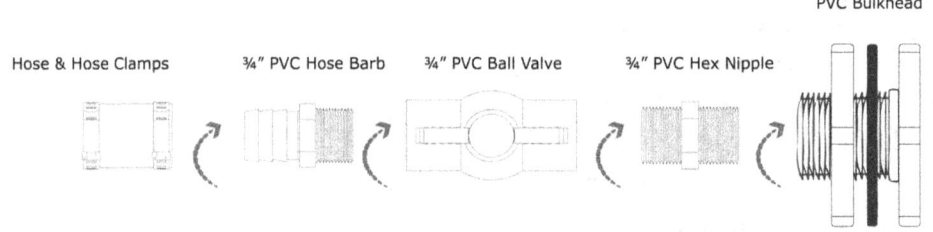

How To Add A PVC Ball Valve To A PVC Pipe

Let's use 2 inch PVC pipe as an example:

Option A

Connect a 2" PVC ball valve with a slip socket, to the 2" PVC pipe. The PVC pipe fits inside the ball valve socket and it is secured with PVC primer and cement.

Option B

Add a male PVC slip adapter to the PVC pipe. This adapter has a socket end that fits your PVC pipe, and a threaded end which screws into your PVC ball valve.

Later on you can add a garden hose adapter to the PVC ball valve if you plan on having a tap or a spigot. Secure the ball valve with Plumber's tape.

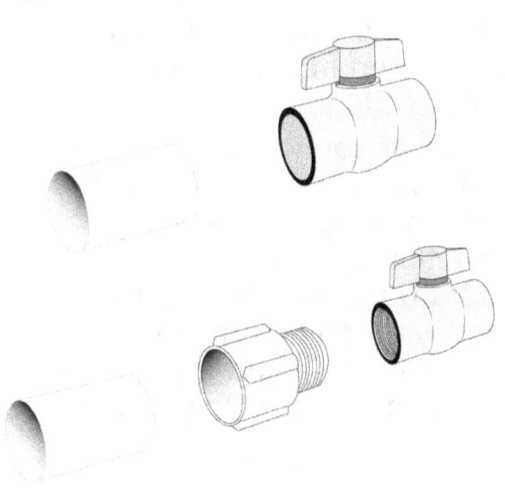

How To Add A Faucet To PVC Pipe

In this example, the faucet is ¾ inch and the PVC pipe is 2 inch. This means that we will have to use a reducer to go from 2 inches to ¾ inches.

1. First, insert the 2" x 1-½" PVC Flush Bushing - Spigot x Hub, into the 2" PVC pipe. This allows us to reduce the size from 2" to 1-½".
2. Next, fit the 1-½" x ¾" PVC Bushing Spigot x FIP into the flush bushing.
3. A spigot can fit directly into the ¾" PVC Bushing Spigot x FIP.

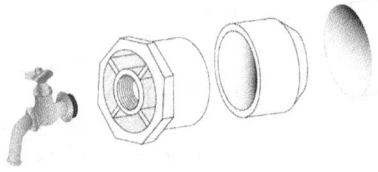

4. Or, insert a ¾" hex nipple and then attach a ¾" ball valve.

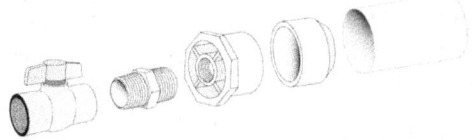

Secure all fittings with PVC primer and cement.

Resources

Water. Why It's An Issue.
http://www.worldbank.org/en/topic/water/publication/high-and-dry-climate-change-water-and-the-economy

Understanding The Origins Of Water
http://en.unesco.org/news/water-action-cop22-address-climate-challenges

http://environment.nationalgeographic.com/environment/freshwater/water-conservation-tips/

Weather Patterns And Systems
http://www.ucmp.berkeley.edu/education/dynamic/session4/sess4_hydroatmo3.htm

http://earthobservatory.nasa.gov/Features/LaNina/la_nina_2.php

https://en.wikipedia.org/wiki/Water#Law.2C_politics.2C_and_crisis

https://en.wikipedia.org/wiki/Water_right

Storing Water For An Emergency
http://naturalsociety.com/recycling-symbols-numbers-plastic-bottles-meaning/

http://edition.cnn.com/2016/02/01/health/bpa-free-alternatives-may-not-be-safe/

Collection And Storage Of Rainwater
https://en.wikipedia.org/wiki/Rainwater_harvesting

https://en.wikipedia.org/wiki/Pump

http://www.s1expeditions.com/2014/08/161-solar-power-off-grid-on-grid.html

Extracting Water From The Ground
http://drillyourownwell.com/index.htm

http://www.waterencyclopedia.com

http://www.waterencyclopedia.com/Ge-Hy/Groundwater.html

http://www.nzdl.org/gsdlmod?a=p&p=home&l=en&w=utf-8

http://siteresources.worldbank.org/INTPHILIPPINES/Resources/RWSVolIDesignManual.pdf

http://siteresources.worldbank.org/INTWRD/864188-1171045933145/21215368/33b.pdf

http://www.sswm.info/sites/default/files/reference_attachments/ABBOT%204000%20Hand%20Dug%20Well%20Manual.pdf

U.S.E.P. Agency. "EPA Region & Drinking Water Unit Tech Tips" Epa.gov. [Online]. Available: https://19january2017snapshot.epa.gov/sites/production/files/2016-06/documents/tech_tip_tank_vents.pdf

Washington State Department of Health. "Drinking Water Tech Tips, Sanitary Protection of Reservoirs: Vents" Wa.gov. [Online]. Available: https://doh.wa.gov/sites/default/files/legacy/Documents/Pubs//331-250.pdf

Oklahoma State University. "Design of Rainwater Harvesting Systems in Oklahoma" Okstate.edu. [Online]. Available: https://extension.okstate.edu/fact-sheets/design-of-rainwater-harvesting-systems-in-oklahoma.html

Rainwater Harvesting Australia. "Sustainable Buildings Save Billons - An Alternative Water Strategy for Sydney" Org.au. [Online]. Available: https://rainwaterharvesting.org.au/

Texas A&M AgriLife Extension Service. "Rainwater Harvesting" Tamu.edu. [Online]. Available: https://rainwaterharvesting.tamu.edu/catchment-area/

Texas Water Development Board. "The Texas Manual On Rainwater Harvesting" Tamu.edu. [Online]. Available: https://texashelp.tamu.edu/wp-content/uploads/2018/08/rainwater-harvesting-manual.pdf

World Water Reserve. "Is it Illegal to Collect Rainwater: 2023 Complete State Guide" Worldwaterreserve.com [Online]. Available: https://worldwaterreserve.com/is-it-illegal-to-collect-rainwater/

Federal Energy Management Program. "Rainwater Harvesting Regulations Map" Energy.gov. [Online]. Available: https://www.energy.gov/eere/femp/rainwater-harvesting-regulations-map

Federal Energy Management Program. "Rainwater Harvesting Tool " Energy.gov.

[Online]. Available: https://www.energy.gov/femp/rainwater-harvesting-tool

Federal Energy Management Program. "Water-Efficient Technology Opportunity: Rainwater Harvesting Systems" Energy.gov. [Online]. Available: https://www.energy.gov/femp/water-efficient-technology-opportunity-rainwater-harvesting-systems

Environment Agency. "Rainwater Harvesting: Regulatory Position Statement" Gov.uk. [Online]. Available: https://www.gov.uk/government/publications/rainwater-harvesting-regulatory-position-statement/rainwater-harvesting-regulatory-position-statement

Environment Agency. "Harvesting Rainwater For Domestic Uses: An Information Guide" Gov.uk. [Online]. Available: http://planning.highpeak.gov.uk/portal/servlets/

International Journal of Science and Research (IJSR). "Rain Water Harvesting - A Case Study " Ijsr.net. [Online]. Available: https://www.ijsr.net/archive/v8i5/ART20197767.pdf

Warwick School of engineering. "Sizing the DRWH system" Ac.uk. [Online]. Available: https://warwick.ac.uk/fac/sci/eng/research/grouplist/structural/dtu/rwh/sizing/

Designing Your Own Filtration System

https://en.wikipedia.org/wiki/Biofilm

https://en.wikipedia.org/wiki/Biosand_filter

http://www.shared-source-initiative.com/biosand_filter/complete_biosand_new.html

http://ojs.library.queensu.ca/index.php/ijsle/article/viewFile/4244/4344

Cleaning Water For Survival

https://en.wikipedia.org/wiki/Portable_water_purification

https://www.brita.com/?locale=us

http://www.bigberkeywaterfilters.com/berkey-water-filters.html?___SID=U

http://doultonusa.com/Pop%20up%20products%20descriptions/Ultracarb%20Water%20Filter%20Elements.htm

http://doultonusa.com/doulton_water_filters/Whole-house-ceramic-water-filter.php

https://www.espwaterproducts.com/about-reverse-osmosis/

https://www.espwaterproducts.com/uv-water-purification/

http://www.ozonepurewater.com

http://www.pelicanwater.com/whole-house-water-filter.php

http://www.culligan.com/home/water-softening

http://www.h2olabs.com

http://www.echotecwatermakers.com/beach_house_desalination_systems.htm

How To Manage Waste

https://en.wikipedia.org/wiki/Pit_latrine

https://www.unicef.org/cholera/Annexes/Supporting_Resources/Annex_9/WHO-tn7_waste_mangt_en.pdf

Centers for Disease Control and Prevention

http://www.cdc.gov/healthywater/disease/az.html

http://wwwnc.cdc.gov/travel/yellowbook/2016/the-pre-travel-consultation/water-disinfection-for-travelers

http://www.cdc.gov/healthywater/drinking/travel/backcountry_water_treatment.html

http://www.cdc.gov/ExtremeHeat/

http://www.cdc.gov/healthywater/emergency/drinking/making-water-safe.html

http://www.cdc.gov/healthywater/drinking/private/wells/maintenance.html

United States Environmental Protection Agency

https://www.epa.gov/privatewells/protect-your-homes-water

www.epa.gov/safewater/mcl.html

www.epa.gov/safewater/sdwa/sdwa.htm

World Health Organization

http://www.who.int/water_sanitation_health/emergencies/fs3_2/en/

http://www.who.int/water_sanitation_health/diseases-risks/diseases/diseasefact/en/

Index

A

A Blue Or Green Color 364
Absolute Pore Size Filters 305
Absorption 305
Acid Rain 60
Ac Pump 196
Add Extra Pipe 240
Adsorption 305
Adsorption 302
Aerobic Zone 102
A Homemade Filtration System 332
Amebiasis (Amoebic Dysentery) 383
Anaerobic Fermentation 99 100
Anaerobic Zone 102 145
Annual Water Well Checkup 292
Apply PVC Cement 396
Apron 258
Aquifer 224
A Red Or Brown Color 364
Arsenic 367
Asphalt Shingles 68
Auger 224
Auger Drilling 246
Automatic Pump 196

B

Backflow 105 222
Backflow (Air Gap) 105 203 204 222
Backup Pumps 197
Bacteria 318
Bags 372
Ball Valve 158 164 181 196 402
Basic Home Purifying System 287
Basic Hygiene When Handling Waste 371
Benefits Of Using Precast Concrete Rings 268
Biodegradable Bags 372
Bisphenol A 41
Boiling 59 213
Boiling Water 307
Booster Pump 196
Bracket 74
Brine 342
Buckets 373
Bulkhead Fitting 170
Bull Float 120
Bungs 176
Bushing 169 170 176 182 183 188 190 403

C

Caisson Ring 258
Calculations 54
Call An Expert 290
Camlock 127 151 153 160
Campylobacter 365
Carbon Filters 312
Catchment Area 54
Cavitation 197 274
Centers For Disease Control And Prevention 59
Centrifugal Pump 197
Ceramic Multi Candle Filter 311
Check Dams 64
Check Valve 200 204 274
Check Valves And Foot Valves 272
Chemical, Chlorine Odor 363
Chlorine 59 212 213 215
Chlorine Based Tablets 329
Chlorine Bleach 325
Chlorine Dioxide Based Tablets 329
Chlorine Dioxide Tablets 325
Cholera 380
Clay 67
Cleaning Water For Survival
  Bacteria 318
  Boiling Water 307
  Carbon Filters 312
  Ceramic Multi Candle Filter 311
  Distillation Systems 313
  Faucet/ Sink/ Desktop/ Pitcher Filters 308
  Filters 304
  Land-Based Marine Filters 317

Ozone Purification 315
Pathogen Distribution 312
Pathogens 306
Protozoa 316
Purifiers 306
Reservoirs & Removable Gravity-Fed Filters 310
Reverse Osmosis Filters 309
Solar Disinfection 318
Ultraviolet Treatment Systems 314
Viruses 314
Water Softeners 317
Cleaning Your Water 302
Closed Hydrological System 25
Cloudburst 148 156 175 185
Cloudy, Milky White Water 364
Communities 15
Comparing Disinfection Options 307
Comparing Treatment Options 307
Concrete 53 67 85 110 116 119 120 258
Concrete Slab 120
Concrete Tiles 67
Condensation 342
Condenses 25
Conduit 176
Conduit Locknut 180
Connecting An Electric Pump 272
Connecting A Solar Water Pump 280
Construct Your Drill 234
Contour Bunds 64
Copper Tubing 344
Corrugated Copper Tubing 345
Cost Estimation 57
Coupler 207
Course Thread 176
Cryptosporidium And Salmonella 367

D

Daisy Chain 177
DC Pump 196
Desalination 302
De-Sludge 215
Digging An Emergency Trench Latrine 374

Digging An Outhouse 374
Digging Your Own Well 224
Discharge Pipe 197
Discussing Waterborne Diseases 377
Disinfecting A Drilled Or Driven Well (Imperial) 296
Disinfecting A Drilled Or Driven Well (Metric) 297
Disinfecting A Dug Well (Imperial) 298
Disinfecting A Dug Well (Metric) 299
Disinfecting A Well 293
Disinfecting Water With Chemicals
  Chlorine Bleach 325
  Disinfection Tablets 327
  Iodine 324
Disinfection Tablets 327
Disinfect Water Using Household Bleach 326
Distillation 342
Distillation Systems 313
Distillation With A Heat Source 343
Dolly 41
Downspout 74 84 85 87 170 183
Drain Plug 354
Draw Down 275
Draw Down Water Level 275
Drilling A Well Yourself 224
Drill Pipe 231
Drill Tip 231
Driving A Well Point 225
Dug Wells 258
Dug Wells And Safety Concerns 267

E

E.coli 367 384
PVC Elbow 134 138 139 147 164 166 169 170 173 174 176 180 182 183
Electrical Hazards 16
El Nino 390
Erosion 21 62 64 108 116 137 143 154 161 193
Evaporation 25
Evaporation 342

Index

Extracting Water From The Ground
  Clay 226
  Get Legal 229
  Gravel 226
  Plan The Type Of Well You Want 229
  Research 228
  Rock 226
  Sand 226
  The Soil 225
  The Water Table 226

F

Fascia 72
Faucet/ Sink/ Desktop/ Pitcher Filters 308
Fecal Coliforms 359
Female And Male Threaded 401
Female Mender Kits 185
Filters 321
Filters 304
Filtration 59 96 97 222 302
Filtration Vs.purification 304
Fine Thread 176
First Flush 102 129 139 142 169 221
Fishy, Earthy Or Moldy Odors 363
Flexible Coupling 237
Flexible Plumbing 149
Float Level Switch 201
Flow Rate 56
Fluoride 369
Foam In Your Water 364
Foot Valve 274
Formwork 120
Foundation 115

G

Galvanized 53 67 70 84 109
Get Legal 229
Getting Your Water Tested 358
Giardia 365
Giardiasis 381
Glass 40
Gravel 120
Gravity-Fed

Filters 337
Gravity-Fed Filtration System 337
Groundwater 224
Groundwater Aquifers 63
Gutter 69 71 73 74 76 77 79 80 81 82 139 147 158
Gutter End Caps 76
Gutter Outlet 147

H

Halogens 302
Having Access To The Inside Of Your Still 354
Hdpe 98 112 149 152
Heavy Duty Home Refuse Bags 372
Hepatitis A 365 386
Hidden Water Sources 50
Hire A Professional 225
Hole Saw 139 151 158 169 176
How To Design And Build A Distillation System 342
How To Design And Build A Filtration System 332
How To Dig And Construct A Well 258
How To Dig And Construct A Well Using An Auger 246
How To Dig And Construct A Well Using A Well Point 252
How To Disinfect A Well 294
How To Drill And Construct A Well Using PVC Pipe 65 230
How To Lower Concrete Rings In A Level Position 266
How To Make Activated Charcoal 339
How To Make A Well-Screen 250
How To Manage Waste 371
How To Pour Pea Gravel Around The Well Screen 245
How Will You Know When You've Hit Water? 241
Humidity 118

I

Ibc Tote 114 169 171 172 174
Inline Pump Controllers 196
Inspection 66

Install A Gutter 73
Install An Irrigation Pump 207
Install An Overflow 135
Install A Pump 205
Installation And Priming 277
Install Chicken Waterer 188
Install Multiple Rain Barrels With Flexible Hose 185
Install Multiple Tanks From The Top 149
Install Multiple Tanks With PVC 164
Install PVC Union Couplings 397
Install Tank Fittings 121
Insulate Your System 354
Insulation 217
Iodine 59
Iodine 324
Iodine Based Crystals 328
Iodine Based Tablets 328
Iodine Crystals 324
Iodine Tablets 324
Ions 302
Irrigation Installation 207

L

Ladder 16 75 89
Land-Based Marine Filters 317
La Nina 392
Lead 368
Leptospirosis 387
Locknut 124

M

Maintenance 211
Make A Mosquito Screen 220
Make A PVC Bail Bucket 249
Manual Drilling 230
Mesh Rating 305
Metal 67
Methane Gas 364
Metric 399
Micro-Filtration 302 321
Micron System 305
Mildew 118
Modular Tanks 191

Mold 118
Mosquito Control Measures 217
Mosquito Screens 219
Municipal Water 31
Municipal Water 197
Musty, Unnatural Odor 363
My Checklist 270

N

Nano-Filtration 303 321
Nitrate 367
Noise Reduction 209
Noroviruses 366

O

Outlet 170
Outlet-To-Outlet Method 157
Overflow 64 104 135 147 217 219 221
Ozone Purification 315

P

Pathogen 303
Pathogen Distribution 312
Pea Gravel 116 171
Permeable Pavement 63
Pest Infestation 118
Petroleum Odor 363
Ph Level 359
Pipe Thread Acronyms 399
Pitless Adapters 274
Plan The Type Of Well You Want 229
Plastic Containers 41
Plumber's (Teflon) Tape 108 132
Point Wells 252
Polishing Filter 303
Poultry Water Cups 188
Pressure Switch 274
Pressure Tank 198 200 274
Pressure Tanks 273
Preventative Maintenance 16
Priming 198
Protective Gear 16 215
Protect Your Well 290
Protozoa 316

Psi 198
Public Water 361
Pump Basics 195 272
Pump Switch 198
Purification 303
Purifiers 306
PVC Board 342 347
PVC Cross Tee 236
PVC Valve 236

Q

Questions To Ask A Licensed Well Driller 36

R

Radon 369
Rain Barrel 58 112 182
Recharge Groundwater 63
Recharge Well 63
Reducer Bushings 236
Reflective Material 354
Rehydration 378
Research 15 228
Reservoirs & Removable Gravity-Fed Filters 310
Reverse Osmosis Filters 309
Rigid Plumbing 149
Roof Material 53
Roof Pitch 66 74
Rotaviruses 366

S

Safety 15
Saving Water 29
Self-Cleaning 66
Self-Priming 197
Self-Priming Pumps 197
Shigella 365
Shigellosis (Bacillary Dysentery) 382
Shock Chlorine Disinfection 212
Silicone Caulk 347
Slate 68
Slimline Tanks 191
Slip Socket 400
Slope 53

Sludge 215
Socket 167 400
Sockets And Spigots 401
Soft, Sandy Areas 60 61
Soft Water 60
Solar Disinfection 318
Solar Pump Basics 1 281
Solar Pump Basics 2 282
Solar Pump Basics 3 282 283
Solar Pump Basics 4 284
Solar Still 342
Solar Still Distillation 346
Spigot 176 181 182 398 403
Spirit Level 74 176
Spring Water 35
Stainless Steel Containers 41
Standing Water Level 224
Storing Water For An Emergency
  Dolly 41
  Glass 40
  Hidden Water Sources 50
    House Pipelines 50
    Pressure Tanks 50
    Swimming Pools 50
    Toilet Top Tanks 50
    Water Heater 50
  Plastic Containers 41
    Hdpe 41
    Ldpe 41
  Stainless Steel Containers 41
  Water In Containers 40
Stormwater 21 57 102 104 105 135 163
Suction Pipe 199
Sulfurous Odor 362
Swales 63

T

Tank Covers 171
Tank Foundation 114
Tank Paint 171
Tank Wraps 171
Tempered Glass 342
Tempered Glass Pane 347
Testing Water: Disease And Contam-

ination
Fecal Coliforms 359
Testing Water: Disease And Contamination
  Ph Level 359
  Public Water 361
  Testing For Odors Or Colors In Your Water
    Foam In Your Water 364
  Testing Odors And Colors In Your Water
    Sulfurous Odor 362
  Testing Odors Or Colors In Your Water
    Chemical, Chlorine Odor 363
  Testing Odors Or Colors In Your Water
    Cloudy, Milky White Water 364
  Testing Odors Or Colors In Your Water
    A Blue Or Green Color 364
    A Red Or Brown Color 364
    Fishy, Earthy Or Moldy Odors 363
    Methane Gas 364
    Musty, Unnatural Odor 363
    Petroleum Odor 363
  Total Coliforms 358
  Wells 359
  When To Test Your Water (EPA Guideline) 360
The Basics Of A Deep Well Jet Pump 277
The Basics Of A Shallow Well Jet Pump 276
The Basics Of A
Submersible Deep Well Pump 278
The Basics Of Managing A Well
  Call An Expert 290
  Visual
  Inspections 291
The Basics Of Wells And Water Pumps
  Check Valves And Foot Valves 272
  Connecting An Electric Pump 272
  Connecting A Solar Water Pump 280
  Pitless Adapters 274
  Pressure Tanks 273

Pump Basics 272
Solar Pump Basics 1 281
Solar Pump Basics 2 282
Solar Pump Basics 3 283
Solar Pump Basics 4 284
The Basics Of A Deep Well Jet Pump 277
The Basics Of A Shallow Well Jet Pump 276
The Basics Of A Submersible Deep Well Pump 278
The Chemical Treatment Of Water 324
The Drill Bit 238
The Drill Handle Bar 239
The Drill Top 234
The Need For Latrines And Toilets 372
The Origins Of Water 23
  Condenses 25
  Global Water Crisis 25
  Saving Water 29
  The Water Cycle 25
  Water And Droughts 27
  Water Vapor 25
The Soil 225
The Water Cycle 25
The Water Table 226
The Well Screen 242
Threaded Socket 400
Thread Plumber's Tape 396
Throttling Valve 201
Tincture 324
To Clean Clear Water From A Tap. 42
To Clean Cloudy Tap Water. 42
Torque Arrestor 280
Total Coliforms 358
Transfer Pumps 196
Treating Waterborne Diseases 379
  Amebiasis (Amoebic Dysentery) 383
  Cholera 380
  E.coli 384
  Giardiasis 381
  Hepatitis A 386
  Leptospirosis 387
  Shigellosis (Bacillary Dysentery) 382

Typhoid 385
Trough 343
Typhoid 385
Typical Symptoms Of Heat Stroke 48
Typical Symptoms Of
    Mild Dehydration 48
Typical Symptoms Of Severe Dehydration 48

U

Ultra-Filtration 302 321
Ultraviolet Treatment Systems 314
Under Deck Tanks 191
Underground Tanks 192
Union Couplings 204
Using An Auger To Drill Your Well 225

V

Vapor 343
Vent 128 132 134 400
Viruses 314
Visual
Inspections 291
Vocs 343

W

Wall-Mounted Tanks 191
Water And Droughts 27
Water Collection Efficiency 66
Water
Disinfection 324
Water Filtration 199
Water In Containers 40
Watering Nipples 188
Water Meter 199
Waterproof Aluminum Duct Tape 354
Water Quality 62 107
Water Related Diseases And Contaminants In Private Wells 365
    Arsenic 367
    Campylobacter 365
    Cryptosporidium And Salmonella 367
    E.coli 367
    Fluoride 369
    Giardia 365
    Hepatitis A 365
    Lead 368
    Nitrate 367
    Noroviruses 366
    Radon 369
    Rotaviruses 366
    Shigella 365
Water Softeners 317
Water Table 225
Water Vapor 25
Weather Patterns And Systems
    El Nino 390
    La Nina 392
Well Basics 224
Well Capacity 275
Well Casing 231
Well Casing 225
Well Drilling Basics 232
Well Point 225
Wells 63
Well Screen 225
Well Total Depth 275
Wetlands 64
What To Test For 358
When To Test 359
When To Test Your Water (EPA Guideline) 360
Where To Test 361
Wildfires 21 107 108 109 110 111
Wildlife 21 62
Windlass 258
Wood Shingles 69

Y

Yield 343
Your Water Rights 37
Y-splitter 185

About The Author

Daniel Schoeman is a scholar and researcher of languages, cultures, carpentry, plumbing, construction, agriculture and how to live off the land. He first started traveling the planet in 1997 and worked as a willing worker on organic farms in Australia. His journey has since taken him to a variety of locales all over the world and given him the privilege to see how different cultures live and deal with their respective environments.

His motivation stems from an aspiration to create a world where self-sufficient, off the grid living is endorsed, and where housing can be affordable to all humans with the will and determination to work for it.

His latest endeavor is researching and writing how-to articles on hydroponics and cob house building techniques.

This is his fourth resource guide on water management.

Dear Reader
"Thanks for purchasing this book.
If you found the information in this book helpful, then please consider leaving an honest review on your favorite store.
It would be greatly appreciated."

www.ingramcontent.com/pod-product-compliance
Lightning Source LLC
Chambersburg PA
CBHW051206290426
44109CB00021B/2363